Parasitic helminths and zoonoses in Africa

Titles of Related Interest

Biology of blood sucking insects
M. Lehane

The biology of Echinococcus and hydatid disease
R. C. A. Thompson (ed.)

A biologist's advanced mathematics
D. R. Causton

Cell movement and cell behaviour
J. M. Lackie

Kalahari hyaenas
M. G. L. Mills

Parasitic protozoa
J. B. Kreier & J. R. Baker

Parasitic helminths and zoonoses in Africa

Edited by

C. N. L. Macpherson
Swiss Tropical Institute Field Laboratory, Tanzania

P. S. Craig
School of Tropical Medicine, Liverpool

London
UNWIN HYMAN
Boston   Sydney   Wellington

Published by the Academic Division of
Unwin Hyman Ltd
15/17 Broadwick Street, London W1V 1FP, UK

Unwin Hyman Inc.,
955 Massachusetts Avenue, Cambridge, Mass. 02139, USA

Allen & Unwin (Australia) Ltd,
8 Napier Street, North Sydney, NSW 2060, Australia

Allen & Unwin (New Zealand) Ltd in association with the
Port Nicholson Press Ltd, Compusales Building, 75 Ghuznee Street,
Wellington 1, New Zealand

First published in 1991

British Library Cataloguing in Publication Data

Parasitic helminths and zoonoses in Africa.
1. Man. Zoonoses
I. Macpherson, C. N. L. (Calum N. L.)   II. Craig, P. S. (Philip S.)
616.959

ISBN 0–04–445565–8

Library of Congress Cataloging-in-Publication Data

Zoonoses and human health: the African perspective/edited by
C. N. L. Macpherson, P. S. Craig.
    p.   cm.
Includes bibliographical references.
Includes index.
ISBN 0-04-445565-8 (hb: alk. paper)
1.   Helminthiasis—Africa. 2.   Zoonoses—Africa. I.   Macpherson,
C. N. L. (Calum N. L.)   II.   Craig, P. S. (Philip S.)
[DNLM: 1.   Helminthiasis—epidemiology—Africa. WC 800 Z87]
RA644.H38Z66 1991
616.9′62′0096—dc20
DNLM/DLC
for Library of Congress   90–12703
                                                              CIP

Typeset in 10 on 12 pt Times by Columns Design and Production
Services Ltd., Reading, UK and printed in Great Britain by Cambridge
University Press

Preface

Helminths include one of the most diverse and geographically widespread groups of parasites which infect humans and animals. About 100 species have been reported from humans, usually producing asymptomatic infection or mild symptoms. However, about 20 species are of public health importance causing severe or even fatal infections. In many parts of Africa parasitic helminths are responsible for enormous economic losses, hampering rural development programmes and reducing the pace of economic growth.

Many parasitic helminths are either zoonoses (diseases naturally transmitted between vertebrate animals and man) or have evolved from animal parasites. The modification of the environment through wars, famine and the ever expanding and increasingly mobile human population brings people into close contact with new environments and wildlife species which makes the study and control of zoonoses of special interest and complexity. In Africa, the transmission of helminth parasites is highly influenced by the ever changing social and cultural differences between diverse groups of peoples and their interaction with wild and domestic animals. It is not surprising, therefore, that approaches to the study and control of parasitic zoonoses require intersectoral cooperation between physicians, veterinarians, parasitologists, zoologists, demographers, anthropologists, engineers and economists to provide the breadth of knowledge and expertise required to develop our understanding of these diseases and to devise methods for their control.

This book provides a selective compilation of parasitic helminths, many of which are zoonoses which create important economic and public health problems in Africa. Methods used to elucidate parasite life cycles, transmission and epidemiology and the investigation of potential animal models, as well as attempts at their control, have been highlighted.

The subjects covered have all been influenced by the remarkable and original studies of a single individual, Professor George Stanley Nelson, who, during a period spanning more than 35 years, made a major impact on our knowledge of helminth diseases. The significance of many of Professor Nelson's contributions can be found in the various chapters and is clear from the Appreciation by Professor Kinoti. All the contributors to this book have been influenced by Professor Nelson and we have had the enjoyable task of editing this *Festschrift* in his honour. We feel that the contributions by these colleagues of George Nelson make this a fitting tribute to his inspiration, dedication and the intellectual stimulation he has given to the study of helminth diseases, an endowment that will be inherited by generations to come.

Calum N. L. Macpherson
Ifakara, Tanzania

Philip S. Craig
Liverpool, England

Professor George Nelson

PROFESSOR GEORGE S. NELSON
An appreciation

by

George Kinoti
Professor of Zoology, University of Nairobi, Nairobi

The occasion of the retirement of Professor George Nelson from the Walter Myers Chair of Parasitology at the Liverpool School of Tropical Medicine gives his many old students, colleagues and admirers in many parts of the world an opportunity to pay tribute to one who over a period of 40 years has made many important contributions to medical parasitology. It is a great honour for me, his first official PhD student I believe, to write this appreciation on their behalf.

Career

George Stanley Nelson was born in Cumbria, England on 27 November 1923. He studied medicine in the University of St Andrews in Scotland where he had a distinguished career, winning medals in anatomy, physiology, medicine and pathology. Graduating in 1948, he worked for a year as a house physician at the Dundee Royal Infirmary and then returned to his old university as a Junior Lecturer in Pathology. His interest in pathology took him to Kampala, Uganda in 1950 where he found the opportunities for learning pathology were rather limited and changed to parasitology. From 1950 to 1956 he worked for the Uganda Medical Service as Medical Officer in charge of the West Nile District. With characteristic energy, he not only looked after the health of about 400 000 people but also carried out research on leprosy, onchocerciasis and schistosomiasis. And he found the time to write a doctor of medicine thesis for which the University of St Andrews awarded him a gold medal in 1956.

Later that year Dr Nelson transferred to the Kenya Medical Service and joined the Division of Vector Borne Disease, which Dr P. C. C. Garnham and other British workers had made famous as a centre for research in parasitic diseases. Here George Nelson commenced the work on filariasis, trichinosis and other zoonoses which rapidly established him as one of the world's leading parasitologists.

His association with the London School of Hygiene and Tropical Medicine started in 1959 when he joined the postgraduate course for the Diploma in Applied Parasitology and Entomology (DAP&E). The School must have formed a very high opinion of him because three years later they appointed him Reader in Medical Parasitology. He took up the appointment in early 1963. Within the short period of three years the University of London appointed him Professor of Helminthology. During the same period his old University awarded him the degree of doctor of science. When that most agreeable professor John Buckley retired Professor Nelson took over headship of the Department of Helminthology. Building on the excellent foundations laid by his predecessors, Professor R. T. Leiper FRS and Professor J. J. C. Buckley and with the outstanding staff team he put together, he made his Department a major centre for studies in schistosomiasis, filariasis, trichinosis, onchocerciasis, hydatid disease and dracunculiasis.

In 1980 he surprised some people by moving from the London School to the Liverpool School of Tropical Medicine. But no one should have been surprised for the Liverpool School was home to him as he had studied there in 1952, taking the Diploma in Tropical Medicine and Hygiene and winning the Milne medal. His appointment as Walter Myers Professor of Parasitology gave him a wider scope and new challenges, which his restless mind and boundless enthusiasm required at that point in his career. Most notable was his contribution to the development of the Wolfson Units of Immunology and Molecular Genetics within the School and his development of the Hydatid Research Group with special collaboration in Kenya.

I believe that of the many honours and awards that George Nelson has received he values most highly the recognition accorded his work by the Royal Society of Tropical Medicine and Hygiene by the award of the Sir Rickard Christopher medal and the election as a Vice-President (1987–89) and as the President of the Society (1989–91).

The scientist

Professor Nelson is a true parasitologist, his interests recognizing no taxonomic boundaries. His researches include the major helminth infections of man – schistosomiasis, filariasis, onchocerciasis, trichinosis, hookworm and hydatid disease as well as leprosy, plague and trypanosomiasis. He is internationally respected as an authority on several parasitic infections and he has served on many WHO committees, including Expert Committees on onchocerciasis, filariasis, schistosomiasis, and zoonoses; the scientific working groups on schistosomiasis and filariasis; the Steering Committee on filariasis; the Scientific and Technical Advisory Committee on the Onchocerciasis Control Programme.

I will not attempt a review of the extensive scientific work done by

Professor Nelson alone or with research associates which has been published in over 180 papers. It is only possible to comment on some aspects of the work that are particularly relevant to this book. Early in his career he developed an interest in the transmission of infections from animals to man. He tackled the problem with vigour and courage. Surely Sir Philip Manson-Bahr was right when, after Dr Nelson's lecture 'Schistosome infections as zoonoses in Africa' to the Royal Society of Tropical Medicine and Hygiene, he commented, 'It is quite evident that on this notable occasion another Nelson has been peering through his microscope upon the battlefield of zoonosis and can be congratulated on the result'.[1] He and Forrester and Sander were the first to show that, contrary to the general belief, trichinosis is present in Africa south of the Sahara. His subsequent studies, each involving hundreds of animals, on the role of wild animals as reservoirs of helminth infections in Kenya are classic. The work which he started in Kenya and continued in London on trichinosis greatly clarified the epidemiology of this infection and led to the classification by Russian workers of *Trichinella spiralis* into four species – *T. nelsoni*, *T. nativa*, *T. spiralis* and *T. pseudospiralis*.[2]

Professor Nelson became interested in schistosomiasis while working in the West Nile, where probably the severest form of the intestinal disease in East Africa occurs. Here he carried out probably the most extensive survey ever made in Africa, examining over 10 000 people for *Schistosoma mansoni* and 2000 for *S. haematobium*. He claims that it was easy to obtain all these samples because all he had to do was get up early in the morning, shoot a hippopotamus on the Nile and trade chunks of its meat for stool specimens! Nearly thirty years later I carried out a much more limited survey using what is generally regarded as a more accurate technique (the Kato thick smear) and was impressed with the accuracy of his estimates of the prevalence of *S. mansoni* infection in the various localities. While at the London School, George Nelson and his colleagues developed the Winches Farm Field Station into one of the most comprehensive centres for schistosomiasis research which attracted high calibre staff for studies on immunopathology, chemotherapy and protective immunization. The important work based at the Winches includes the development and field testing of the first vaccine against schistosomiasis in cattle and the application of the technique for the control of *S. japonica* in the animal reservoirs of the disease in China. Work on heterologous immunity contributed new insights into the epidemiological importance of cross-immunity and led to the recent isolation in other laboratories of protective antigens which are common to parasites of man and animals. Studies with *S. haematobium* showed the baboon to be a good model for studies of cancer of the bladder. Chemotherapy studies contributed to the launching by Bayer and WHO of praziquantel, currently the 'magic bullet' against schistosomes and some other parasitic platyhelminths.

The filariasis unit which George Nelson established in London was also very successful. It attracted funds from the Medical Research Council, WHO and industry in support of studies on immunology and chemotherapy of filariasis and onchocerciasis.

Another infection of special interest to Professor Nelson is hydatid disease. While working in Kenya in the late 1950s and early 1960s he carried out a study which confirmed the identity of the parasite occurring in Kenya and showed that although the parasite occurred in wild carnivores (hunting dogs, jackal, hyena) the domestic dog was the main host of *Echinococcus granulosus* in the Turkana district. In more recent years he has been involved in a major effort to unravel the epidemiology of hydatid disease in Turkana, an enormous semi-desert district which has the doubtful distinction of having the highest prevalence of the disease in the world. Although the picture is not complete yet, the results have contributed new insights into the model of transmission from dogs to man and the role of wild animals as maintenance hosts. New immunological methods have been developed for diagnosis of the infection and for distinguishing *Echinococcus* eggs which are morphologically identical to other taeniids that occur widely in the Turkana environment and complicate epidemiological studies. The Turkana studies have also shown that chemotherapy can often replace surgery in the treatment of the infection.

In addition to making many original contributions to knowledge, George Nelson has contributed to science significantly in two other ways. First, his masterly reviews are mines of information and ideas for young scientists. Some of them, e.g. *Onchocerciasis, Pathology of filariasis,*[3] are still very valuable twenty years after they were written. Second, wherever he is, whether in his laboratory, at a meeting or visiting another country, George Nelson is, as an old student of his puts it, always 'bubbling with ideas'. In this way he stimulates research by others which might otherwise not be done.

Although as a naturalist George Nelson finds parasites fascinating, his researches are primarily motivated by the desire to control human parasitic disease. This is evident from his choice of research projects which aim at solving practical problems. When he has studied parasites of animals it has been either in search of a model for a human infection or in order to better understand the epidemiology of a medically important infection. He keeps in constant touch with endemic areas, carrying out field projects on schistosomiasis in Sudan, Egypt and Kenya; on filariasis in Tanzania, Trinidad, Egypt and India; on onchocerciasis in Sudan, Cameroon and Mexico; and on hydatid disease in Kenya. I think his concern with the practical problems of parasitic disease in poor countries is nicely brought about by remarks he made when opening a symposium of the British Society for Parasitology on parasites and molecular biology. He said, in part, 'Molecular biologists and immunologists working with parasites may have the satisfaction of winning a scientific race but, unless they are in close contact with the problems in the field, they will become frustrated and disillusioned if their elegant diagnostic tools and synthesized vaccines remain on the shelf whilst the parasites continue to take their toll in the countries which cannot afford or are not prepared to use the new tools'.[4]

The teacher

Professor Nelson is known internationally not only as a scientist but also as a teacher. He attracted large numbers of students to both the London School and the Liverpool School. Both in the classroom and at scientific meetings he is a most stimulating lecturer. He has the ability to make even the most boring subject come alive as many former students know from his lectures on cestodes. He has been an invited speaker on medical and veterinary parasitology in many parts of the world including Australia, Brazil, China, Europe, India, Japan, Trinidad, the USA and USSR, in addition to numerous meetings in Great Britain. And he is a regular lecturer on the course 'The biology of parasitism' held at Woods Hole in the USA.

A few years ago a young Kenyan protozoologist travelled thousands of miles to attend a major international conference on parasitology. When she returned I asked her what she had found most interesting at the conference and her answer was, 'Professor Nelson's lecture on worms'. What is the secret of his success? I think it is a remarkable combination of several elements: a wide knowledge of his subject, a fresh approach, concentration on the 'picture' rather than on detail, genuine enthusiasm for his subject, clarity of expression, a warm personality and a touch of dogmatism. Yes, dogmatism! Years ago he admitted that he could be deliberately dogmatic in order to 'entertain the intellect by arousing interest' and to 'challenge by stimulating others to individual thought based on aroused curiosity'.[1]

Through research training Professor Nelson has also made a most valuable contribution. He has supervised the research projects of numerous MSc students and over 20 PhD students from many countries. Several of his former students and research assistants are now professors in their own right. They in turn are producing other parasitologists who can be truly regarded as George Nelson's intellectual grandchildren. In addition he has done much to promote the teaching of medical parasitology in Thailand, Kenya and other countries.

The person

Prior to his entry into the scientific world George Nelson distinguished himself on the rugby field, playing for England schoolboys against Wales at Cardiff Arms Park. Whilst receiving his medical education at St Andrews he could not have avoided playing golf, which remains among his favourite hobbies, along with gardening, music and natural history – particularly ornithology. His interest in the latter was no doubt stimulated by spending his schooldays at Haversham Grammar School in the Lake District and later at St Andrews through the influence of D'Arcy Wentworth Thompson, who was at that time in his 63rd year as a professor. Whilst in Dundee, George Nelson met an attractive young nurse, Sheila, to whom he was married in the Nairobi Cathedral before taking up his appointment in Arua in western Uganda on the Zaire border. Sheila has been and still is a pillar of strength, supporting George in the remote areas in which they have lived and travelled. They have four children, three of whom were born in East Africa.

Though now officially retired, George is as active as ever in the service of parasitology and human health in the tropics. 'Retirement' to him means a new phase in his career. May he and Sheila have many more years of enjoyable work and usefulness.

Notes

1 *Trans. R. Soc. Trop. Med. Hyg.* **54,** 301–24 (1960).
2 See W. C. Campbell 'Trichinella in Africa and the *nelsoni* affair' Chapter 4 of this book.
3 Nelson, G. S. (1966). The pathology of filarial infections. *Helminthological Abstracts* **35,** 311–36.
   Nelson, G. S. (1970). Onchocerciasis. In: *Advances in parasitology,* Vol. 8, 172–224.
4 Nelson, G. S. (1986). Parasites and molecular biology: applications of new techniques. *Parasitology* **91,** 83–5.

Contents

List of contributors

Albert E. Bianco, Research Fellow, Department of Pure and Applied Biology, Imperial College of Science and Technology, Prince Consort Road, London SW7 2BB, UK

William C. Campbell, Dana Fellow, Department of Biology, College of Liberal Arts, Drew University, Madison, New Jersey, USA

Philip S. Craig, Lecturer in Parasitology, Department of Parasitology, Liverpool School of Tropical Medicine, Pembroke Place, Liverpool, L3 5QA, UK

Alan Fenwick, Project Leader, Schistosomiasis Control Unit, Gezeira Project, Sudan

John M. Goldsmid, Department of Pathology, Clinical School, Royal Hobart Hospital, 43 Collins Street, Hobart 7000, Tasmania, Australia

Leslie J. S. Harrison, Research Fellow, Centre for Tropical Veterinary Medicine, University of Edinburgh, Royal (Dick) School of Veterinary Studies, Easter Bush, Roslin, Midlothian EH25 9RG, UK

Robert A. Harrison, United States Naval Medical Research, Unit No. 3, FPO New York 09527, USA

Mansour F. Hussein, Department of Animal Production, College of Agriculture, King Saud University, PO Box 2460, Riyadh 11451, Saudi Arabia

Lotfi F. Khalil, Lecturer, C.A.B. International Institute for Parasitology, 395a Hatfield Road, St Albans AL4 0XU, UK

George K. Kinoti, Professor, Department of Zoology, University of Nairobi, Chiromo Campus, PO Box 30192, Nairobi, Kenya

Calum N. L. Macpherson, Head, Swiss Tropical Institute Field Laboratory, PO Box 53, Ifakara, Tanzania and Honorary Lecturer, Department of Parasitology, Liverpool School of Tropical Medicine, Pembroke Place, Liverpool L3 5QA, UK

Ralph Muller, Director, C.A.B. International Institute for Parasitology, 395a Hatfield Road, St Albans AL4 0XU, UK

John H. Ouma, Research Scientist, Division of Vector Borne Diseases, PO Box 20750, Nairobi, Kenya

Calvin W. Schwabe, Professor of Epidemiology, Department of Epidemiology and Preventive Medicine, University of California, Davis, California 95616, USA

Morley M. H. Sewell, Professor, Centre for Tropical Veterinary Medicine, University of Edinburgh, Royal (Dick) School of Veterinary Studies, Easter Bush, Roslin, Midlothian EH25 9RG, UK

Martin G. Taylor, Reader in Medical Helminthology, Department of Medical Parasitology, London School of Hygiene and Tropical Medicine, Keppel Street, London WC1E 7HT, UK

List of tables

1 Helminth zoonoses in African perspective

'About the only genuine sporting proposition that remains [in medicine] . . . is the war against . . . [those] ferocious fellow creatures, which stalk us in the bodies of rats, mice and all kinds of domestic animals'. Hans Zinnser 1935.

Introduction

Before the 'Western' mind's eye, the very notion of diseases in Africa projects a kaleidoscopic montage of the unfamiliar, the mysterious, the romantic. Practitioners of tropical medicine have not been totally immune to such perceptions. The extent to which these views reflect reality is related to the particular prominence in Africa of 'exotic' parasites, infections transmitted by arthropod and snail vectors, and zoonoses.

Mere mention of zoonoses to many physicians elicits a clouded vision of poorly differentiated diseases and obscure man–animal relationships, many of which seem beyond the influence of the medical establishment. Helminthic diseases reflect another atypical image of human medical problems, forcing attention upon bio-ecological considerations peripheral to medicine's main visual fields. Combine all three of these elements – Africa, zoonoses, helminths – and it is probably impossible to project a more polyopic or astigmatic view of medicine, one more demanding of the broadest range of intellectual complements to conventional approaches.

The scientific challenge

This special challenge to medicine reflects, first of all, the unusual biological complexity of helminths themselves. The most anatomically and physiologically advanced animals broadly adapted to causing disease in other species, parasitic nematodes, cestodes and trematodes undergo successive stages of development, often as quite dissimilar forms, including in their feeding and other habits. Most have free-living, as well as parasitic, stages, often within a succession of different environments. Different developmental stages may exhibit alternating sexual and asexual patterns of reproduction. As a result of all this, biologists with other than conventional medical training have played more creative, and broader, roles within parasitology in the past than within any other

1

branch of medicine (Schwabe 1981a).

Besides such complexities common to all parasitic helminths, zoonotic helminths present a whole additional dimension of biological intricacy. This general phenomenon of zoonoses defines one of the broadest and most important sectors for interaction between human and veterinary medicine. Zoonoses comprise, in total, about four-fifths of all described human infections[1] and provide socially important examples of the biological principle of the relative lack of host specificity among infectious agents. Most new infections of people described during the last 60 years have been zoonoses, the vast majority of them caused either by animal parasites or viruses. The card file at Beltsville (USA) of the Index-Catalog of Medical and Veterinary Zoology currently references well over 100 different parasitic animal species which have been reported to naturally infect both man and other vertebrate animals, at least occasionally. Most of these are helminths, some causing prevalent human infections which frequently are misdiagnosed.[2] Of this total, the World Health Organization (WHO) Expert Committee on Parasitic Zoonoses (1979) identified 17 nematodes (or groups of closely related nematodes), five cestodes (or groups) and 12 trematodes (or groups) as causing important human infections in which other vertebrate animal hosts play epidemiologically significant roles.

Living agents which cause zoonotic infections are shared in nature by people with many other species of domestic and wild vertebrates, and are often transmitted naturally from one to the other.[3] This may occur by all known routes for transmission of infectious agents, including via invertebrate animal vectors, which adds still another level of biological complication to some helminthic zoonoses. For many zoonoses, their reservoir hosts, the animals in which the parasite is maintained naturally (i.e. depends for its own survival as a species), are one or more species of domestic or wild vertebrate animals. In most zoonoses of this type human infection is a biological dead-end and person to person, or person to animal, transmission never takes place. In a few zoonoses, this reservoir relationship is reversed, the parasite's reservoir host being man, with frequent or occasional transmission to other vertebrate animal species, for example *Entamoeba histolytica* transmission to the dog. In a few other zoonotic infections, transmission may occur in both directions (see discussion in Nelson 1960), including the special case of human taeniasis where cyclical transmission between man and other vertebrates is obligatory.

Four principal types of zoonoses cycles exist depending upon the nature of the reservoir(s) required (WHO 1979, Schwabe 1984a). The biologically simplest type, direct zoonoses, can be perpetuated in nature by a single vertebrate reservoir species. Trichinosis (Chapter 4) is a helminthic example. Cyclo-zoonoses, on the other hand, require more than one vertebrate species for their maintenance cycles, with echinococcosis

(Chapter 2) an example.[4] In the meta-zoonoses, both vertebrate and invertebrate host species comprise the maintaining cycle, as in certain schistosomiases (Chapters 8 and 9) or filariases (Chapter 6). Finally sapro-zoonoses (see note 2) depend also upon some non-animal reservoir or developmental site, including soil, decayed organic matter or plants. The nematodal larva migrans are examples.

Among direct zoonoses (Schwabe 1984a) there are at least one common nematodiasis and one cestodiasis; among cyclo-zoonoses, two nematodiases and three cestodiases; among meta-zoonoses, 12 nematodiases, eight cestodiases and 16 trematodiases; among sapro-zoonoses, 11 nematodiases.[5] That is, helminthic zoonoses tend to have epidemiologically complicated cycles.

It is noteworthy that three of the seven human diseases singled out by WHO as in urgent need of special programme emphases are helminthic zoonoses prevalent on the African continent (and that one additional helminthiasis also important in Africa is among five zoonoses given highest programme priority by the Pan American Health Organization). Of the 35 zoonoses given special attention in 1979 by the WHO Expert Committee, eight were caused by nematodes, four by cestodes and seven by trematodes.

Some zoonotic infections are the cause of very important human diseases and major public health problems over large areas,[6] while others are economically important diseases of domestic animals, threatening man's food supply, and some may assume both roles. Particular zoonoses also reflect situations in which ecologically and aesthetically important wildlife species pose actual or potential hazards to people directly or through their livestock. Helminthic zoonoses of this type are discussed in Chapters 2, 5 and 8 (also with reference to non-human primates) and in Chapters 2, 4 and 10 (referring to a variety of other wild vertebrates). For these reasons and others, helminthic zoonoses as a group are not only of medical and veterinary importance, but of ecological importance too.

New zoonotic relationships continue to be recognized:

1 as people in ever larger numbers enter new geographical areas and ecologic regions to open up lands for cultivation, mining, logging or similar exploitation and to construct roads, pipelines, dams and the like;
2 as new animal species are disseminated widely among human populations as pets, zoo specimens, laboratory species or food animals;
3 with changes in uses of particular animals or animal products and
4 with changes in local food habits or in food technology.

The best illustration of the extent of biological complexity possible in parasitic zoonoses is provided by *Nanophyetus salmincola*, a small intestinal fluke found in the Pacific northwest of North America, which is associated with an often fatal bloody diarrhoea in dogs. Its observed association in dogs with their eating of raw salmon resulted in the classic disclosures, first by Donham, Simms and colleagues, that this fluke . . . whose first intermediate host is a freshwater snail, *Oxytrema silicula*: with metacercarial encystment in marine salmonid fish which spend three years at sea and then return to freshwater streams to spawn . . . is itself a vector of a rickettsia, *Neorickettsia helminthoeca*, the actual agent of the fatal disease in dogs popularly called 'salmon poisoning' (a summary of these findings may be found in Schwabe 1984a, p. 346). Possible zoonotic implications of this most complex of infectious cycles are suggested in the recent report by Eastburn *et al.* (1987) of human intestinal infections with this same *Nanophyetus salmincola* caused by consumption of raw salmon. The eating of raw fish in the form of sashimi is becoming a popular custom in several parts of the world.

After Hans Zinnser wrote his classic 'biography' of some prominent zoonoses in history in 1935, other students *par excellence* of microbial zoonotic infections like K. F. Meyer and Eugene Pavlovsky 'fleshed out' the theoretical bases and research methodologies of Zinnser's medical ecology of infections to define a broad biological, or qualitative, basis for epidemiology, one which embraced both its human and veterinary aspects, including concepts such as 'natural foci of infections' and 'webs of causation'.

In retrospect, we can appreciate that many of the diverse biological phenomena these and other microbiologists described and interpreted as microbes had already been observed among some parasitic infections, but were initially regarded by most physicians and veterinarians as 'biological curiosities'. Underestimation of the commonness of some of these phenomena simply reflected parasitology's relative historic isolation from the mainstreams of western medicine. Thus, while Meyer was to document the generality within infectious disease epidemiology of the phenomenon of inapparent infections, the earliest instances had been recognized by parasitologists. For example, *Trichinella spiralis* infection, even with heavy parasite burdens, was well known to occur without any clinical signs in domestic swine. And, as has been learned since from the work of George Nelson and his colleagues (see Chapter 4) in sub-Saharan Africa, inapparent trichinosis also exists there as a prevalent and almost totally silent infection not only among wild relatives of domestic swine, but in other carnivores and scavengers as well.

The related phenomenon of latency also was recognized early by helminthologists in equine strongylosis caused by *Strongylus vulgaris*, subsequently in several gastro-intestinal nematodiases of ruminants, but only much later as a fairly frequently encountered phenomenon in viral and other microbial infections.[7] Similarly, no better example can be provided of prolonged clinical incubation of an infection (a 'slow infection') than the long-recognized one of hydatid disease. Or, as George Nelson is fond of pointing out, the first indications of arthropod-borne diseases of vertebrates were provided by disclosures of the role of cyclops in guineaworm transmission (1869) and of evidence for filarial development in mosquitoes (1877); and so on.

With full development of this medical ecology perception of epidemiology – in a way the extension throughout all of infectious diseases epidemiology of the broad biological perspective inherent to parasitology – environmental factors once again assumed scientific respectability as possible direct or indirect determinants of infectious disease patterns, determinants which, along with specific agents, could be shown to also interact with other essential variables characteristic of the host itself, to possibly define specific webs of causality for each and every infection.[8] Concurrent with such advances, a few parasitologists like Theodore von Brand and Ernest Bueding,[9] observing how progress in diagnostic bacteriology and in understanding bacterial pathogenesis followed quickly upon fundamental investigations of the *in vitro* physiology of these simpler organisms, commenced the process of exposing among more complex animal parasites also some of the chemical and other bases of their physiologies *per se*, or in interaction with their hosts.

This new physiological and biochemical route for parasite study accelerated after World War II as Desmond Smyth and others began to succeed in the *in vitro* cultivation of higher animal parasites and these two research areas merged with the slowly accelerating progress of parasite immunology. Together these activities have ushered in what now is referred to as molecular parasitology. Prospects for such a more reductionistic parasitology are now well understood (Wood 1978, Warren & Purcell 1981).

Partially, one begins to fear, however, that these newer shifts in research directions, in which this author participated enthusiastically, may be taking place at the expense of appreciating the equally vital importance of continuing to carry out high quality field-based research on such biologically complex problems, especially of improving the scientific value of field research by better utilizing rapidly developing methodologies of quantitative analytical epidemiology. As George Nelson's career exemplifies, tropical medicine cannot afford to shift all of its support base to vaccine development, or to any other single line of research, no matter how promising it may be of rich rewards. Yet to be realized, in this whole field, is meaningful juxtaposition of these modern reductionistic and holistic strains of scholarship either in the purely parasitological, or the disease control spheres.[10]

The generally diverse characteristics of helminthic zoonoses and the multiple relationships represented among them, take on even more added dimensions of complexity and importance when considered within the, in many ways, unique environment of Africa. The variety of African climatic and topographical conditions – ranging from desert to tropical rain forest, from high mountains, prominent geological fissures and large lakes, to enormous grassy savannah, riverine flood plains and huge swamps (Phillips 1959), affords the widest imaginable range of ecological niches for both parasites and hosts. This is especially true considering that the African continent contains the world's largest and most diverse populations of wild mammals, an ancient and nearly symbiotic relationship between man and domestic cattle over vast areas of the continent[11] and the perpetuation virtually side by side of interspersed hunter-gatherer, nomadic or transhumant pastoral, settled agrarian and urban economies. As a result, social and cultural, as well as biological, complications can be visualized that seldom, if ever, have been experienced in medicine.

Unprecedentedly broad scientific approaches are required in response. Cross-professional and cross-disciplinary research approaches already have characterized past efforts directed toward understanding helminthic zoonoses in Africa to an unusual degree. George Nelson has been the major catalyst of such efforts. But this past record of co-operation is only a beginning compared to what may be required ultimately. Study of such complex interactions must henceforth involve not only the broadest spectrum of natural scientists, but also specialists in other diverse disciplines, such as social scientists and humanists who possess a vital 'insider's understanding of highly valued aspects of traditional cultures (values that need not necessarily be sacrificed totally in the future to either western or Marxist notions of human progress). Especially do such considerations need to be weighed more realistically against the increasingly recognized socially or psychologically negative accompaniments of uncritically received 'material progress'.[12]

The urgency of this situation is indicated by the fact that, while reductionistic studies (high technology science) can be expected ultimately to explain the actual mechanisms by which particular helminths (or other living agents) infect, develop and survive within, and produce disease in individual animals, the means by which their hosts, as well as vaccines, drugs and other products of technology, might thwart them. Reductionistic science runs the risk of setting premature research priorities, encouraging unrealistic optimism about wide applicability of simple 'cures', of confounding value-free objectivity in scientific approach with the value-sensitive purposes and applications of science, or otherwise of missing the forest for the trees. It becomes clearer that considerations such as these are overlooked only at great risk when contemplating transfers of scientific technology from one part of the world to another, or in anticipating the likely consequences of doing so. Failure to take social and cultural factors into sufficient account with respect to disease control can have far-reaching consequences.

More holistic research approaches, incorporating epidemiological, anthropological, economic and other social science (and humanities) perspectives – could be expected to yield not only important new hypotheses of predisposing, or other indirect causes of occurrences, or patterns of occurrence (components of the actual webs of causation) of particular diseases. They could also identify those causal variables which might most readily be manipulated to greatest advantage, or which otherwise might serve as essential keys to planning, execution and evaluation of socially meaningful programmes of disease control, within the overall context of appropriate development.

As attention necessarily turns to a greater extent in the future from purely scientific aspects of understanding complex diseases like helminthic zoonoses – to documenting more adequately the social implications of some such diseases, and of their control, in Africa – better understanding must be realized of the social and cultural milieus in which particular diseases occur than has been true in the past. Throughout Africa deeply rooted and highly valued patterns of life exist that are totally unfamiliar to most scientists, or economists, in countries from which most outside medical 'experts' still come to Africa.[13] Some such considerations are virtually unprecedented outside Africa, and unquestionably, others of importance remain to be recognized, or to be made known at all to the medical community. Here again, helminthic zoonoses under African circumstances afford exciting, socially significant challenges – for new ideas and new actions, for visualizing and practically implementing new intervention tactics. For, within the context of prospects for implementing optimal development strategies within the African continent, helminthic zoonoses are seen to exert negative social influences in a variety of ways, and their control will necessitate a high degree of community under-standing and cooperation[14] – and of inter-sectoral co-operation within government itself. To realize these will involve practical interactions between the public and bearers of professional and disciplinary skills. These requirements have seldom been demonstrated elsewhere and practical intersectoral co-operation in government still seems little more than occasional lip-service. Yet nowhere are cross-professional and interagency efforts more obviously essential than within the poorest countries of the Third World (Schwabe 1981b).

The social challenge

Another way to state this is that the whole panorama of human experience still may be observed as living systems on the African continent, evolutionary site of our species and its immediate predecessors, the 'naked apes'. Speaking numerous varieties of four major phyla[13] of human languages, contemporary African society today ranges from small numbers of hunter-gatherer Khoisan ('click') speaking Bushmen and their relatives to exploding populations of urban peoples living traditional as well as recently imported life styles in some of the world's largest and fastest-growing cities, like Lagos with its 9–11 million inhabitants. These urban populations speak not only almost the full range of African languages, but increasingly, major lingua francas, long-established Kiswahili and Arabic, more recently adopted English and French.

Large and diverse groups between these two extremes include transhumant and nomadic pastoralists, ranging from Niger-Kordofanian-speaking Fulani (numbering over six million) and even larger numbers of pastoral Bantu peoples; through almost all speakers of Nilo-Saharan languages. The latter include peoples such as Sudan's over two million Dinka and related East African herd-keepers like the Turkana and Maasai; plus other sizable Afro-Asiatic-speaking pastoralist populations like the some two million Somalis and even more numerous desert-inhabiting Berbers, Arabs and others.

Exceeding in total numbers these diverse pastoralists, with their most intimate man–domestic animal contacts, are many other sedentary agricultural peoples speaking a very wide variety of languages of the second, and fourth-mentioned, phyla. These include most of the Niger-Kordofanian-speaking peoples, such as the numerous Kikuyu of Kenya and Yoruba of Nigeria, as well as Afro-Asiatic Semitic language-speakers like the Ethiopian Amhara. Some of these peoples have utilized domestic work animals in support of plant agriculture and traditional economy, while others have not. These rural sedentary peoples include the majority of formally educated Africans (in the western sense) and are usually the peoples in power in existing African states. It becomes quite obvious that, out of this unusual diversity, a wider range of important choices may still be available to mankind in Africa than in most other parts of the world.

Not only does such cultural diversity, often within single African countries, make all communications, including health education unusually difficult, but it embraces an infinite variety of man–animal relationships, and other patterns of habits, which may increase, or decrease the risk of acquiring particular diseases. It also may mask various beliefs and practices which could facilitate or deter particular approaches to disease control. In the case of the numerous pastoral peoples, it makes the delivery of any kinds of governmental services, including health, unbelievably difficult. To further illustrate this social challenge to combat helminthic zoonoses in Africa, let us consider briefly two important man–animal relationships that exist there.

Dog and man in Africa

Despite increasingly restrictive Western views today of dogs' roles only as man's closest companion, we can observe within the African continent much more complex relationships of dog to man, most of which probably evolved first in prehistoric times but which are being perpetuated in great variety today among Africa's diverse peoples. According to Simoons (1961), the dog's most widespread roles in Africa are as a scavenger of refuse and guardian of people from predators and intruders. Herding and hunting uses occur among some African peoples, but the most strikingly different canine functions perpetuated there, from a present-day western perspective, would be those the dog and its relatives perform in connection with death.[16] In some places this function seems gradually to have evolved also into beliefs that the dog is able to help ward off (prevent) death, or undesirable consequences of death – or otherwise to assure health.[17]

On the African continent this spectrum of death (funerary) to protecting and healing roles for dogs and their relatives must have evolved quite early as evidenced by the very prominent funerary roles of the Egyptian canine god Anubis who acted as guide for the dead, thence evolving protective, even healing roles as evidenced during Ptolemaic times by his Greek-adopted equivalent Hermanubis. It is clear, however, from their very earliest funerary literature that ancient Egyptians expected (and feared) dismemberment after death, with the scattering of their bones by canine and other scavengers.[18] And, with respect to their chief or pharaoh, they attempted to prevent this (even by building stone burial pyramids). Hence Anubis' original roles and the need to mollify him.

Similar dog–man beliefs persist in parts of Africa today. As early as 1911 Beech recorded practices among the Pokot (Suk), Nilotic neighbours of the Turkana in northwestern Kenya, which are strikingly similar to some of those just mentioned. The Pokot ordinarily lay their dead out on the surface of the ground in order that scavenging carnivores (dog, jackal, hyena) can eat them. The only exception Beech noted was in the case of cattle-rich Pokot elders whose bodies were buried instead (protected from scavengers?) under mounds of cattle dung. Subsequently, Huntingford (1953a) reported similar practices, and associated beliefs, among Kenya's Nilotic Nandi. For all but especially aged individuals (or deceased infants) among them, referred to as 'toothless ones', Nandi believe that, if a hyena (or other carnivore?) does not eat the deceased's body, his *mukuleldo* (the personality aspect of his soul) cannot reach spirit-land and is destined to roam forever as a ghost on earth. Evolution of similar Turkana practices with respect to disposal of their dead may be assumed to have had related origins.

It becomes apparent from contemplation of such beliefs and practices with respect to zoonoses transmission, that a human dead-end to transmission of certain zoonoses in most societies elsewhere results from cultural (i.e. burial, cremation) practices rather than from biological causes. This would obviously be so for hydatid disease, where a biological cycle of transmission incorporating man can take place where canid hosts of *Echinococcus granulosus* feed on human dead as in Kenya's

Turkanaland. It is against a background of such prevalent man–dog relationships in Africa that the unusual hyperendemic occurrence of echinococcosis among the northernmost of Kenya's Nilotic Turkana (Chapter 2), as contrasted to its much lower endemicity among even the southern Turkana, or other closely related Nilotes, like the Maasai, must be considered. The possibility also exists (Chapter 4) that uncommonly reported human *Trichinella spiralis* in sub-Saharan Africa also could be transmitted to scavenger species by this route.

At the same time, however, we see a close companion role for dogs among diverse African peoples today. Some of these relationships apparently derived from co-operation in hunting, a role dogs continue to fulfill among certain ethnic groups. Huntingford (1953b) recorded that the Nilotic Kuku of southern Sudan especially honour dogs and treat them as tribal benefactors because dogs are believed to have explained childbirth to them, as well as to have taught them to use fire and herbs in cooking, and to grind grains. Among the Koma of southern Ethiopia reddish-coloured dogs are venerated and are unusually well cared for in connection with yearly extensions of their ruler's reign marked by dog sacrifice at a time of the New Moon (Simoons 1961).

Among the Turkana, this companion function assumes a near-healing (i.e. nursing) dimension among women, in that the dog owned by each Turkana woman not only is depended upon to clean her babies after they defaecate or vomit, but also to clean up her own menses. Whether these diverse death–nursing roles were somehow connected originally, we can only guess. In any event, such very close relations as these can result not only in heavy contamination of pastoralists' huts with *E. granulosus* eggs (Chapter 2), but unusually favourable opportunities for exposures of infants to hydatid infection.[19]

Such varied cultural roles for the dog may exist side by side with others in which that species enjoys equally high regard as a food animal.[20] According to Simoons (1961), the main African centre for dog meat eating embraces the Congo Basin and the rain forests and adjacent savannahs of West Africa. Peoples such as the Talensi of Ghana and Poto of Zaire are said to prefer eating dog meat above all other flesh. Some such peoples castrate dogs intended for food, or otherwise fatten them deliberately or, in various ways, tenderize their flesh. The numerous Azande of southern Sudan, Zaire and the Central African Republic are an example of a dog-eating people within this African centre for dog meat-eating who also regard the dog very highly as a companion and see no inconsistency in its important dual roles.

Elsewhere in Africa the dog may be consumed locally only by particular ethnic groups. For example special regard for dog flesh has been observed among the Mittu of southern Sudan, while the Gogo and Rangi of Tanzania are said to eat it only under conditions of dire need. In some other instances described by Simoons, dog sacrifice is an accompaniment of its consumption, or sacrifice may be associated with other beliefs, as with respect to the iron-working occupation. Hambley (1935) has described the latter among the Nigerian Yoruba where twice a year a dog is sacrificed to the god Ogun, patron of blacksmiths and god of war. Its head then is hung conspicuously in the blacksmith's shop.

Other recorded reasons for dog sacrifice are to cure or keep away diseases. Beyond that, Simoons indicates that flesh of a sacrificed dog is thought by some African peoples to have medicinal or spiritual value, while other peoples like the Yaka of Zaire fear that eating dog meat actually will cause disease. Kenya's Turkana, with their unusually high rate of hydatid infection (Chapter 2), generally deny eating dog or related carnivores, but there are some indications to the contrary, at least with respect to jackals. Serious consideration in all such unusual situations of hyperendemicity must be given also to knowledge that, in many societies, folk medicines sometimes may include 'vile' substances such as faeces.[21]

Similarly, other African ethnic groups observe strong general taboos against dog meat eating. Different bases for these are described. Early western travellers reported that southern Sudanese Dinka and Bongo would prefer to die of hunger than to eat dog meat. Simoons cites a number of other specific instances, including ones within the West and Central African dog-eating centre, for different kinds of local dog meat taboos.

The breakdown of traditional food taboos, sometimes as the result of rapid and profound cultural disruptions, as has occurred amongst Kenya's Kikuyu, was illustrated well in the totally unexpected disclosure by George Nelson and his associates of the presence, since reported elsewhere in sub-Saharan Africa by others, of silently endemic trichinosis (Forrester *et al.* 1961). In that first instance of human infection (Chapter 4), beside the very serious human illnesses which resulted from non-observance by a group of young Kikuyu men of their tribal taboos against eating wild pigs, the dog companion participating with them in this tragic adventure also acquired the parasite.

Cattle and man in Africa

No other vertebrate species, however, not even the dog, has been as intimately associated with people over broad geographical areas and extensive periods of time as have cattle (see von Lengerken & von Lengerken 1955, Conrad 1957, Epstein 1971, Schwabe 1978b). Nowhere has this been more true than in Africa, where in prehistoric times hunting magic associated with wild ancestors of one of the largest, strongest, bravest and most libidinous animals man encountered there prompted his domestication of cattle, precipitating a peculiarly embracing 'cattle-culture' which still survives very broadly throughout that continent today, among unrelated language groups, and which is based upon all-encompassing covenants between man and cattle. The ox's eventual yoking to the digging stick in the lower Nile Valley provided a minority of early Africans with the first harnessable source of power greater than their own muscles. That, in turn, provided the key to local grain production beyond subsistence needs, hence sufficient leisure for a talented and powerful few to assure emergence of the other requisites for an ancient civilization (Lobban 1989).

Today we find as still vital systems for millions of diverse Africans almost every aspect of the whole chronology of man–cattle relationship since the earliest of these developments. Among East Africa's Nilotes, for example, some of people's deepest historic perceptions of man–cattle interdependence still are maintained. Thus as recorded consistently by diverse observers: the Suk (Pokot), whose name for his tribe is *pipatic*, literally means 'cattle people . . . lives for his cattle, and everything is done to make them an object of reverence' (Beech 1911); 'the Nandi view of nature is largely a matter of how, in its various aspects, it is friendly or hostile to cattle' (Huntingford 1953a); among the Dinka, 'cattle are the medium of reconciliation between man confronting God and ancestral spirits [and they] often apply the word *aciek*, "creator", to cattle' (Deng 1971); in fact these people 'explicitly conceive their lives and the lives of cattle . . . on the same model. Men imitate cattle . . .; the [uncastrated] bull, the centre and sire of the herd, is associated with the . . . senior man of the camp; . . . cattle are integrally part of human social life . . .; men are thought of as bulls, begetters and fighters . . . . The bull represents virility for the Dinka' (Lienhardt 1961); 'it is no exaggeration to say that the [Dinka] youth attaches himself so strongly to this animal that the process called by psychologists "identification" takes place' (Seligman 1959); 'start on whatever subject I would, and approach it from whatever angle, we would soon be speaking of cows and oxen, heifers and steers; *cherchez la vache* is the best advice that can be given to those who desire to understand Nuer behaviour' (Evans-Pritchard 1940).

For such numerous peoples, cattle are not only their wealth, but provide the basis for all social interactions and are the focus for their entire culture, the central feature of their cosmology (Schwabe 1987). In West Africa, too, Lott and Hart (1977) have described how the principal traditional education of boys among the unrelated Fulani is to assume their roles as 'bulls' with respect to cattle, as well as within their own communities.

How the calendars of cattle, ecology and human activities are inextricably linked among such peoples has been sketched by a Dinka former Director of Veterinary Services for southern Sudan:[22]

'There are four seasons. *Keer* (March–May) marks the start of the rainy season. Cattle leave the *toich* (grazing areas in the Sudd swamp) for high ground and permanent settlement areas. Cultivation of sorghum and other crops begins. *Ruer* (June–August) is when rains are heaviest. Crops begin to be harvested and adolescent boys are initiated into manhood. People eat more in order to be physically fit for the difficult dry season ahead. In *Rur* (September–November) harvesting is completed, marriages are settled and cattle sacrifices made to the 'gods' (*jok*) to secure their blessing for the hard-to-manage dry season, especially to avert cattle diseases while people and cattle are in the *toich* seeking grazing and water, to have safe crossings of rivers without crocodiles attacking cattle, etc. During *Mei* (December–February) the people drink much milk (if the *toich* provides good grazing) and eat a lot of fish taken from the shallow flooded Sudd. Marriages are planned as young men and unmarried girls interact actively and future couples identify themselves. Thus, the most important seasons in the lives of Dinka and their cattle are *Rur* and *Mei*, when all the most vital events for both take place.'

Buxton (1987) has described a similarly interactive calendar for the Nilotic Atuot.

It should come as no surprise, therefore, that diseases of cattle have influenced Africa profoundly since at least the times recorded in Exodus. No better documentation of this exists than for the disastrous ecological, economic and political consequences of African peoples of the great pandemic of rinderpest there in the late 19th century (Brannagan & Hammond 1965). Typical of its most immediate consequences were those recorded by a contemporary observer among Kenya's then partly sedentary, partly pastoral Kamba along the Tana river: 'only a herd of about 20 [cattle] had survived out of what we estimated, from a survey of the described carcasses, at an original strength of 40,000 head' (Simon 1963). Indeed, this single cattle disease 'broke the economic backbone of many of the most prosperous and advanced communities [and] . . . initiated the breakdown of a long-established ecological balance . . .' (Kjekshus 1977). Politically, cattle disease hastened European conquest, for as put by Lord Lugard (1893) at the time 'powerful . . . pastoral tribes [had] their pride . . . humbled and our progress [in conquest] was facilitated by this awful visitation [of rinderpest]. The advent of the white man had else not been so peaceful'.

Among sedentary peoples who employed oxen to till their fields, and bear their loads, the total effects of cattle disease were equally disastrous, causing Ethiopia's emperor Menelik to lament in utter despair at the time: 'Oh! how my country has fallen in ruins! My people are finished!' (Pankhurst 1966).

Ecologically, rinderpest's devastation among people, cattle and wildlife alike also was such that the extensive man–cattle–savannah/cultivation ecosystem became physically separated, and remained so for many years, from the game animal–tsetse fly–savannah/woodland ecosystem over extensive areas of Africa (Ford 1971). All man–cattle–wildlife zoonoses were significantly affected.[23]

The effects of most helminthic zoonoses prevalent in Africa have been far less dramatic than rinderpest's and in some of the most dramatic instances of a helminthic disease's effects, as in onchocerciasis, animal reservoirs of infection are far less important than that provided by man himself. Yet it was considerations of great importance, such as these, that caused Julian Huxley to note sagely as early as 1931 that 'to be a good veterinary officer . . . [in Africa] you must be a first-class biologist and you must be a knowledgeable and sympathetic anthropologist as well'. This advice is no less true for physicians and zoologists, especially in their considerations of possibilities for control of such biologically and culturally complex phenomena as helminthic zoonoses in African settings.

Despite such admonitions, superficial comparisons by outsiders between cattle numbers and current cattle productivity in Africa, in relation to those of other areas such as Europe and North America, frequently prompt beliefs that this single 'untapped' cattle resource offers a simple key to rapid economic development over large areas of Africa. Africa in 1977 had 147 million cattle and 349 million people, as compared to 129 million and 470 million, respectively, for Europe and 134 million and 233 million, respectively, for North America. And it is true that, despite these large numbers of animals, the annual offtake (for meat) from Africa's cattle population is estimated (1981) as only five million metric tons, versus 12.9 million for Europe and 15.1 million for North America. Similarly, Africa's cattle herd yields only an estimated 16.4 million metric tons of milk annually, compared to 193.7 million for Europe and 64.9 million for North America. On the other hand, the realities of close cattle–man relationships have been consistently disregarded or underestimated in almost all development efforts which contemplate economic exploitation of Africa's cattle.

As a result, most cattle development projects for beef production attempted in Africa since that continent's independence have suffered from insufficient attention to Huxley's admonition and have been relative failures in terms of western standards of cattle utility.[24] Nevertheless, surprisingly few studies have endeavoured to identify specific reasons for these many 'failures' and similar 'development' efforts are continually reinitiated. In one rare attempt to assess the probable outcome of such an effort, Doran and associates accompanied a United Nations Development Programme (UNDP) cattle production project undertaken in Swaziland in 1973 by a study of past marketing experiences with Swazi cattle. That new UNDP project was based upon economists' assumptions that as the 'value' (for outside sale as beef) of Swazi cattle was raised through the programme's intended technical efforts, and if Swazis were offered a higher price per animal, they would respond by selling more cattle (i.e.

increase their offtake rate) to earn more money (i.e. there would be a high correlation between price offered and offtake rate). Doran showed this latter had, in fact, been true, but the correlation was just the opposite of that the economists anticipated from experiences in their own countries. For, when the price offered for cattle had gone up in the past, Swazis actually had sold fewer, rather than more, cattle. The explanation, which might have been anticipated by any close observer of the African cattle-culture, was that Swazis had only limited, fixed, non-traditional needs for money (for such purposes as education of sons) and that the fewer cattle (their real wealth) they needed to part with to realize that fixed money need, the better. Facts such as these, and the traditional values that underlie them, have equal pertinence when it comes to considering health education, or other interventions, with respect to control of cattle–man zoonoses (or any other aspects of development which impinge upon cattle).

Such intimate man–animal relationships must assume their obvious importance in even considering any cattle–man zoonosis in such a setting, especially in any situations where interventions are contemplated which might affect cattle or man–cattle interactions. For not only do most African pastoralists deeply cherish their traditional style of life, but in many ecologically marginal areas they inhabit, alternative uses of land to support such sizable human populations are not apparent (although they sometimes are urged by outsiders out of inexcusable ignorance).

Even our still very sketchy knowledge of such relationships, and their very poor communication to persons involved in health or other aspects of development – plus the effects trypanosomiases and other hema-protozoal diseases commonly are believed to exert upon development prospects throughout large areas of Africa – caused Julian Huxley to note, again back in 1931, that 'the prosperity and indeed habitability of enormous areas [in Africa] hangs upon success or failure in research and research along the broadest biological and medical lines'. As has been suggested, nowhere does this remain truer today than in instituting the degree of medical–veterinary–zoological cooperation required of efforts to combat zoonoses.

Future medical–veterinary–zoological co-operation

In contemplating either research progress or control efforts against zoonoses it is obvious that close medical–veterinary cooperation is essential over broad areas. When parasitic (or vector-borne) zoonoses, or those involving wildlife, are taken into account, such co-operation often must extend to embrace participation of other zoologists who may be trained in either medical tradition. Up to now this cross-professional co-operation has been much more satisfactorily realized in research than in planning and execution of control efforts. Various reasons for this latter failure to co-operate sufficiently have been identified (WHO 1975, Schwabe 1984a).

Obvious advantages of intersectoral co-operation in government – economic, scientific, logistic – have been clearly stated, especially for resource-poor countries, yet when, for example, a planning group was appointed to indicate expeditious routes for implementing WHO's goal of primary health care in the largely cattle-culture southern Sudan (Lolik et al. 1976), it stressed the need for intersectoral co-operation between health and educational authorities (as well as some other government sectors), but inexplicably failed to suggest this with the veterinary services, nor even to mention the centrality of all things involving cattle to most of the southern Sudanese population.

On the other hand, subsequent efforts to actually initiate the WHO/UNICEF Expanded Programme of Immunization (EPI) in southern Sudan have indicated the key role there at the local level, in extending any possibility of EPI to millions of pastoral peoples, of merging these human health efforts (especially their cold-chain requirements under extremely difficult circumstances) with already long-established and parallel 'grass-roots' veterinary immunization efforts which actually have extended mass immunization (of cattle) to the transhumant camp level (C. L. Schwabe et al. in preparation). Efforts like those go far beyond much more obvious needs for medical–veterinary co-operation vis-à-vis zoonoses control per se and reflect the practical implications of prior observations by anthropologist Jean Buxton (1973: 318) for the cattle-culture Mandari of southern Sudan, that it is 'animals rather than humans who benefit most [in pastoralist Africa] from any scientific medical treatment. Herders who have themselves never visited a government dressing station, still less a hospital bring their cattle for inoculation'.

In any such specific effort today we should note, too, that current international co-operation in Africa through instruments of the Organization of African Unity (OAU) had its antecedents in colonial times in the programmes of the Commission for Technical Cooperation in Africa South of the Sahara (OCTA). It is salutary in considering zoonoses in Africa within the context of these especially intimate man–animal relationships and the needs for meaningful intersectoral co-operation to note that OCTA's very first president was P. J. Du Toit, former Director of South Africa's Onderstepoort Laboratory for Veterinary Research. His appointment as head of that parent organization then reflected the fact that Pan-African co-operation in development began with formation of an Inter-African Bureau for Epizootic Diseases and a second Bureau for Tsetse and Trypanosomiasis. The priorities for Africa even then were clearly evident.[25]

With reference both to governmental outreach and to community co-operation, it is equally important to note that virtually all African peoples have various kinds of traditional healers – most of whom do not distinguish in practice between human and veterinary medicine – including persons who employ different herbs and other 'medicines', or manual healing skills (like wound and abscess surgery, dentistry, orthopaedics, cautery and obstetrics) or 'powers' to exorcize believed causes of disease or perform related psychotherapy.

There has yet been little systematic exploration of what such healers actually do and do not know, but there is wider appreciation that they often enjoy considerable popular reputations. Some, but not all, of these healers are indigenous priests or persons who also fulfil religious or quasi-religious roles (Schwabe 1978b). Among the Maasai, for example, one class of healers has been the *ol obani* (Merker 1904) and, among the Turkana, the *imuron*. The Dinka recognize several types of healers (Schwabe & Kuojok 1981), a herbalist (*ran wal*), an exorcist (*tiet*) and a practical manual healer (*atet*, which simply means 'specialist'). *Atet* perform wound, abscess, horn, castration and venesection surgery, bonesetting, and obstetrics (the latter, including embryotomy, plus castration and horn surgery, on animals only).

Some traditional African healers have been shown to possess considerable amounts of reasonably correct anatomical and physiological knowledge and ability to identify and differentiate various diseases. For the Dinka *atet*, for example, cattle diseases commonly recognized include anthrax, contagious bovine pleuropneumonia, haemorrhage septicaemia, blackquarter, trypanosomiasis, rinderpest, tuberculosis, foot-and-mouth disease, infectious conjunctivitis, fascioliasis, hydatid cysts, tapeworms, scabies, mange, calfhood diarrhoeal syndrome, and mastitis.

For some diseases, casual theories have also evolved locally, some of which may reflect important epidemiological observations, and are worthy, therefore, of serious consideration. As an example, the Maasai call malignant catarrh of cattle the 'wildebeest disease' because they have associated risk of cattle acquiring it with their coming into contact with wildebeest newborn and placentas during that prevalent wild ruminant's annual calving season. Subsequent veterinary investigations have confirmed that wildebeests are, indeed, the reservoir host for malignant catarrh virus in Kenya. On the other hand, while Turkana recognized hydatid cysts in their livestock before outsiders came there to study that disease, they did not associate these with the prevalent disease of the big belly in themselves, which some Turkana patients indicated to the author in 1961 they believed to be a curse placed upon them by the warlike Murle of southern Sudan.

Merker (1904) described the quite remarkable early Maasai practice of immunizing cattle against contagious bovine pleuropneumonia (said to have been originated by an *ol obani* named Mbatyan) by rubbing a piece of the lung of an animal newly dead from that disease into cuts made on the noses of unaffected cattle. According to Merker, most cattle so treated survived the resultant local infections and became resistant to subsequent natural challenges. The same folk prophylaxis has been referred to by veterinarians working in West Africa.

Among the Bantu-speaking South Kitosh in Kenya, Wagner (1949) noted that it was believed by them that if a *omuloli wemunda* (diviner of internal organs) found small pointed swellings (*tijisala*) on the intestinal wall of cattle (*Oesophagostomum* nematode larvae?), this was a sign that all cattle of the owner would soon develop a disease called *lukata* (which Wagner did not identify further). But he did record another Kitosh belief that if the small intestine of a Kitosh cow contained a large number of 'red worms' (blood-sucking *Haemonchus contortus* or *Ostertagia ostertagi* normally seen in the abomasum rather than intestine?), a serious epidemic was imminent, and if these were seen in several sacrificial cattle in succession, the Kitosh would break down the village and move their cattle quickly to a safer place. Dinka healers also described to the author the presence of such red worms in the abomasum of their cattle that 'bleed' when they are cut.

Trichostrongylus spp., also common nematode parasites of ruminants, may infect people (see Chapter 5). Additional human zoonotic helminthiases known to be acquired from cattle reservoirs in Africa include fascioliasis (which, as mentioned, Dinka recognize in cattle), hydatid disease (Chapter 2) and taeniasis (Chapter 3). The situation *vis-à-vis* any epidemiological role for African cattle in schistosomiasis is unclear (Chapters 8 and 9).

The value of ascertaining the beliefs and practices concerning diseases of local traditional healers and other local observers is obvious. Not only have many efficacious drugs been discovered elsewhere in that way, but also valuable epidemiological information. Beyond such needs to know, are possibilities in some places and circumstances that ways can be envisaged for different classes of traditional healers (or of individuals among them) to be offered various elements of training and be enlisted as co-operators with 'official medicine' (rather than antagonistic competitors) not only in selected phases of primary health care, but also at the truly grass-roots level of disease surveillance (Schwabe 1980, Schwabe & Kuojok 1981).

Conclusions

In summary, helminthic zoonoses in Africa represent a range of medical, economic and ecological problems of enormous biological and social complexity. Meaningful progress in research presents an array of exciting scientific challenges demanding participation of an unusual spectrum of natural scientists – as well as of social scientists and other scholars who too rarely participate fully in such practical activities. Control efforts present additional social challenges of realizing community understanding and participation and, especially, the degree of intersectoral co-operation in government essential for accomplishing even the technical aspects of effective zoonoses control.

Notes

1 Actual numbers of zoonoses vary in different tallies (Hubbert *et al.* 1975, Steele 1981, Schnurrenberger & Hubbert 1981, Schwabe 1984a, Acha & Szyfries 1987) depending on whether or not one lumps, or splits into different zoonoses, infections caused by agents closely related taxonomically. Useful book-length works on parasitic zoonoses specifically include Garnham (1971) and Soulsby (1974).

2 For example, zoonotic possibilities for human hookworm infections (Chapter 5) reflect two quite different circumstances. One is the straightforward one of cutaneous larva migrans, partial development with lesions in the human skin by *Ancylostoma* spp. (and more rarely other hookworm species) which normally mature in the intestines of dogs or other vertebrate animals. But in Africa especially, a second problem is posed by misdiagnoses as human intestinal hookworm diseases of zoonotic infections with other strongylid nematodes which have wild and domestic mammalian reservoirs. This represents a quite different type of medical problem accounted for by a general unfamiliarity of medical technicians and physicians with egg (and larval) morphology upon stool examination (or even with the existence!) of common helminths of animals like *Trichostrongylus colubriformis* (and other species in this genus) which usually are parasites of ruminants, and *Oesophagostomum* spp. and *Ternidens deminutus*, normally parasites of other primates. (Less excusable are misdiagnoses as hookworm disease of human infections with *Strongyloides fuelleborni* of primates). In Chapter 5, these aspects of hookworm disease illustrate some of the most immediate problems posed by the relative inattention to, and confused perceptions within human medicine of, zoonotic possibilities.

3 Some difference in numbers of infections classified as zoonoses depend, too, upon just how zoonoses are defined, for example whether systemic mycoses shared in nature by man and other vertebrates, but probably not directly transmitted often or ever from any infected host to another, are zoonoses (for definitions, classifications and discussion, see Schwabe (1984a: Chapter 10)).

4 Man may be required as one of these maintaining vertebrates, as indicated, for bovine and porcine taeniasis.

5 During many years of zoonoses study, the author has gained the impression that British physicians were more reluctant historically than continental or American physicians to recognize the possibilities of zoonotic relationships for newly disclosed agents of human disease. Perhaps this is but one reflection of a lesser British than continental or American interest in, or recognition of, the closeness of human and veterinary medicine and the importance to human health of the relationships represented. If these perceptions are true, George Nelson has proven one of several notable exceptions, John Hunter and Edward Jenner being predecessors that leap immediately to mind.

6 The WHO Expert Committee on Parasitic Zoonoses (1979) lists as nematodal diseases in this major public health category (and in which non-human vertebrates' roles are important) trichinosis and dracunculiasis; as cestodal diseases diphyllobothriasis, echinococcosis and the taeniasis; as trematodal diseases clonorchiasis, fasciolopsiasis, paragonimiasis and schistosomiasis japonica.

7 A number of very interesting 'immune' phenomena have been observed, too, in helminthiases which yet lack disclosed counterparts in most other types of infections. Examples include decreases in an individual parasite's size, or in its rate, or extent of

development (in addition to diminutions in parasite numbers or the parasite population's rate of multiplication) in resistant host animals.

8 Harvard's John E. Gordon, with others, showed the frequent inadequacy of simplistic perceptions of the 'necessary and sufficient' roles of specific agents in disease causality by developing an orderly body of theory encompassing the full ranges of possible interactions among agent, host and environment variables in infectious disease epidemiology. He then took the critically important further step which assured epidemiology's emergence as the basic science of population medicine of extending this 'new' more holistic perspective on causes of disease to the investigation and understanding also of observable patterns of occurrence of non-infectious diseases (with instruction of a post-World War II generation of epidemiologists, both physicians and veterinarians, in this research approach to an understanding of disease causality).

9 See as the pioneering compilation, Von Brand (1952).

10 Among cestode parasites viewed from a broad combined reductionistic–holistic perspective by Schwabe and Kilejian in 1968 were several which cause zoonotic infections. For some efforts to incorporate aspects of a modern epidemiological approach into zoonotic helminth control, see Burridge and Schwabe 1977, Schwabe 1979a.

11 First described as the 'cattle complex' by Melville Herskovits (1926: 650): 'The cattle complex of East Africa [broadly defined geographically] includes a number of definite traits . . . . In all of East Africa, cattle constitute wealth. A man may have iron, produce from his fields, sheep and goats, implements of all kinds, but unless he has cattle, his wealth amounts to nothing'.

12 Need for such a broad future juxtaposition of perspectives and knowledge (e.g. *vis-à-vis* cattle and man) has been elaborated elsewhere (Schwabe 1978a). For more general discussion of cross-professional and cross-disciplinary initiatives in the interest of stimulating discoveries of new knowledge, see Schwabe (1984b). At the time early works on development in Africa like Phillips' (1959) were written, there was no inclination for authorities to seek or be guided by the 'insider' views of indigenous Africans, such as pioneering African anthropologist Jomo Kenyatta already had provided other scholars on his own Kikuyu culture (*Facing Mount Kenya, the tribal life of the Gikuyu*) as early as 1938.

13 Few, if any, persons have possessed the variety and depth of background knowledge and experience to meaningfully conceptualize anything approaching the full range of problems and interactions understanding, and control, of helminthic zoonoses in Africa will entail. But pre-eminent in this effort, at almost all levels of its pursuit, has been the unusually prolific scientist to whom this volume is dedicated. Even to scan the range of his publications on a variety of helminthic diseases: effects of economic development on disease risk; field epidemiology, including medical geography, natural host ranges and human behaviour; parasite development and host–parasite relations, including *in vitro* cultivation, experimental infections in new domestic, wild and laboratory hosts, mechanisms of pathogenesis, pathology, nutrition and parasitism, homologous and heterologous immune mechanisms; parasite taxonomy and morphology, including parasite hybridization and ultrastructure; isoenzyme and immunological techniques to differentiate similar parasites and applications of other molecular level technology; vector biology and control; immunization against parasites; experimental therapy – none of it very far removed from clinical medicine (diagnosis and pathology, as well as treatment), or its teaching – indeed runs the practical gamut of medicine.

14 The socio-developmental classification of F. W. Riggs (1973) is especially useful in envisaging some of the complications inherent in undertaking societal modifications from a less to a more differentiated mode. For inclusion of animals' roles in Riggs' classification, see Table 1.3 and relevant discussions in Schwabe (1984a). In the author's view, more realistic consideration of actual development alternatives areas by area, especially for present-day pastoralists, must occur in Africa. This will necessitate

thoughtful challenges to widely held western social Darwinian assumptions that pastoralism is inherently better than hunting-gathering, sedentary agriculture better than pastoralism and industrialization better than sedentary agriculture (i.e. the objective of development is 'to be perfect as we are perfect'). Thoughtful people, like Francis Mading Deng (especially see Deng 1971, 1985) are attempting to define optimal local balances between 'tradition' and 'modernization', i.e. appropriate development, in Africa in terms broader and more satisfactory than the purely material bases of virtually all past Western conceptions of development.

15 Phyla are the largest groupings of languages, e.g. the Indo-European phylum, regarded by linguists as being related to one another (see Greenberg 1963).

16 These death connections resemble those which existed over quite broad geographical areas of the world at history's dawn. Among Biblical expressions, see Psalm xxii, 15–16 'Thou hast brought me into the dust of death, for dogs have compassed me,' and v. 20, 'Deliver . . . my darling from the power of the dog'. More specific are 1st Kings (xiv, 11), 'Him that dieth of Jeroboam in the city shall the dogs eat' and, in xxi, 23, 'the dogs shall eat Jezebel by the wall . . .'. Similar beliefs, as well as ones about the dog as messenger or forewarner of death, and use of dog-shaped amulets to ward off death, and the like, were recorded by the Babylonians. Some of these fears survive in statements about the dog's 'uncleanliness' found in the Muslim *Hadith*, or Sayings of the Prophet. Useful compendia of historical lore about dog and man which indicate needs for more detailed African scholarship on this subject in the future include Vesey-Fitzgerald (1957), Leach (1961) and Howey (1972).

17 The evolution of early Zoroastrian beliefs that a dog must consume, or rend, the body of a deceased, to later beliefs that a dog must only gaze upon the deceased after death, was reflected in unusual regard for, and protection of, dogs by ancient Persians, including directives that the dog itself must be provided a quality of medical care equal to that for people. Perhaps similar evolutions of early ideas about dogs' associations with death (and afterlife?) resulted in their becoming the healing companions of the healing god Asklepios in ancient Greece and also of the Babylonians' healing goddess Gula. In ancient Greece, dogs (together with snakes, also animals with dual life–death associations) actually were kept as co-therapists in the healing temples.

18 During personal exploration of the extensive ancient necropolis adjacent to the Hawara Pyramid of Amenemhat III in the Egyptian Fayoum in 1984, an area with numerous open or partially open human graves, some with deep shafts, and strewn all over with human bones and mummy wrappings, the author was startled by the sudden emergence from a tomb he had just climbed down into of a jackal which almost knocked him over in its eagerness to leave.

19 Other commonly diagnosed helminthic zoonoses with canine reservoirs in Africa include heterophyidiasis (Chapter 10).

20 In Chapter 6, on dog and cat meat, of *Unmentionable cuisine* (Schwabe 1979a) the author discusses dog-eating worldwide; also see Simoons (1961) for further details.

21 It is salutary in such regards to recall, as an example, that the feeding of mice to children as a preventive of tuberculosis was commonly practised in parts of Europe in the not very distant past.

22 Dr Aggrey Ayuen Majok, personal communication, 1988.

23 One notable consequence was that the epidemiology of the trypanosomiases was atypically altered during the period these diseases first began to be studied by Europeans in man, cattle and wildlife. As a result many observers reached very misleading conclusions about what we now recognise as profound zoonotic relationships within this disease complex.

24 Despite the facts that ancestors of Indo-European language-speakers once held strikingly similar beliefs, as stated in both the ancient Hindu and Zoroastrian scriptures, perpetuated in modern Hindu beliefs and practices, and clearly evident in the original cattle-culture meanings of the antecedents of such English-language words as pecuniary

(cf. Spanish *pecuaria*), chattel and capital (i.e. originally 'head of cattle').

25 It is not surprising, therefore, given the importance of such animal questions in Africa that significant African political leadership has arisen out of this situation. Examples include assumption by veterinarian Fanuel Walter Odede of leadership of the Kenya Union immediately after British arrest of KAU's founder Jomo Kenyatta, that Luo leader and former Kenya vice-president Oginga Odinga once had headed a school for veterinary assistants, that another veterinarian Kabwimukya Babiiha became Uganda's first vice-president or that Daoud Kairaba Jawara, formerly principal veterinary officer of The Gambia, became his country's first prime minister and president (Schwabe 1976).

References

Acha, P. & B. Szyfries 1987. *Zoonoses and communicable diseases common to man and animals*, 2nd edn. Pan American Health Organization Scientific Publication 503.

Beech, M. W. H. 1911. *The Suk, their language and folklore*. Oxford: Oxford University Press.

Brannagan, D. & J. A. Hammond 1965. Rinderpest in Tanganyika: a review. *Bull. Epiz. Dis. Afr.* **13**, 225–46.

Burridge, M. J. & C. W. Schwabe 1977. Epidemiological analysis of factors influencing rate of progress in *Echinococcus granulosus* control in New Zealand. *J. Hyg.* **78**, 151–63.

Burton, J. 1973. *Religions and healing in Mandari*. Oxford: Oxford University Press.

Buxton, J. 1987. *A Nilotic world. The Atuot-speaking peoples of the Southern Sudan*. New York: Greenwood Press.

Conrad, J. R. 1957. *The bull and the sword. The history of the bull as symbol of power and fertility*. New York: Dutton.

Deng, F. M. 1971. *Tradition and modernization: a challenge for law among the Dinka of the Sudan*. New Haven and Oxford: Yale University Press.

Deng, F. M. 1985. Development in context. In *Modernization in the Sudan*. M. W. Daly (ed.), New York: Lilian Barber Press.

Eastburn, R. L., T. R. Fritsche & C. A. Terhune 1987. Human intestinal infection with *Nanophyetus salminocola* from salmonid fish. *Am. J. Trop. Med. Hyg.* **36**, 586–91.

Epstein, H. 1971. *The origin of the domestic animals of Africa*, Vols I and II. New York, London, Munich: Africana Publishing Corporation.

Evans-Pritchard, E. E. 1940. *The Nuer*. London: Oxford University Press.

Ford, J. 1971. *The role of trypanosomiasis in African ecology, A study of the tsetse fly problem*. Oxford: Clarendon Press.

Forrester, A. T. T., G. S. Nelson & G. Sander. 1961. The first record of an outbreak of trichinosis in Africa south of the Sahara. *Trans. R. Soc. Trop. Med. Hyg.* **55**, 503–13.

Garnham, P. C. C. 1971. *Progress in parasitology*. University of London, London: Athlone Press.

Greenberg, J. H. 1963. *The languages of Africa*. Bloomington: Indiana University Press.

Hambley, W. D. 1935. *Culture areas of Nigeria*. Field Museum of Natural History, Anthropological Series, 21 (No. 3), Chicago.

Herskovits, M. J. 1926. The cattle complex of East Africa. *Am. Anthropol.* **28**, 230–72, 361–88, 494–528, 633–64.

Howey, M. O. 1972. *The cults of the dog.* Essex: C. W. Daniel.

Hubbert, W. T., W. F. McCulloch & P. R. Schnurrenberger (eds) 1975. *Diseases transmitted from animals to man*, 6th edn Springfield: Charles C Thomas.

Huntingford, G. W. B. 1953a. *The Nandi of Kenya.* London: Routledge & Kegan Paul.

Huntingford, G. W. B. 1953b. *The Northern Nilo-Hamites.* East Central Africa, Pt. 6. *Ethnographic Survey of Africa.* London: International African Institute.

Huxley, J. 1931. *Africa View.* New York: Harper & Brothers.

Kenyatta, J. 1938. *Facing Mount Kenya, the tribal life of the Gikuyu.* London: Secker & Warburg.

Kjekshus, H. 1977. *Ecology control and economic development in East African history.* Berkeley: University of California Press.

Leach, M. 1961. *God had a dog. Folklore of the dog.* New Brunswick: Rutgers University Press.

Lienhardt, G. 1961. *Divinity and experience, the religion of the Dinka.* London: Oxford University Press.

Lobban, R. A. 1989. Cattle and the rise of the Egyptian state. *Anthropozoos.* **2**, 194–201.

Lolik, P. L. (chairman) 1976. *Primary Health Programme, Southern Region, Sudan, 1977/78.* Juba: The Democratic Republic of the Sudan.

Lott, D. F. & B. L. Hart 1977. Aggressive domination of cattle by Fulani herdsmen and its relation to aggression in Fulani culture and personality. *Ethos.* **5**, 174–186.

Lugard, F. D. 1893. *The rise of our East African empire*, Vol. 1. Edinburgh: Blackwood.

Merker, M. 1904. *Die Maasai. Ethnographische Monographie einer Ostafrikanischen Semitenvolkes.* Berlin: Dietrich Reimer.

Nelson, G. S. 1960. Schistosome infections as zoonoses in Africa. *Trans. R. Soc. Trop. Med. Hyg.* **54**, 301–24.

Pankhurst, R. 1966. The great Ethiopian famine of 1888–92: a new assessment. *J. Hist. Med.* **21**, 95–124.

Phillips, J. 1959. *Agriculture and ecology in Africa. A study of actual and potential development south of the Sahara.* London: Faber & Faber.

Riggs, F. W. 1973. *Prismatic society revisited.* Morristown: General Learning Press.

Schnurrenberger, P. R. & W. T. Hubbert. 1981. *An outline of the zoonoses.* Ames: Iowa State University Press.

Schwabe, C. L., C. W. Schwabe & S. S. Basta. Childhood immunization: practical intersectoral cooperation in pastoral Africa (manuscript in preparation).

Schwabe, C. W. 1976. On treating political animals. *J. Am. Vet. Med. Assoc.* **168**, 329–34.

Schwabe, C. W. 1978a. The holy cow – provider or parasite? A problem for humanists. *Southern Humanit. Rev.* **13**, 251–78.

Schwabe, C. W. 1978b. *Cattle, priests and progress in medicine.* Fourth Wesley W. Spink Lecture in Comparative Medicine. Minneapolis: University of Minnesota Press.

Schwabe, C. W. 1979a. *Unmentionable Cuisine.* Charlottesville: University Press of Virginia.

Schwabe, C. W. 1979b. Epidemiological aspects of the planning and evaluation of hydatid disease control. *Aust. Vet. J.* **55**, 109–17.

Schwabe, C.W. 1980. Animal disease control. Part II. Newer methods, with possibility for their application in the Sudan. *Sudan J. Vet. Sci. Ani. Husb.* **21**, 55–65.

Schwabe, C. W. 1981a. A brief history of American parasitology – the veterinary connection between medicine and zoology. In *The current status and future of parasitology.* K. S.

Warren & E. F. Purcell (eds), 21–43. New York: Josiah Macy Jr. Foundation.

Schwabe, C. W. 1981b. Animal diseases and primary health care: intersectoral challenges. *Wld Hlth Org. Chron.* **35**, 227–32.

Schwabe, C. W. 1982. Epidemiology of parasitic zoonoses. In *Parasites – their world and ours.* D. F. Mettrick & S. S. Desser (eds), 357–62. Proceedings of the Fifth International Congress of Parasitology, Toronto. Amsterdam: Elsevier Biomedical Press.

Schwabe, C. W. 1984a. *Veterinary Medicine and Human Health*, 3rd edn. Baltimore: Williams and Wilkins

Schwabe, C. W. 1984b. *Knot-tying, bridge-building, chance-taking – the art of discovery.* University of California, Davis, Library, Chapbook number 9.

Schwabe, C. W. 1987. Dinka 'spirits', cattle and communion. *J. Cultur. Geogr.* **7**, 117–26.

Schwabe, C. W. & A. Kilejian. 1968. Chemical aspects of the ecology of platyhelminths. In *Chemical zoology* M. Florkin & B. T. Scheer (eds), New York and London: Academic Press.

Schwabe, C. W. & I. M. Kuojok. 1981. Practices and beliefs of the traditional Dinka healer in relation to provision of modern medical and veterinary services in the southern Sudan. *Human Org.* **40**, 231–8.

Seligman, C. 1959. *Races of Africa*, 3rd edn. London: Oxford University Press.

Simon, N. 1963. *Between the sunlight and the thunder, the wildlife of Kenya.* Boston: Houghton Mifflin.

Simoons, F. 1961. *Eat not this flesh: food avoidances in the Old World.* Madison: University of Wisconsin Press.

Soulsby, E. J. L. (ed.) 1974. *Parasitic zoonoses.* New York and London: Academic Press.

Steele, J. H. (ed.) 1981. *CRC handbook series in zoonoses*, Boca Raton: CRC Press.

Vesey-Fitzgerald, B. 1957. *The domestic dog. An introduction to its history.* London: Routledge & Kegan Paul.

Von Brand, T. 1952. *Chemical physiology of endoparasitic animals.* New York: Academic Press.

Von Lengerken, H. & E. Von Lengerken. 1955. *Ur, Hausrind und Mensch.* Berlin: Deutsche Academie Landwirtschafts Wissenschaften.

Wagner, G. 1949. *The Bantu of North Kavirondo.* London: Oxford University Press.

Warren, K. S. and E. F. Purcell (eds) 1981. *The current status and future of parasitology.* New York: Josiah Macy Jr Foundation.

Wood, C. (ed.) 1978. *Tropical medicine: from romance to reality*, Proceedings of an Anglo-American Conference, Royal Society of Medicine and others. London and New York: Academic Press.

WHO 1975. *The veterinary contribution to public health practice.* Report of the World Health Organization Expert Committee on Veterinary Public Health, WHO Tech. Report Series 573.

WHO 1979. *Parasitic zoonoses.* Report of a World Health Organization Expert Committee. WHO Tech. Report Series 637.

Zinnser, H. 1935. *Rats, lice and history.* Boston: Little, Brown.

2 Echinococcosis – a plague on pastoralists

Introduction

Jenner was perhaps one of the first scientists to realize the danger of keeping pets, and he warned 'The deviation of man from the state in which he was placed by nature seems to have proven to him a prolific source of diseases . . . . The wolf disarmed of its ferocity is now pillowed in the Lady's lap' (Jenner 1798). It is uncertain why dogs were domesticated but it may have been due to their value in transportation, hunting, herding, guarding, fighting, scavenging, and as food, bedwarmers and pets (Manwell & Ann Baker 1984). Since the domestication of the dog approximately 14 000 years ago (Schwabe 1984) the relationship now facilitates the transmission of about 50 diseases (Baxter & Leck 1984), perhaps the most important of which is hydatid disease, caused by the cestode, *Echinococcus granulosus*. The adaptation of the parasite to a wide variety of different hosts made possible the broad geographical distribution and the parasite has now been recorded from all the inhabited continents.

Life cycle

Echinococcus granulosus like all taeniid cestodes is transmitted cyclically between carnivore and herbivorous hosts (see Thompson 1986) (Fig. 2.1). The minute adult tapeworms are non-pathogenic and occur in the small intestine of the definitive host (e.g. dog). Intermediate hosts, including man, become infected through oral ingestion of the microscopic eggs which are passed in carnivore faeces. Following hatching and activation in the gut of the intermediate host, the mobile embryo (oncosphere) is released from the egg, penetrates the epithelium of the intestinal villi, and enters the circulation and vesiculates within 1–2 days, commonly in the liver or lungs. The larval cystic stage (metacestode) develops over months or years into the mature hydatid cyst. The life cycle is completed when protoscoleces present in the hydatid are ingested following predation or scavenging (or deliberate feeding) by the carnivore host, and develop into the adult tapeworms, which may number many thousand in one dog.

In domestic livestock, hydatid cysts appear to be well tolerated with no general indication of infection. However, in man cysts growing deep in

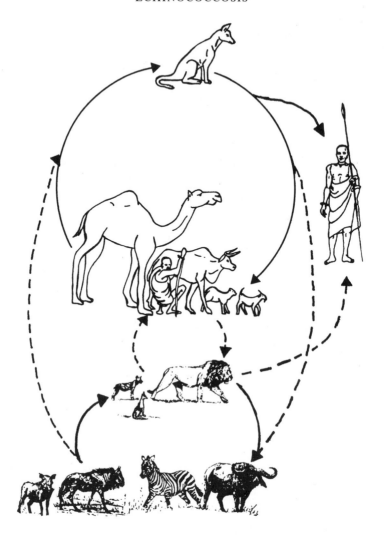

Figure 2.1 Life cycle of *Echinococcus granulosus* in sub-Saharan Africa.

organs or tissues may result in morbidity and sometimes mortality, primarily due to pressure effects (see Diagnosis and treatment).

Current status of human hydatidosis in Africa

Until as recently as thirty years ago, the disease was thought to be rare in Africa south of the Sahara. It was not until then that the studies of Nelson and Rausch (1963), which were stimulated by the publication of Wray (1958), revealed the extent of the hydatid problem amongst the

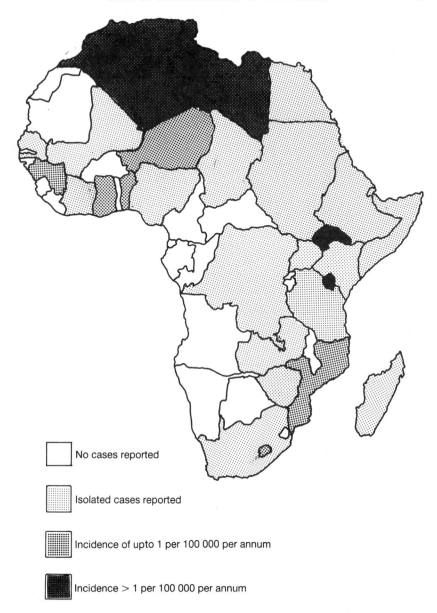

No cases reported

Isolated cases reported

Incidence of upto 1 per 100 000 per annum

Incidence > 1 per 100 000 per annum

Figure 2.2   Reported distribution and importance of human hydatidosis in Africa.

Turkana pastoralists of northwestern Kenya. With the development of health services the disease has now been recorded from over half of African countries and is endemic in many of them (Fig. 2.2). There are two major foci of hydatidosis in Africa, one extending along all the countries bordering the Mediterranean coast including Morocco (Chenebault 1967,

Pandey *et al.* 1988), Algeria (Cherid & Nosny 1972, Larbaoui & Alloula 1979, Abada *et al.* 1977), Tunisia (Ben Rachid *et al.* 1984, Jaiem 1984, Mlika *et al.* 1984, Mlika *et al.* 1986, Bchir, Jemni *et al.* 1985, Gharbi *et al.* 1985), Libya (Gebreel *et al.* 1983, Aboudaya 1985) and to a lesser extent Egypt (Cahill *et al.* 1965, Hegazi *et al.* 1986). The other major focus occurs among the nomadic pastoralists of Eastern Africa which includes Kenya, Uganda, Tanzania, Sudan and Ethiopia (Macpherson, Spoerry *et al.* 1989). Only sporadic human cases are reported elsewhere in the continent but without mass prevalence surveys the true extent of the disease problem is unknown for any country.

From their studies, Nelson and Rausch (1963) stated that hydatid disease was more prevalent in man and livestock in Africa than in many other parts of the world, an argument which has recently been proved by mass ultrasound surveys which demonstrated that approximately 9% of the Turkana in northwestern Turkana District, Kenya, had the disease, a figure that is many times higher than anywhere else in the world (Macpherson, Zeyhle *et al.* 1987). The hyperendemic focus of the disease among the Turkana extends into the pastoralists living in neighbouring countries i.e. the Toposa in southern Sudan (Eisa *et al.* 1962); the Karamajong in eastern Uganda (Owor & Bitakaramire 1975); and the Nyangatom and Dassanetch in south-western Ethiopia (Fuller & Fuller 1981, Lindtjorn *et al.* 1982, Macpherson, Spoerry *et al.* 1989).

Public health importance of *E. granulosus* in Africa

With improved medical and diagnostic capabilities the importance of hydatid disease in many countries in Africa has only recently been recognized. For example, in the Turkana District of Kenya, Schwabe (1964) estimated the minimum surgical incidence rate of 40 cases per 100000 Turkana per annum at a time when there were no operating facilities in the District and patients were treated elsewhere. Once operations began being performed within the District the figure was revised to 96 per 100000 per annum (O'Leary 1976) and more recently to 220 per 100000 per annum for the north of the District (French & Nelson 1982). In Tunisia, the annual surgical incidence rate varies from 11 to 30 per 100000 per annum (Mlika *et al.* 1984, Bchir, Jemni *et al.* 1985) and an ultrasound prevalence rate of 0.4–3.6% has been reported in two different communities (Mlika *et al.* 1986, Bchir, Larouze *et al.* 1987, 1988). It is estimated that in Tunisia, between 800 and 1200 new cases of hydatidosis are diagnosed every year (Gharbi *et al.* 1985). A retrospective study in Morocco indicated a national incidence of 7.8 operated cases per 100000 inhabitants per annum (Pandey *et al.* 1988). In the Central Hospital in Tripoli, Libya, 111 cases of hydatidosis were confirmed surgically from 1972 to 1979 representing an incidence of 0.5% of all admissions (Aboudaya 1985). Over a ten-year period in Algeria, 5305 cases of hydatid disease were treated giving an incidence of between 3.4 and 4.6 per 100 000 inhabitants annually (Larbaoui & Alloula 1979).

Details of the public health importance of the disease including the cost of hospitalization, diagnosis, treatment, time lost from work, etc. is as yet unavailable for any African country. In the University Hospital of Sousse, Tunisia, an area endemic for hydatid disease (Bchir, Jemni *et al.* 1985), 8.3% of the total hospital annual budget for 1983 was spent on hydatid patients (Khlifa *et al.* 1985). To these direct costs must be added the immeasurable suffering caused by the disease, for example, one girl with multiple hydatidosis since the age of 11 was operated on 14 times in 5 years (Djilali *et al.* 1983). It would be impossible to calculate the real cost to persons who undergo surgery, for instance the 69 children under 15 years of age operated on for cerebral hydatidosis (Abada *et al.* 1977), or the young Turkana girl with a uterine cyst who later became an outcast in society because she could not bear children. There is also the less tangible factor of anxiety related to knowledge about, and justifiable fear of, the disease among certain heavily infected populations (Schwabe 1986).

In addition to the human costs most African countries suffer economically from hydatid infections among domestic animals. In spite of careful prevalence studies very little attention has been paid to this aspect and the economic loss is unknown but must be considerable given the high prevalences in livestock in several countries.

Diagnosis of human hydatidosis

It is frequently difficult to establish a specific clinical diagnosis of human hydatidosis, as symptoms are often non-specific. In addition, unlike many other human parasitic helminths like *Schistosoma* or *Onchocerca*, there is invariably no direct parasitological evidence of hydatid disease and cysts are usually located deep in the internal organs. Hydatid cysts may grow to large sizes in man, sometimes >20 cm, and contain many litres of fluid (Fig. 2.3a & 2.3b). Organs affected are most often the liver, 65–70% of cases, followed by the lungs, 10–20%, and to a lesser extent, the spleen, kidney, bones, orbit, brain and almost any other site, including 'free' cysts in the peritoneal cavity. Clinical symptoms are generally associated with pressure effects of the relatively slow-growing cysts, though occasionally cyst rupture with release of excess foreign protein may precipitate fatal anaphylaxis in the sensitized patient.

In remote areas, like the Turkana district of Kenya with limited medical facilities, self-diagnosis of the 'swollen belly' disease is common and, consequently, advanced cases are frequently presented. Where facilities exist, the methods for the detection of space-occupying lesions are very useful.

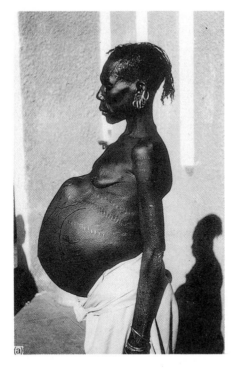

Figure 2.3a Turkana woman with large secondary hydatidosis of the abdomen.

Figure 2.3b Successful removal of complete hydatid cyst. The prognosis such an operation is favourable.

Conventional X-rays can be used for detecting pulmonary cysts (Fossati 1970) and can be used as a guide to further investigations of osseous hydatidosis (Tazi *et al.* 1985). Angiography, scintography (Mechmeche *et al.* 1985), computerized axial tomography (Ozgen *et al.* 1979) and ultrasound scanning (Fig. 2.3c) (Vicary *et al.* 1977, Gharbi *et al.* 1986, Gharbi *et al.* 1985) are well established techniques for diagnosing cysts in many organs. Computerized tomography and ultrasound scanning have also been used for investigating changes in cystic appearance during chemotherapy (Morris *et al.* 1984) and with time when treatment was unavailable (Romig *et al.* 1986).

The recent development of portable, generator-powered ultrasound scanners has facilitated their use in the field and mass prevalence studies have been carried out in a number of countries in Africa including Tunisia (Gharbi *et al.* 1986, Mlika *et al.* 1986), Ethiopia, Sudan, Kenya and Tanzania (Fig. 2.5a) (Macpherson, Craig *et al.* 1989, Macpherson, Spoerry *et al.* 1989). The technique offers many advantages for mass screening of hydatidosis in Africa (Macpherson, Zeyhle *et al.* 1987). However, usually only abdominal cysts can be detected.

Serology

In most endemic areas of the world, including Africa, serology has proved useful in the diagnosis of cystic hydatid disease particularly in support of X-ray, ultrasound or CT findings (Schantz & Gottstein, 1986). Seroepidemiological surveys for human hydatidosis have indicated the prevalence of the disease in several endemic communities including: South America (Varela-Diaz *et al.* 1983), Northern Europe (Moir *et al.* 1986), China (Chai Jun Jie *et al.* 1986) and Africa (Mlika *et al.* 1984, Bchir, Hamdi *et al.* 1988), and asymptomatic cases have been identified after follow-up (Coltorti *et al.* 1988).

In Africa serologic mass surveys have resulted in human seropositive rates of 2.7% of 147 rural Egyptians (Botros *et al.* 1973), 1.3% and 2.2% of 670 and 355 rural Tunisians (Mlika *et al.* 1984, Bchir, Hamdi *et al.* 1988), 0.7% of 273 Maasai in northern Tanzania (Macpherson, Craig *et al.* 1989), 4.2% of 400 Somali including nomads (Kagan & Cahill 1968), 6.4% of 327 Nilotic pastoralists in south-west Ethiopia (Fuller & Fuller 1981) and 2.2–9.4% of Turkana nomads in north-west Kenya (French &

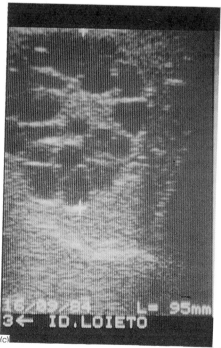

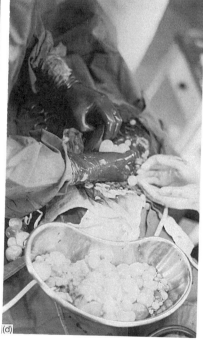

Figure 2.3c Ultrasound scan of a hepatic hydatid cyst full of daughter cysts.

Figure 2.3d Surgical removal of the cyst shown in Figure 2.3c. Failure to remove all cysts can result in a recurrence of the disease.

Ingera 1984; Craig, Zeyhle *et al.* 1986). In all these serologic surveys, crude hydatid cyst fluid preparations were employed as antigens for the detection of anti-hydatid antibodies. The sensitivity and specificity of tests are by no means one hundred per cent, and false positive reactions, through cross-reactions with other parasitic helminth infections such as *Schistosoma*, did occur (Botros *et al.* 1973). In contrast, a relatively high degree of false negativity was observed when sensitive serological tests, such as the enzyme linked immunosorbent assay (ELISA) (Craig, Zeyhle *et al.* 1986) and the indirect hemagglutination test (IHAT) (Njeruh *et al.* 1986) were used to test sera from Turkana hydatid patients. Lack of sero-reactivity in Turkana patients compared to antibody activity in hydatid patients from Europe has been suggested to relate to parasite strain differences (Chemtai *et al.* 1981), and to high levels of circulating parasite antigen and immune complexes with resultant lack of detectability and also possible immunosuppression (Craig 1986, Craig, Zeyhle *et al.* 1986). Recent studies using immunoblotting techniques have identified putative specific antigens in hydatid cyst fluid with approximate molecular weights between 8–16 KDa (Shepherd & McManus 1987, Maddison *et al.* 1989). However, at present there is still no serodiagnostic test for detection of anti-*E. granulosus* antibodies which has >90% sensitivity and specificity. The combination of existing serodiagnostic tests (with crude cyst fluid antigen) such as IHAT or ELISA with portable ultrasound screening for hydatidosis has been used effectively in north-west Kenya (Macpherson, Zeyhle *et al.* 1987) and Tunisia (Mlika *et al.* 1986), and currently probably provides the best mass screening approach for human hydatidosis (Macpherson, Zeyhle *et al.* 1987, Bchir, Larouze *et al.* 1987).

Circulating specific antibodies do not appear to have any effect on established hydatid cysts in both humans and animals. However, the infective oncosphere stage is highly susceptible *in vitro* to antibody-mediated complement dependent killing (Rickard & Williams 1982), and furthermore, antibody dependent cellular cytoxicity has been observed against oncospheres of *T. hydatigena* (Beardsell & Howell 1984). In ovine hydatidosis continuous egg challenge is probably most effective in stimulating acquired immunity (Gemmell *et al.* 1987a) (see also Chapter 3). Whether humans develop resistance to infection with *Echinococcus* is not known but is likely, and will probably also be dependent on anti-oncosphere antibodies (Craig 1988) (see also Conclusions and future prospects).

Treatment of human hydatidosis

Surgery (Fig. 2.3b & 2.3d), often carried out using suitable scolicidal adjuncts such as hydrogen peroxide (Djilali *et al.* 1983) or Cetrimide

(Frayha *et al.* 1981, Macpherson, Wood *et al.* 1982) to prevent secondary hydatidosis, remains the most important form of treatment. However, there is growing optimism for the use of chemotherapy both for simple and inoperable cases (Eckert 1986, Morris *et al.* 1986, Mansueto *et al.* 1987). The first compounds shown to be effective against hydatidosis were the benzimidazole carbamates including mebendazole. However, although this drug was beneficial to many inoperable patients treated in Turkana, Kenya (French 1984), a few side-effects were noted and the drug was subsequently withdrawn from use in Kenya. More recently, albendazole, at a dosage of 10 mg/kg body-weight has been used in treating Kenyan patients with encouraging results (Okelo 1986) and further studies are in progress. However, multi-centre clinical trials indicate only a 30% success rate in albendazole treatment for cystic hydatid disease (Davis *et al.* 1986). Recent *in vitro* studies using praziquantel, an isoquinoline compound, both by itself and in combination with albendazole sulphoxide have shown it to be highly effective against protoscoleces (Taylor *et al.* 1988). However, the action of praziquantel on the germinal layer of the parasite appears to be less effective *in vitro* than albendazole (Taylor *et al.* 1989). It has been shown in animal models that albendazole and praziquantel may be effective for preventing recurrences (Morris *et al.* 1986, Taylor & Morris 1988), which in Turkana, can account for up to 22% of patients seen (Macpherson, Zeyhle *et al.* 1987). The possibility of treatment of large numbers of hydatid cases within hyperendemic zones rather than after transport to distant hospitals in major cities like Nairobi, is preferable, and has been realized in northern Turkana where an 'albendazole-village' has been constructed for this purpose (Macpherson, Wachira *et al.* 1986).

Domestic intermediate hosts

In Africa hydatidosis is propagated primarily by a domestic cycle involving the ubiquitous domestic dog as the definitive host and the many species of livestock including camels, cattle, sheep, goats, pigs, horses, donkeys and buffaloes as intermediate hosts. The local importance of each intermediate host species in maintaining the cycle varies in different regions but this factor is often difficult to determine since usually many different domestic species from the same country harbour hydatid infections. Camels are found all over northern Africa extending southward as far as Kenya in the east and Nigeria in the west. In all countries where the camel has been reported as an intermediate host (Chad, Djibouti, Ethiopia, Egypt, Kenya, Libya, Mauritania, Morocco, Nigeria, Somalia, Sudan, Tunisia, Western Sahara) it is generally considered to be an important host for the local maintenance of the life cycle of the parasite (Macpherson 1981). In the hyperendemic regions of north Africa sheep and cattle, and buffaloes in Egypt (El-Kordy 1946), in addition to camels, appear to be important intermediate hosts (Faure 1949, Gusbi *et al.* 1987, Jore d'Arces 1953, Larbaui *et al.* 1980, Jaiem 1984, Dar & Taguri 1978, Senevet 1951, Pandey *et al.* 1988). Sheep and goats are the main intermediate hosts in Maasailand in Kenya (Macpherson 1985) and probably in Ethiopia (Bekele *et al.* 1988), and goats and camels are thought to be the main hosts in Turkana (Macpherson 1981). In Sudan (El-Badawi *et al.* 1979), South Africa (Verster & Collins 1966) and Nigeria (Dada 1980) cattle are also thought to be important sources of infection for local dogs along with sheep and goats. Donkeys are poor intermediate hosts (Pandey 1980) and in most countries little has been reported on the role of pigs although they are known to be intermediate hosts in many African countries (Macpherson 1981).

Role of domestic dogs

High prevalences of *Echinococcus* infections have been reported in dogs from areas where human hydatidosis is endemic. In north Africa, *E. granulosus* infections have been reported in dogs in five countries – 20% of dogs in Algeria (Senevet 1951), 20–51% of dogs in Morocco (Sirol & Lefevre 1971, Pandey *et al.* 1987, 1988), 22–43% of dogs in Tunisia (Deve 1923, Jaiem 1984, Bchir, Jemni *et al.* 1985), 11–27% in Libya (Gusbi 1987, Packer & Ali 1986) and 1–10% in Egypt (Abdel Azim 1938, El-Garhy & Salim 1958, Moch *et al.* 1974, Abo-Shady 1980, Abo-Shady *et al.* 1982, Hegazi *et al.* 1986). In the East African focus *E. granulosus* infections were found in 3–87% of dogs in Sudan (Eisa *et al.* 1962, El-Badawi *et al.* 1979, Saad & Magzoub 1986), 39–70% of dogs in Turkana, Kenya (Macpherson, French *et al.* 1985, Nelson & Rausch 1963), 27–50% of dogs in Maasailand, Kenya (Eugster 1978, Macpherson, Craig *et al.* 1989a, Ngunzi 1986, Nelson & Rausch 1963) and 23% of stray dogs in Somalia (Macchioni *et al.* 1985). Away from these hydatid endemic foci *E. granulosus* prevalences have been much lower, being 0.9% in South Africa (Verster 1979), 0% in 643 and 76 dogs examined respectively in Mozambique (Pires Ferreira 1980) and Niger (Develoux *et al.* 1985), 0.6–6.2% in Nigeria (Dada *et al.* 1979a, 1979b, Dada *et al.* 1980, Arene & Nweke 1985, Ugochukwu & Ejimadu 1985) and 3.0% in Chad (Troncy & Graber 1969).

The epidemiology of hydatid disease is closely related to the prevalence of *E. granulosus* in the domestic dog, and to the behaviour of man in relation to dogs (Nelson 1972, French & Macpherson 1983). In the hot, dry environment of north Africa and the regions occupied by the nomadic pastoralists in East Africa, *Echinococcus* eggs can only survive for a short period of time and it is likely that people must come into frequent contact with freshly deposited eggs (Fig. 2.4a) for successful transmission to occur. In Tunisia, a man : dog ratio of 6 : 1 was reported (Ben Rachid *et al.* 1984), this is despite the high Islamic religious influence in North Africa, whose laws forbid the close association of man and dogs. Elsewhere in north Africa large numbers of dogs are kept and tolerated in the houses and at religious and marriage feasts (Dew 1928, Ben Rachid *et al.* 1984, Gusbi 1987). In Libya stray dogs are common particularly in urban and semi-urban areas and a significant percentage (40%) have been found infected (Gusbi 1987). The Somalis of East Africa, who are also Moslems do not keep many dogs and despite a significant incidence of hydatidosis in camels (McManus *et al.* 1987) and probable human exposure (Kagan & Cahill 1968) no human infections have as yet been reported in these pastoralists (Macpherson, Spoerry *et al.* 1989). However, all other pastoral groups keep large numbers of dogs, particularly those groups living along the international borders.

The possibility of variable susceptibility to infection with *E. granulosus* within and between domestic dog breeds could be a factor affecting the intensity of infection in dogs and therefore the transmission dynamics. In most endemic areas of the world worm burdens in the domestic dog are usually less than 200 worms (Gemmell *et al.* 1987), but in Turkana, Kenya, this figure is much higher and worm burdens of up to 30 000 worms are not unusual (Macpherson *et al.* 1985).

It is the concentration of the dogs in the home with the women and children which is thought to be responsible for the increased incidence of hydatidosis amongst Turkana women (Macpherson 1983). The close association with the dogs is heightened by allowing dogs to clean babies (Fig. 2.4b) and also allowing dogs to lick clean inanimate objects around the house (Nelson & Rausch 1963, Schwabe 1964, Fuller & Fuller 1981, French & Nelson 1982, French *et al.* 1982, Watson-Jones & Macpherson 1988). *E. granulosus* eggs (Fig. 2.4c) have been isolated from soil samples and utensils found within Turkana and Maasai households (Craig, Macpherson *et al.* 1988, Macpherson, Craig *et al.* 1989) and the increased period women and children spend in the houses may also be partly responsible for a higher infection rate than in men (Watson-Jones & Macpherson 1988). Drinking water samples from wells dug in the dry Turkana river beds (Fig. 2.4d) have also been shown to be contaminated with *Taenia* sp. eggs (Stevenson & Macpherson 1982), and the presence of *Echinococcus* eggs has been confirmed by immunofluorescence using a specific anti-*Echinococcus* oncosphere monoclonal antibody (Craig, Macpherson *et al.* 1988).

The Maasai have just as close an association with their dogs as do the Turkana (Macpherson, Craig *et al.* 1989) and Nelson (1972, 1987) has suggested that the lower infection rate amongst the Maasai may be due to the much higher incidence of *Taenia saginata* amongst the Maasai, affording a degree of cross protection or resistance to infection with

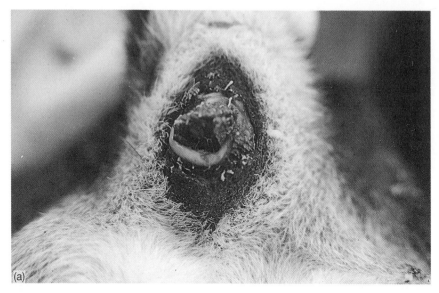

Figure 2.4a Anus of a dog showing *Echinococcus granulosus* proglottides being expelled.

Figure 2.4b Turkana child being 'cleaned up' by a dog. Such close contact must facilitate the rapid transmission of *Echinococcus* eggs.

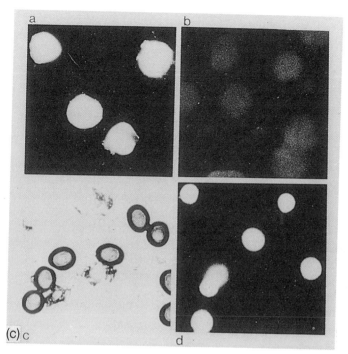

Figure 2.4c Immunofluorescence of oncospheres from dog tapeworms using an anti-*Echinococcus* oncosphere monoclonal antibody (4E5). (a) Positive fluorescence with known *Echinococcus* oncospheres. (b) Negative fluorescence with known *T. hydatigena* oncospheres. (c) *Echinococcus* eggs on a Scotch tape perianal swab from a naturally infected dog. (d) Oncospheres hatched from the Scotch tape showing positive fluorescence.

Figure 2.4d Turkana pastoralists drawing water from a water hole in a dry river bed.

E. granulosus. Cross-immunity between taeniid larval cestodes has been demonstrated in experimental animals (Rickard & Williams 1982). Many examples of zooprophylaxis or as Nelson has recently renamed it 'the Jennerian principle of cross protection' are now recognized and have recently been reviewed (Nelson 1987).

Wildlife cycles in Africa

It is likely that before the introduction and domestication of livestock in Africa, wildlife cycles of *E. granulosus* existed, as they still do in isolated areas, such as in the National Parks of many countries in central, east and southern Africa where large concentrations of wild carnivores and their prey species still exist (Nelson 1982, Macpherson 1986).

Some of the most extensive records of *Echinococcus* in wildlife come from Kenya, where in 1958 Ginsberg reported an infection in a jackal (*Canis* sp.) which led Wray (1958) to speculate that jackals and hyenas were the main definitive hosts of *Echinococcus* there, thus explaining the distribution of hydatidosis amongst the nomadic pastoralists. Nelson subsequently examined 143 carnivores representing 23 species and found *Echinococcus* adults in 27 out of 43 domestic dogs, in 3 out of 4 Cape hunting dogs (*Lycaon pictus*), in 1 out of 7 silver-backed jackals (*Canis mesomelas*) and in 3 out of 19 hyenas (*Crocuta crocuta*) (Nelson & Rausch 1963). They concluded that the most important definitive host in Kenya was the domestic dog and the main reason for the high prevalence in Turkana was the close relationship between women and children and dogs.

Nelson's search for wild intermediate hosts was less productive and only 1 wildebeest (*Connochaetes taurinus*) of 92 ungulates representing 18 species was found infected (Nelson & Rausch 1963). At the same time none of 2000 rodents or 271 monkeys were found infected (Nelson 1983). Although there was little evidence for a secondary wildlife cycle in Kenya Nelson suggested that this may not be true elsewhere in the continent (Nelson *et al.* 1965), and that further studies were required.

Today more than 18 different wild intermediate host species have been reported including the most common prey species of the lion (*Panthera leo*) which has been implicated as the main definitive host involving a wildlife cycle in South Africa (Ortlepp 1937, Verster & Collins 1966, Young 1975), Kenya (Eugster 1978), Tanzania (Dinnik & Sachs 1972, Rodgers 1974) and the Central African Republic (Graber & Thal 1980). The most important wild intermediate host species in South Africa appear to be the zebra (*Equus burchelli*), of which 60% were found infected in the Kruger Park (Young 1975) and the buffalo (*Syncerus caffer*) (Young & Van den Heever 1969, Basson *et al.* 1970). The warthog (*Phacochoerus aethiopicus*) appears to be the most important host in the

Central African Republic (Graber & Thal 1980). In East Africa, warthogs, buffalo and wildebeest have all been found to frequently harbour hydatid infections (Woodford & Sachs 1973, Eugster 1978, Dinnik & Sachs 1969, Sachs & Sachs 1968, Macpherson, Karstad *et al.* 1983, Macpherson 1986).

Rausch & Nelson (1963) found that morphologically the material they collected from the wild hosts was within the normal range of *E. granulosus*. However, the parasite in the lion is morphologically distinct from *E. granulosus* and warthog protoscoleces appear to be not infective to dogs (Graber & Thal 1980). Felids are not normally definitive hosts of *E. granulosus* (Thompson 1979) and since a discrete predator–prey parasite transmission could occur between the lion and its main food animals, the taxonomic status of the parasite in the lion needs to be re-evaluated and specific status (*Echinococcus felidis* Ortlepp 1937) may be applicable (Macpherson 1986). Although many other wild felids have been examined for *E. granulosus* infections, only the African wild cat (*Felis lybica*) examined in South Africa has been found to harbour the adult parasite (Verster & Collins 1966).

Hyenas are still relatively abundant and may appear to be more suitable as maintenance hosts than lions but only a few lightly infected hyenas have been reported (Nelson & Rausch 1963, Schwabe 1984, Young 1975). These infections may have been with a wildlife strain of the parasite, for hyenas scavenge from lion kills, they are phylogenetically closer to cats than dogs and are refractory to experimental infection with hydatid material from Turkana which was highly infective to jackals and dogs (Macpherson & Karstad 1981, Macpherson, Karstad *et al.* 1983). This factor requires further study as does the role of jackals and Cape hunting dogs in the epidemiology of hydatidosis in Africa. The infectivity of the wildlife strain to man is unknown but in practical terms it is unlikely that the wildlife cycle could be responsible for many human infections. The responsibility for this must be reserved for the domestic dog, which has been described as the host *par excellence* for *Echinococcus* (Nelson 1972).

Species and strain differences of *Echinococcus* in Africa

A single case of *Echinococcus multilocularis* has been reported in man in northern Tunisia (Robbana *et al.* 1981). However, the main causative agent of human hydatidosis in Africa is *E. granulosus* although there is increasing evidence that the speciation of the parasite is complicated by the occurrence of strains (McManus & Macpherson 1984). The presence of unusual strains in East Africa was suggested by Nelson & Rausch (1963), who speculated that in most parts of Kenya the parasite was poorly adapted to man, but that in Turkana a strain of higher infectivity existed, thus explaining the existence of the parasite in the hot, arid areas. This suggestion was supported by Eugster (1978) who found a much lower incidence of the disease among a similar pastoral group in southern Kenya, the Maasai. Besides a rapid growth rate and the usual development of a single, fertile cyst in man (Macpherson 1983) no evidence for a highly infective strain in Turkana has emerged despite the examination of a number of extrinsic and intrinsic differential criteria (Macpherson 1981, Macpherson & McManus 1982, McManus 1981, Macpherson & Smyth 1985, Macpherson, French *et al.* 1985). Isoenzyme studies in particular indicated that nevertheless there may be three strains of *E. granulosus* in Kenya, one involving sheep, goats and humans, another in camels and goats and the third in cattle (McManus & Macpherson 1984, Wachira 1989). More recent studies, using DNA hybridization analysis to characterize *E. granulosus* protoscoleces from Kenyan isolates indicate that camel material is genetically distinct from that obtained from sheep, humans, goats and cattle types which were more similar to each other (McManus *et al.* 1987). McManus *et al.* (1987) also speculate that the virtual absence of human hydatidosis in Somalia may be due to the lack of infectivity to man by the predominant camel/dog parasite.

Comparative *in vitro* studies, however, indicate that all the Kenyan isolates studied share similar physiological/nutritional requirements (Macpherson & Smyth 1985), and camel/dog, sheep/dog, goat/dog and cattle/dog *E. granulosus* eggs were experimentally infective to baboons (*Papio cynocephalus*) suggesting that the strains in Kenya may be more uniform than indicated by biochemical or DNA criteria (Macpherson, Wachira *et al.* 1986). Elsewhere in Africa, there is further evidence of a cattle strain. In Egypt (El-Kordy 1946), Nigeria (Dada 1980), Kenya (Macpherson 1985, Eugster 1978), and in many other countries in the world (see Thompson *et al.* 1984), cattle cysts are predominantly sterile. However, in South Africa (Verster 1962) and Sudan (Eisa *et al.* 1962) as in Belgium, Sri Lanka, Switzerland, Germany and India (Thompson & Lymbery 1988) high fertility rates have been reported suggesting a strain of *E. granulosus* adapted to cattle. Morphologically the South African cattle material described by Verster (1965) resembles that of Swiss cattle origin (Thompson *et al.* 1984), and the latter authors have suggested that the status of this strain needs consideration, and suggest that specific status (*Echinococcus ortleppi*) which was originally described for the South African cattle material (Lopez-Neyra & Solar Planas 1943) may be applicable.

Control

The first programme for controlling hydatid disease was started over 120 years ago in Iceland where an estimated one in six people had the disease (Schwabe 1984); today the disease has not been seen there for over a quarter of a century. Other island control programmes began in the 1960s in New Zealand and Tasmania, and in the 1970s in Cyprus and the Falklands (Gemmell et al. 1986). Spectacular reductions in the incidence of the disease were achieved in man, sheep and dogs and today the programmes have entered the maintenance phase (Gemmell et al. 1986). Education, administration of the drug arecoline hydrobromide for surveillance by examining the purges from dogs, and strict meat inspection, have all contributed to these programmes. Since they began, praziquantel (Droncit, Bayer, Leverkusen, Germany) a taeniacide which is 100% effective in removing the adult parasite (Thakur et al. 1978), has been introduced, thus providing an effective control measure against the adult worm population in dogs. As praziqantel has no ovicidal properties extreme care should be exercised especially at the start of any dog control programme. A suitable ovicide such as sodium hypochlorite should be used to sterilize instruments etc. (Craig & Macpherson 1988). Evidence that control was possible encouraged the formation of other control programmes and in Africa one such programme began in Turkana, Kenya in 1983 (Macpherson, Zeyhle et al. 1984, Macpherson, Wood et al. 1984). The programme was introduced only once the baseline data had been collected by a research team under the guidance of Professor Nelson and set up by the African Medical and Research Foundation in 1976 (French et al. 1982, Nelson 1986).

Because it was realized that human behaviour is largely responsible for the infection pressure from dogs to man (Nelson & Rausch 1963, Nelson 1972), the control programme in Turkana began with an educational campaign (Fig. 2.5b). Since changes in human behaviour take a long time, strict dog control measures were also introduced (Macpherson, Wachira et al. 1986) and evidence for control could be shown by the vastly reduced number of infected dogs, as determined by six-weekly purging of all dogs in the control area (Wachira et al. 1990). Surveillance was carried out in the people themselves using initially sero-epidemiological surveys which were subsequently replaced with the cheaper, more sensitive and more appropriate ultrasound scanning surveys (Macpherson, Zeyhle et al. 1987). Many of the recent advances in research on hydatid disease made in Kenya and in other parts of the world were assimilated into the Turkana programme which is being developed for an area inhabited by nomads and where abattoirs would serve little purpose. It is envisioned that the pilot control area will be expanded to cover the whole of Turkana district and also possibly into neighbouring countries where the disease is also hyperendemic.

Figure 2.5a Ultrasound scanning survey in the field.

Figure 2.5b Education *baraza* being conducted in Turkana.

Conclusions and future prospects

Hydatid disease is one of the world's most important and geographically widespread parasitic zoonoses (Matossian *et al.* 1977). Its chronic nature and primary association with pastoralists, dogs, and livestock make it difficult to study and control. Although a great deal of information is now known about the biology of the causative organism, *E. granulosus*, and the public health importance of human hydatid disease (Thompson 1986), there is still a lack of knowledge about basic aspects of parasite transmission and host–parasite interaction, and still a need to further improve diagnosis, treatment and control.

In Africa cystic hydatid disease is essentially a problem for arid-land pastoralists of the northern Sahara and south-eastern Sahel belts. In these harsh, and sometimes remote areas, traditional nomadic and semi-nomadic lifestyles still persist with often intimate cultural associations between man, livestock and dogs (see also Chapter 1).

Adaptation of the tapeworm to particular predator–prey cycles, involving both domestic or wildlife species, has resulted in a tendency for the parasite to form infra-specific variants or strains which may differ in morphological and biochemical, as well as biological characteristics (McManus & Smyth 1986). The recent application of recombinant DNA techniques has enabled the start of detailed analysis of the *E. granulosus* species complex at the gene level, rather than the more variable level of expressed products such as enzymes. DNA hybridization studies have shown distinct differences between UK horse and sheep strains of *E. granulosus* (Rishi & McManus 1987, McManus & Rishi 1989), and have already begun to help unravel the potentially more complex sub-species variant picture in parts of East Africa (McManus *et al.* 1987).

The production of cDNA expression libraries from mRNA extracted from *Echinococcus* protoscoleces or oncospheres, should enhance the ultimate goal of development of species-specific and pure antigens for diagnosis and protection in both intermediate and definitive hosts. The recent report of a >90% protection against experimental ovine cysticercosis, using a recombinant peptide antigen vaccine derived *in vitro* from a *Taenia ovis* oncosphere cDNA library (Johnson *et al.* 1989) is very encouraging and augurs well for development of a defined anti-*Echinococcus* vaccine.

The use of improved immunoserologic methods, such as ELISA, in mass screening for human hydatidosis has proved beneficial in a number of endemic areas including regions in North and East Africa (Mlika *et al.* 1984, Craig, Zeyhle *et al.* 1986). As a result of research associated with the Turkana Hydatid Project, an ELISA for the detection of circulating parasite antigens, rather than antibodies, was developed (Craig 1986), and may have additional application in post-treatment (surgery or chemotherapy) surveillance.

The possibility for early diagnosis by ultrasound now requires similar advances in the development of more active and specific organ targeted anti-hydatid drugs, particularly those which have a major effect against the germinal epithelium as well as the protoscolex, such as albendazole.

New tools are becoming available which should enable *Echinococcus* transmission to be more effectively studied. For the first time, the ability to identify the infective egg stage of *Echinococcus*, using a specific oncosphere monoclonal antibody (Craig, Macpherson *et al.* 1986), has already been applied to pinpoint environmental contamination sites (Craig, Macpherson, Watson-Jones *et al.* 1988, Macpherson, Craig *et al.* 1989). In addition, the development of tests for human exposure based on detection of anti-oncosphere antibodies (Craig, Zeyhle *et al.* 1986), and for immunodiagnosis of canine echinococcosis (Jenkins & Rickard 1986) will greatly improve our ability to study transmission and microepidemiology patterns in detail, and also to help monitor hydatid control programmes, especially where livestock infection data cannot be collected or is insufficient.

Echinococcus control programmes, including those first undertaken in Africa (Macpherson, Zeyhle *et al.* 1984), primarily rely on reducing the adult worm population in the dog definitive host, and in providing relevant health education to the communities at risk. The creation of computer-assisted mathematical models for the study of the dynamics of interaction of extrinsic (e.g. egg survival), intrinsic (e.g. biotic potential of the parasite and host acquired immunity) and socioecological (e.g. meat inspection and husbandry practices) factors associated with transmission of echinococcosis, can provide further valuable insights into collection of baseline data and optimal strategies for control in a given region (Gemmell *et al.* 1987b).

It is now a quarter of a century since Professor Nelson first published a paper on hydatid disease in Turkana, Kenya and it is a testimony to his character that he is still involved with the control programme. His enthusiasm for the project and appreciation of the potential and the application of 'new technologies' in research and control provide a powerful stimulus for scientists to endure the hardships of working in such a remote place, which must surely be one of the hardest places in the world to live. Hopefully, for the Turkana, it will be a better place without the constant worry of *epespes*, hydatid disease.

References

Abada, M., I. Galli, A. Busallah & G. Lehmann 1977. Kystes hydatiques du cerveau. Problèmes diagnostiques et thérapeutiques à propos de 100 cas. *Neuro-Chirurgie* **23**, 195–204.

Abdel Azim, M. 1938. On the intestinal helminths of dogs in Egypt. *J. R. Egypt. Med. Assoc.* **21**, 118–22.

Abo-Shady, A. F. 1980. Intestinal helminths among stray dogs in Mansoura city, Egypt. *J. Egypt. Soc. Parasitol.* **10**, 289–94.

Abo-Shady, A. F., E. S. I. El Kholy, M. M. Ali & A. A. Abouzkham 1982. The prevalence of canine echinococcal infection in Dakahalia Governorate, Egypt. *J. Egypt. Soc. Parasitol.* **12**, 453–57.

Aboudaya, M. A. 1985. Prevalence of human hydatidosis in Tripoli region of Libya. *Int. J. Zoonosis.* **12**, 304–7.

Arene, F. O. I. & O. O. Nweke 1985. The epidemiology of *Echinococcus* infection in the Niger Delta. *Pub. Hlth* **99**, 30–2.

Basson, P. A., R. M. McCully, S. P. Kruger, J. W. Van Niekerk *et al.* 1970. Parasitic and other diseases of the African buffalo in the Kruger National Park. *Onders. J. Vet. Res.* **37**, 11–28.

Baxter, D. N. & I. Leck 1984. The deleterious effects of dogs on human health: 2. Canine zoonoses. *Comm. Med.* **6**, 185–97.

Bchir, A., L. Jemni, M. Allegue, A. Hamdi *et al.* 1985. Epidémiologie de l'hydatidose dans le Sahel et le Centre Tunisiens. *Bull. Soc. Path. Exotique.* **78**, 687–90.

Bchir, A., A. Hamdi, L. Jemni, M. C. Dazza *et al.* 1988. Serological screening for hydatidosis in households of surgical cases in central Tunisia. *Ann. Trop. Med. Parasitol.* **82**, 271–3.

Bchir, A., B. Larouze, H. Bouhaoula, L. Bouden *et al.* 1987. Echtomographic evidence for a highly endemic focus of hydatidosis in Central Tunisia. *Lancet.* **ii**, 684.

Beardsell, P. L. & M. J. Howell 1984. Killing of *Taenia hydatigena* oncospheres by sheep neutrophils. *Zeit. Parasitkd.* **70**, 337–44.

Bekele, T., E. Mukasa-Mugerwa & O. B. Kasali 1988. The prevalence of cysticercosis and hydatidosis in Ethiopian sheep. *Vet. Parasitol.* **28**, 267–70.

Ben Rachid, M. S., R. Ben Ammar, T. Redissi, M. Ben Said *et al.* 1984. Géographie des parasitoses majeures en Tunisie. *Arch. Inst. Pasteur Tunis*, **61**, 17–41.

Botros, B. A. M., R. W. Moch & I. S. Barsoum 1973. Echinococcosis in Egypt: evaluation of the indirect hemagglutination and latex agglutination tests for echinococcal serologic surveys. *J. Trop. Med. Hyg.* **76**, 243–7.

Cahill, K. M., W. Attala, & R. D. Johnson 1965. An echinococcal survey in Egypt and Sudan. *J. Egypt. Pub. Hlth Assoc.* **40**, 293–6.

Chai Jun Jie *et al.* 1986. Epidemiological significance of antibody detection in hydatidosis. *End. Dis. Bull.* **1**, 197–200.

Chemtai A. K., T. R. Bowry & Z. Ahmed 1981. Evaluation of five immunodiagnostic techniques in echinococcosis patients. *Bull. Wld Hlth Org.* **59**, 767–72.

Chenebault, J. 1967. L'échinococcose pulmonaire. *Bull. Acade. Nat. Med.* **151**, 230–7.

Cherid, A. & Y. Nosny 1972. Considérations anatomo-pathologiques sur l'hydatidose en Algérie. *Bull. Soc. Path. Exotique.* **65**, 128–38.

Coltorti E, E. Fernandez, E. Guarnera, J. Lago *et al.* 1988. Field evaluation of an enzyme immunoassay for detection of asymptomatic patients in a hydatid control program. *Am. J. Trop. Med. Hyg.* **38**, 603–7.

Craig, P. S. 1986. Detection of specific circulating antigen, immune complexes, and antibodies in human hydatidosis from Turkana (Kenya) and Great Britain by enzyme-immunoassay. *Parasite Immunol.* **8**, 171–81.

Craig, P. S. 1988. Immunology of human hydatid disease. *ISI Atlas of Science: Immunology* Vol. 1, 95–100.

Craig, P. S. & C. N. L. Macpherson 1988. Sodium hypochlorite as an ovicide for *Echinococcus*. *Ann. Trop. Med. Parasitol.* **82**, 211–3.

Craig, P. S. & G. S. Nelson 1984. The detection of circulating antigen in human hydatid disease. *Ann. Trop. Med. Parasit.* **78**, 219–27.

Craig, P. S., C. N. L. Macpherson & G. S. Nelson 1986. The identification of eggs of

Echinococcus by immunofluorescence using a specific monoclonal antibody. *Am. J. Trop. Med. Hyg.* **35**, 152–8.

Craig, P. S., E. Zeyhle & T. Romig 1986. Hydatid disease: research and control in Turkana. II. The role of immunological techniques for the diagnosis of hydatid disease. *Trans. R. Soc. Trop. Med. Hyg.* **80**, 183–92.

Craig, P. S., C. N. L. Macpherson, D. L. Watson-Jones & G. S. Nelson 1988. Immunodetection of *Echinococcus* eggs from naturally infected dogs and from environmental contamination sites in settlements in Turkana, Kenya. *Trans. R. Soc. Trop. Med. Hyg.* **82**, 268–74.

Dada, B. J. O. 1980. Taeniasis, cysticercosis and echinococcosis/hydatidosis in Nigeria: I – prevalence of human taeniasis, cysticercosis and hydatidosis based on a retrospective analysis of hospital records. *J. Helminthol.* **54**, 281–6.

Dada, B. J. O., D. S. Adegboye & A. N. Mohammed 1979a. A survey of gastro-intestinal helminth parasites of stray dogs in Zaria, Nigeria. *Vet. Rec.* **104**, 145–6.

Dada, B. J. O., D. S. Adegboye & A. N. Mohammed 1979b. The epidemiology of *Echinococcus* infection in Kaduna State, Nigeria. *Vet. Rec.* **104**, 312–3.

Dada, B. J. O., D. S. Adegboye & A. N. Mohammed 1980. The epidemiology of *Echinococcus* infection in Kano State, Nigeria. *Ann. Trop. Med. Parasitol.* **74**, 515–7.

Dar, F. K. & S. Taguri 1978. Human hydatid disease in Eastern Libya. *Trans. R. Soc. Trop. Med. Hyg.* **72**, 313–4.

Davis, A., Z. S. Pawlowski & H. Dixon 1986. Multicentre clinical trials of benzimidazole-carbamates in human echinococcosis. *Bull. Wld Hlth Org.* **64**, 383–8.

Deve, F. 1923. Enquête étiologique sur l'échinococcose en Tunisie. *Rev. Vet. Milit.* 133–63.

Develoux, M., F. Lamothe, A. Sako, J. Landois *et al.* 1985. Deux cas d'hydatidose en République du Niger. *Bull. Soc. Path. Exotique.* **78**, 216–20.

Dew, H. R. 1928. Historical. In *Hydatid disease, its pathology, diagnosis and treatment.* 9–45. Sidney: Aust. Med. J.

Dinnik, J. A. & R. Sachs 1969. Cysticercosis, echinococcosis and sparganosis in East Africa. *Vet. Med. Rev.* **2**, 104–14.

Dinnik, J. A. & R. Sachs 1972. Taeniidae of lions in East Africa. *Z. Tropenmed. Parasitol.* **23**, 197–210.

Djilali, G., A. Mahrour, T. Oussedik, M. Abad *et al.* 1983. L'eau oxygénée dans la chirurgie du kyste hydatique. *Presse Med.* **12**, 235–7.

Eckert, J. 1986. Prospects for treatment of the metacestode stage of *Echinococcus*. In *The biology of Echinococcus and hydatid disease*. R.C.A. Thompson (ed.), 250–84. London: Allen & Unwin.

Eisa, A. M., A. A. Mustafa & K. N. Soliman 1962. Preliminary report on cysticercosis and hydatidosis in southern Sudan. *Sudan J. Vet. Sci.* **3**, 97–108.

El-Badawi, E. K. S., A. M. Eisa, N. B. K. Slepenez & M. B. A. Saad 1979. Hydatidosis of domestic animals in the central region of the Sudan. *Bull. Ani. Hlth Prod. Afr.* **27**, 249–51.

El-Garhy, M. T. & M. K. Selim 1958. Incidence of echinococcosis in camels slaughtered for meat production in Egypt. *Vet. Med. J. Giza.* **4**, 191–200.

El-Kordy, M. I. 1946. On the incidence of hydatid disease in domestic animals in Egypt. *J. R. Egypt. Med. Ass.* **29**, 265–79.

Eugster, R. O. 1978. A contribution to the epidemiology of echinococcosis/hydatidosis in Kenya (East Africa) with special reference to Kajiado District. DVM thesis, University of Zurich.

Faure, J. 1949. Contribution a l'étude de l'échinococcose dans la région de Marrakech. *Bull. l'Inst. Hyg. Maroc.* **2**, 211–32.

Fossati, C. J. 1970. Las parasitosis respiratorias halladas en pacientes arabolibicos de Cirenaica (Libya) en los ultimos diez anos. Nota II. Hidatidosis toracica. *Rev. Iber Parasitol.* **30,** 587–647.

Frayha, G. J., K. J. Bikhazi & T. A. Kachachi 1981. Treatment of hydatid cysts *Echinococcus granulosus* by Cetrimide®. *Trans. R. Soc. Trop. Med. Hyg.* **75,** 447–50.

French, C. M. 1984. Mebendazole and surgery for human hydatid disease in Turkana. *E. Afr. Med. J.* **61,** 113–9.

French, C. M. & E. W. Ingera 1984. Hydatid disease in the Turkana District of Kenya. V. Problems of interpretation of data from a mass serological survey. *Ann. Trop. Med. Parasitol.* **78,** 213–8.

French, C. M. & C. N. L. Macpherson 1983. The man–dog relationship in the culture of the Turkana and its significance in the transmission of human hydatid disease. In *Current public health research in the tropics with a special session on communicable diseases*, P. M. Tukei & A. R. Ngogu (eds), 121–6. Nairobi: AMREF.

French, C. M. & G. S. Nelson 1982. Hydatid disease in the Turkana District of Kenya. II. A Study in medical geography. *Ann. Trop. Med. Parasitol.* **76,** 439–57.

French, C. M., G. S. Nelson & M. Wood 1982. Hydatid disease in the Turkana District of Kenya. I. The background to the problem with hypotheses to account for the remarkably high prevalence of the disease in man. *Ann. Trop. Med. Parasitol.* **76,** 425–37.

Fuller, G. K. & D. C. Fuller 1981. Hydatid disease in Ethiopia: clinical survey with some immunodiagnostic test results. *Am. J. Trop. Med. Hyg.* **30,** 645–52.

Gebreel, A. O., H. M. Gilles & J. E. Prescott 1983. Studies on the sero–epidemiology of endemic diseases in Libya. 1. Echinococcosis in Libya. *Ann. Trop Med. Parasitol.* **77,** 391–3.

Gemmell, M. A., J. R. Lawson & M. G. Roberts 1986. Control of echinococcosis/hydatidosis: present status of worldwide progress. *Bull. Wld Hlth Org.* **64,** 333–9.

Gemmell, M. A., J. R. Lawson & M. G. Roberts 1987a. Population dynamics in echinococcosis and cysticercosis: evaluation of the biological parameters of *Taenia hydatigena* and *T. ovis* and comparisons with those of *Echinococcus granulosus*. *Parasitology* **94,** 161–80.

Gemmell, M. A., J. R. Lawson & M. G. Roberts 1987b. Towards global control of cystic and alveolar hydatid diseases. *Parasitol. Today* **3,** 144–51.

Gharbi, H. A., W. Hassine & K. Abdesselom 1985. L'hydatidose abdominale a l'échographie. Réflexions et aspects particuliers *Echinococcus granulosus*. *Ann. Radiol.* **28,** 31–4.

Gharbi, H. A., J. Mrif, M. Ben Abdallah, B. Abdelmoula *et al.* 1986. Epidémiologie du kyste hydatique en Tunisie. I. Résultats de l'enquête par échographie abdominale partant sur 3116 sujets dans la région de Menzel Bouruiba. *Med. Mal. Infect.* **16,** 151–6.

Ginsberg, A. 1958. Helminthic zoonoses in meat inspection. *Bull. Epiz. Dis. Africa.* **6,** 141–9.

Graber, M. & J. Thal 1980. L'échincoccose des artiodactyles sauvages de la République Centrafricaine: existence probable d'un cycle lionphacochere. *Rev. d'Elevage Med. Vet. Trop.* **33,** 51–9.

Gusbi, A. M. 1987. Echinococcosis in Libya, I. Prevalence of *Echinococcus granulosus* in dogs with particular reference to the role of the dog in Libyan society. *Ann. Trop. Med. Parasitol.* **81,** 29–34.

Gusbi, A. M., M. A. Q. Awan & W. N. Beesley 1987. Echinococcosis in Libya, II. Prevalence of hydatidosis (*Echinococcus granulosus*) in sheep. *Ann. Trop. Med. Parasitol.* **81,** 35–41.

Hegazi, M. M., S. A. Abdel–Magied, F. M. Abdel–Wahab & R. A. Atia 1986.

Epidemiological study of echinococcosis in Dakahlia governorate, Egypt. *J. Egypt. Soc. Parasitol.* **16**, 541–8.

Jaiem, A. 1984. L'échinococcose hydatique dans la région de Sousse: enquête épidémiologique. *Maghreb Vet.* **1**, 9–12.

Jenkins, D. J. & M. D. Rickard 1986. Specific antibody responses in dogs experimentally infected with *Echinococcus granulosus*. *Am. J. Trop. Med Hyg.* **35**, 345–9.

Jenner, E. 1798. *An enquiry into the causes and effects on the veriolae vaccinae, a disease discovered in some of the western counties of England, particularly Gloucestershire, and known by the name of cow pox.* London: Sampson Low.

Johnson, K. S., G. B. L. Harrison, M. W. Lightowlers, K. L. O. Hoy *et al.* 1989. Vaccination against ovine cysticercosis using a defined recombinant antigen. *Nature.* **338**, 585–7.

Jore d'Arces, P. 1953. L'échinococcose en Algérie. *Bull. Off. Int. Epiz.* **40**, 45–51.

Kagan, I. G. & K. M. Cahill 1968. Parasitic serologic studies in Somaliland. *Am. J. Trop Med. Hyg.* **17**, 392–6.

Khlifa, K. G. Foulon, A. Bchir, M. Jeddi *et al.* 1985. Cout de l'hydatidose chirurgicale à l'Hôpital Universitaire de Sousse (Tunisie). *Bull. Soc. Path. Exotique.* **78**, 691–5.

Larbaoui, D., R. Alloula, L. V. Osiiskaya, I. Yu Osiiskii *et al.* 1980. (Hydatidosis in Algiers). *Med. Parasit. Parazit. Bolezni*, **49**, 21–8.

Larbaoui, D. & R. Alloula 1979. Etude épidémiologique de l'hydatidose en Algérie. Résultats de deux enquêtes rétrospectives portant sur 10 ans. *Tunisie Med.* **57**, 318–26.

Lindtjorn, B., T. Kiserud & K. Roth 1982. Hydatid disease in some areas of southern Ethiopia. *Eth. Med. J.* **20**, 185–8.

Lopez–Neyra, C. A. & M. A. Solar Planas 1943. Revision del genero *Echinococcus*. Rud. y descripcion de una especie nueva parasita intestinal del perro en Almoia. *Rev. Iberica Parasitol.* **3**, 169–94.

Macchioni, G., P. Lanfranchi, M. A. Abdallatif & F. Testi 1985. Echinococcosis and hydatidosis in Somalia. *Bull. Sci. Fac. Zoo. Vet.* **5**, 179–89.

Macpherson, C. N. L. 1981. Epidemiology and strain differentiation of *Echinococcus granulosus* in Kenya. PhD thesis, University of London.

Macpherson, C. N. L. 1983. An active intermediate host role for man in the life cycle of *Echinococcus granulosus* in Turkana, Kenya. *Am. J. Trop. Med. Hyg.* **32**, 397–404.

Macpherson, C. N. L. 1985. Epidemiology of hydatid disease in Kenya: a study of the domestic intermediate hosts in Masailand. *Trans. R. Soc. Trop. Med. Hyg.* **79**, 209–17.

Macpherson, C. N. L. 1986. *Echinococcus* infections in wild animals in Africa. In *Wildlife/livestock interfaces on rangelands.* S. MacMillan (ed.), 73–8. Nairobi: Inter–African Bureau for Animal Resources.

Macpherson, C. N. L. & L. Karstad 1981. The role of jackals in the transmission of *Echinococcus granulosus* in the Turkana District of Kenya. In *Wildlife disease research and economic development.* L. Karstad, B. Nestel & M. Graham (eds), 53–6. Ottawa, Canada: International Development Research Centre.

Macpherson, C. N. L. & D. P. McManus 1982. A comparative study of *Echinococcus granulosus* from human and animal hosts in Kenya using isoelectric focusing and isoenzyme analysis. *Int. J. Parasitol.* **12**, 515–21.

Macpherson, C. N. L. & J. D. Smyth 1985. *In vitro* culture of the strobilar stage of *Echinococcus granulosus* from protoscolosus of human, camel, cattle, sheep, goat origin from Kenya and buffalo origin from India. *Int. J. Parasitol.* **15**, 137–40.

Macpherson, C. N. L., J. E. Else & M. Suleman 1986. Experimental infection of the olive baboon (*Papio cynocephalis*) with *Echinococcus granulosus* of camel, cattle, sheep and goat origin from Kenya. *J. Helminthol.* **60**, 213–7.

Macpherson, C. N. L., A. M. Wood & C. M. French 1982. The use of Cetrimide® as a scolicidal adjunct to hydatid surgery. In *Current medical research in East Africa with emphasis on zoonoses and waterborne diseases*, P. M. Tukei & A. R. Njogu (eds), 21–5. Nairobi: Africascience International.

Macpherson, C. N. L., E. Zeyhle & T. Romig 1984. An *Echinococcus* pilot control programme for north–west Turkana, Kenya. *Ann. Trop. Med. Parasitol.* **78**, 188–92.

Macpherson, C. N. L., L. Karstad, P. Stevenson & J. H. Arundel 1983. Hydatid disease in the Turkana District of Kenya (iii). The significance of wild animals in the transmission of *Echinococcus granulosus* with particular reference to Turkana and Maasailand. *Ann. Trop. Med. Parasitol.* **77**, 61–73.

Macpherson, C. N. L., P. S. Craig, T. Romig, E. Zeyhle et al. 1989. Observations on human echinococcosis (hydatidosis) and evaluation of transmission factors in the Maasai of northern Tanzania. *Ann. Trop. Med. Parasitol.* **83**, 489–97.

Macpherson, C. N. L., C. M. French, P. Stevenson, L. Karstad et al. 1985. Hydatid disease in the Turkana District of Kenya. IV. The prevalence of *Echinococcus granulosus* infections in dogs and observations on the role of the dog in the lifestyle of the Turkana. *Ann. Trop. Med. Parasitol.* **79**, 51–61.

Macpherson, C. N. L., A. Spoerry, E. Zeyhle, T. Romig et al. 1989. Pastoralists and hydatid disease: an ultrasound scanning prevalence survey in East Africa. *Trans. R. Soc. Trop. Med. Hyg.* **84**, 243–7.

Macpherson, C. N. L., T. M. Wachira, E. Zeyhle, T. Romig et al. 1986. Hydatid disease: research and control in Turkana, IV. The pilot control programme. *Trans. R. Soc. Trop. Med. Hyg.* **80**, 196–200.

Macpherson, C. N. L., E. Zeyhle, T. Romig, P. H. Rees et al. 1987. Portable ultrasound scanner versus serology in screening for hydatid cysts in a nomadic population. *Lancet,* **ii**, 259–62.

Macpherson, C. N. L., A. M. Wood, C. Wood, C. M. French et al. 1984. Perspective on options for the implementation of a pilot hydatidosis control programme in the Turkana District of Kenya. *E. Afr. Med. J.* **61**, 513–23.

Maddison, S. E., S. B. Siemenda, P.M. Schantz, J. A. Fried et al. 1989. A specific diagnostic antigen of *Echinococcus granulosus* with an apparent molecular weight of 8 KDa. *Am. J. Trop. Med. Hyg.* **40**, 377–83.

Mansueto, S. S., S. Di Rosa, E. Farinella & S. Orsini 1987. Albendazole in the treatment of hydatid disease: more than a hope. *Trans. R. Soc. Trop Med. Hyg.* **81**, 168.

Manwell, C. & C. M. Ann Baker 1984. Domestication of the dog: hunter, food, bed-warmer, or emotional object? *Z. Tierzunctg Zuchtgsbiol.* **101**, 241–56.

Matossian, R. M., M. D. Rickard & J. D. Smyth 1977. Hydatidosis: a global problem of increasing importance. *Bull. Wld Hlth Org.* **55**, 499–507.

McManus, D. P. 1981. A biochemical study of adult and cystic stages of *Echinococcus granulosus* of human and animal origin from Kenya. *J. Helminthol.* **55**, 21–7.

McManus, D. P. & C. N. L. Macpherson 1984. Strain characterisation in the hydatid organism *Echinococcus granulosus*: current status and new perspectives. *Ann. Trop. Med. Parasitol.* **78**, 193–8.

McManus, D. P. & A. K. Rishi 1989. Genetic heterogeneity within *Echinococcus granulosus* isolates from different hosts and geographical areas characterised with DNA probes. *Parasitology,* **99**, 17–29.

McManus, D. P. & Smyth J. D. 1986. Hydatidosis: changing concepts in epidemiology and speciation. *Parasitol. Today* **2**, 163–8.

McManus, D. P., A. J. G. Simpson & A. K. Rishi 1987. Characterisation of the hydatid organism, *Echinococcus granulosus* from Kenya using cloned DNA markers. In *Helminth zoonoses with particular reference to the tropics*. S. Geerts, V. Kumar & J. Brandt (eds), 29–36, Dordrecht: Martinus Nijhoff.

Mechmeche, R., M. Ben Cheikh, A. Bousnina, H. A. Gharbi et al. 1985. Imagerie dans l'échinococcose cardiaque. A propos de 21 cas. Codification d'une stratégie de son exploration. *Ann. Radiol.* **28**, 373–80.

Mlika, N., B. Larouze, M. Dridi, R. Yang *et al.* 1984. Serologic survey of human hydatid disease in high risk populations from central Tunisia. *Am. J. Trop. Med. Hyg.* **33**, 1182–4.

Mlika, N., B. Larouze, C. Gaudebout, B. Braham *et al.* 1986. Echotomographic and serologic screening for hydatidosis in a Tunisian village. *Am. J. Trop. Med. Hyg.* **35**, 815–7.

Moch, R. W., J. B. Cornelius, A. M. Boulas, I. S. Botros *et al.* 1974. Serological detection of echinococcal infection in camels by the indirect hemagglutination (IHA) and latex agglutination (LA) tests. *J. Egypt. Pub. Hlth Assoc.* **49**, 146–55.

Moir, I. L., C. J. Chrystall, A. W. L. Joss & H. Williams 1986. Further investigations of a hydatid focus in north west Scotland. *J. Hyg.* (Cambridge) **96**, 113–9.

Morris, D. L., J. B. Chinnery & J. D. Hardcastle 1986. Can albendazole reduce the risk of implantation of spilled protoscoleces? An animal study. *Trans. R. Soc. Trop. Med. Hyg.* **80**, 481–4.

Morris, D. L., H. Skene-Smith, A. Haynes and F. G. O. Burrows 1984. Abdominal hydatid disease: computed tomographic and ultrasound changes during albendazole therapy. *Clin. Rad.* **35**, 297–300.

Nelson, G. S. 1972. Human behaviour in the transmission of parasitic diseases. In *Behavioural aspects of parasite transmission.* E. U. Canning & C. A. Wright (eds), 109–22. London: *Zool. J. Linnean Soc.*

Nelson, G. S. 1982. Carrion-feeding cannibalistic carnivores and human disease in Africa with special reference to trichinosis and hydatid disease in Kenya. *Symp. Zool. Soc. London.* **50**, 181–98.

Nelson, G. S. 1983. Wild animals as reservoir hosts of parasitic diseases of man in Kenya. In *Tropical parasitoses and parasitic zoonoses*, J. D. Dunsmore (ed.), 59–72. Tenth Meeting of the World Association for the Advancement of Veterinary Parasitology, Australia.

Nelson, G. S. 1986. Hydatid disease: research and control in Turkana, Kenya, I. Epidemiological observations. *Trans. R. Trop. Med. Hyg.* **80**, 177–82.

Nelson, G. S. 1987. More than a hundred years of parasitic zoonoses with special reference to trichinosis and hydatid disease: *J. Comp. Pathol.* **98**, 135–53.

Nelson, G. S. & R. L. Rausch 1963. *Echinococcus* infections in man and animals in Kenya. *Ann. Trop. Med. Parasitol.* **57**, 136–49.

Nelson, G. S., F. R. N. Pester & R. Rickman 1965. The significance of wild animals in the transmission of cestodes of medical importance in Kenya. *Trans. R. Soc. Trop. Med. Hyg.* **59**, 507–24.

Ngunzi M. M. 1986. The effect of prolonged drought on infestation of dogs with *Echinococcus granulosus* in Maasailand, Kenya. In *Recent Advances in the Management and Control of Infections in Eastern Africa.* P. M. Tukei, D. K. Koech & S. N. Kinoti (eds), 231–9. Nairobi: English Press.

Njeruh, F. M., J. M. Gathuma & A. G. Tumboh-Oeri 1986. Diagnosis of human hydatidosis. I. The role of indirect hemagglutination test (IHA) based on a thermo-stable antigen. *E. Afr. Med. J.* **63**, 312–7.

Okelo, G. B. A. 1986. Hydatid disease: research and control in Turkana, Kenya, III. Albendazole in the treatment of inoperable hydatid disease in Kenya; a report on 12 cases. *Trans. R. Soc. Trop. Med. Hyg.* **80**, 193–5.

O'Leary, P. 1976. A five year review of human hydatid cyst disease in Turkana District, Kenya. *E. Afr. Med. J.* **53**, 540–4.

Ortlepp, R. J. 1937. South African helminths. Part I. *Onders. J. Vet. Sci.* **9**, 311–36.

Owor, R. & P. K. Bitakaramire 1975. Hydatid disease in Uganda. *E. Afr. Med. J.* **52**, 700–4.

Ozgen, T., A. ErBengi, V. Bertan, S. Saglam *et al.* 1979. The use of computerized tomography in the diagnosis of cerebral hydatid cysts. *J. Neurosurgery* **50**, 339–42.

Packer, D. E. & T. M. Ali 1986. *Echinococcus granulosus* in dogs in Libya. *Ann. Trop. Med. Hyg.* **80**, 137–9.

Pandey, V. S. 1980. Hydatidosis in donkeys in Morocco. *Ann. Trop. Med. Parasitol.* **74**, 519–21.

Pandey, V. S., A. Dakkak & M. Elmamoune 1987. Parasites of stray dogs in the Rabat region, Morocco. *Ann. Trop. Med. Parasitol.* **81**, 53–5.

Pandey, V. S., H. Ouhelli & A. Moumen 1988. Epidemiology of hydatidosis/echinococcosis in Ouarzazate, the pre–Saharian region of Morocco. *Ann. Trop. Med Parasitol.* **82**, 461–70.

Pires Ferreira, M. L. 1980. Occurrence of the multivesicular form of hydatidosis in cattle in Mozambique. *Bull. Ani. Hlth Prod. Africa* **28**, 254–8.

Rausch, R. L. & G. S. Nelson 1963. A review of the genus *Echinococcus* Rudolphi, 1801. *Ann. Trop. Med. Parasitol.* **57**, 127–35.

Rickard, M. D. & J. F. Williams 1982. Hydatidosis/cysticercosis: immune mechanisms and immunization against infection. *Adv. Parasitol.* **21**, 229–96.

Rishi, A. K. & D. P. McManus 1987. Genomic cloning of human *Echinococcus granulosus* DNA: isolation of recombinant plasmids and their use as genetic markers in strain characterization. *Parasitology* **94**, 369–83.

Robbana, M., M. S. Ben Rachid, M. M. Zittouna, N. Heldt *et al.* 1981. Première observation d'échinococcose alvéolaire autochtone en Tunisie. *Anat. Cytol. Pathol.* **29**, 311–2.

Rodgers, W. A. 1974. Weights, measurements and parasitic infestation of six lions from southern Tanzania. *E. Afr. Med. J.* **12**, 157–8.

Romig, T., E. Zeyhle, C. N. L. Macpherson, P. H. Rees *et al.* 1986. Cyst growth and spontaneous cure in hydatid disease. *Lancet* **ii:** 861.

Saad, M. B. & M. Magzoub 1986. *Echinococcus granulosus* infection in dogs in Tambool, Sudan. *J. Helminthol.* **60**, 299–300.

Sachs, R. & C. Sachs 1968. A survey of parasitic infestation of wild herbivores in the Serengeti region in northern Tanzania and the Lake Rukwa region in southern Tanzania. *Bull. Epizot. Dis. Afr.* **16**, 455–72.

Schwabe, C. W. 1964. *Veterinary medicine and human health*, Baltimore: Williams & Wilkins, pp. 211 and 395.

Schwabe, C. W. 1984. *Veterinary medicine and human health*, 3rd edn. Baltimore: Williams & Wilkins.

Schwabe, C. W. 1986. Current status of hydatid disease: a zoonosis of increasing importance. In *The biology of Echinococcus and hydatid disease*. R. C. A. Thompson (ed.), 81–113. London: Allen & Unwin.

Schantz P. M. & B. Gottstein 1986. *Echinococcus* (hydatidosis) In *Immunoserology of parasitic diseases*. K. F. Walls & P. M. Schantz (eds), 69–107. New York: Academic Press.

Senevet, G. 1951. Epidémiologie du kyste hydatique en Afrique du Nord. *Arch. Int. Hidat.* **12**, 113–20.

Shepherd J. C. & D. P. McManus 1987. Specific and cross-reactive antigens of *Echinococcus granulosus* hydatid cyst fluid. *Mol. Biochem. Parasitol.* **25**, 143–54.

Sirol, J. & M. Lefevre 1971. L'hydatidose humaine au Tchad pays d'élevage. A propos de trois observations colligées à Fort Lamy. Tchad. *Bull. Soc. Pathol. Exotique* **64**, 887–9.

Stevenson, P. & C. N. L. Macpherson 1982. The recovery of cestode eggs from water and soil in the Turkana District of Kenya. In *Current medical research in Eastern Africa with an emphasis on zoonoses and water-borne diseases*. P. M. Tukei & A. R. Njogu (eds), 13–15. Nairobi: Africascience International.

Taylor, D. H. & D. L. Morris 1988. The current management of hydatid disease. *Br. J. Clin. Pract.* **42**, 401–6.

Taylor, D. H., D. L. Morris & K. S. Richards 1988. Combination chemotherapy of *Echinococcus granulosus* – *in vitro* studies. *Trans. R. Soc. Trop. Med. Hyg.* **82**, 263–4.

Taylor, D. H., D. L. Morris & K. S. Richards 1989. A comparison of the effects of albendazole sulphoxide, albendazole sulphane and praziquantel on whole cysts of *Echinococcus granulosus* in an *in vitro* culture system. British Society for Parasitology 20–22 March 1989, University of Southampton.

Tazi, Z., N. Boujida, N. Hamdouch & N. Boukhrissi 1985. Hydatidose vertébro-médullaire. Apport de la radiologie et da la tomodensitométrie. A propos de 36 observations. *J. Radiol.* **66**, 183–8.

Thakur, A. S., U. Prezioso & N. Marchevsky 1978. Efficacy of Droncit against *Echinococcus granulosus* infection in dogs. *Am. J. Vet. Res.* **39**, 859–60.

Thompson, R. C. A. 1979. Biology and speciation of *Echinococcus granulosus*. *Aust. Vet. J.* **55**, 93–8.

Thompson, R. C. A. (ed.) 1986. *The biology of Echinococcus and hydatid disease.* London: Allen & Unwin.

Thompson, R. C. A & A. J. Lymbery 1988. The nature, extent and significance of variation within the genus *Echinococcus*. *Adv. Parasitol.* **27**, 210–63.

Thompson, R. C. A., L. M. Kumaratilake & J. Eckert 1984. Observations on *Echinococcus granulosus* of cattle origin in Switzerland. *Int. J. Parasitol.* **14**, 283–91.

Troncy, P. & M. Graber 1969. L'échinococcose – hydatidose en Afrique centrale. III. Taeniasis des carnivores à *Echinococcus granulosus* (Batsch, 1786. Rudolphi, 1801). *Rev. Med. Vet. Trop.* **22**, 75–84.

Ugochukwu, E. I. & K. N. Ejimadu 1985. Comparative study on the infestations of three different breeds of dogs by gastro-intestinal helminths. *Int. J. Zoonos.* **12**, 318–22.

Varela-Diaz, V. M., E. A. Coltorti, O. De Zavaleta, H. Perez Caviglia *et al.* 1983. Immunodiagnosis of human hydatid disease: application and contributions to a control programme in Argentina. *Am. J. Trop. Med. Hyg.* **32**, 1079–87.

Verster, A. 1962. Hydatidosis in the Republic of South Africa. *S. Afr. J. Sci.* **58**, 71–4.

Verster, A. J. M. 1965. Review of *Echinococcus* species in South Africa. *Onder. J. Vet. Res.* **32**, 7–118.

Verster, A. 1979. Gastro-intestinal helminths of domestic dogs in the Republic of South Africa. *Onder. J. Vet. Res.* **46**, 79–82.

Verster, A. & M. Collins 1966. The incidence of hydatidosis in the Republic of South Africa. *Onder. J. Vet. Res.* **33**, 49–72.

Vicary, F. R., G. Cusick, I. M. Shirley & R. J. Blackwell 1977. Ultrasound and abdominal hydatid disease. *Trans. R. Soc. Trop. Med. Hyg.* **71**, 29–31.

Wachira, T. M. 1989. Studies on the epidemiology and control of *Echinococcus granulosus* in Kenya. PhD thesis. University of Nairobi.

Wachira, T. M., C. N. L. Macpherson & J. M. Gathuma 1990. Hydatid disease in the Turkana District of Kenya. VII. Analysis of the infection pressure between definitive and intermediate hosts of *Echinococcus granulosus*, 1979–88. *Ann. Trop. Med. Parasitol.* **84**.

Watson-Jones, D. L & C. N. L. Macpherson 1988. Hydatid disease in the Turkana district of Kenya. VI. Man–dog contact and its role in the transmission and control of hydatidosis amongst the Turkana. *Ann. Trop. Med. Parasitol.* **82**, 343–56.

Woodford, M. H. & R. Sachs 1973. The incidence of cysticercosis, hydatidosis and sparganosis in wild herbivores of the Queen Elizabeth National Park, Uganda. *Bull. Epizoot. Dis. Afr.* **21**, 265–71.

Wray, J. R. 1958. Note on human hydatid disease in Kenya. *E. Afr. Med. J.* **35**, 37–9.

Young, E. 1975. *Echinococcus* (hydatidosis) in wild animals of the Kruger National Park. *J. S. Afr. Vet. Assoc.* **46,** 285–6.

Young, E. & L. W. Van den Heever 1969. The African buffalo as a source of food and by-products. *J. S. Afr. Vet. Assoc.* **40,** 83–8.

3 The zoonotic *Taeniae* of Africa

Introduction

Human taeniasis is caused by infection with the adult stage of the tapeworms *Taenia saginata* and *T. solium*, while human cysticercosis results from infection with the larvae (cysticerci) of the latter species. Both of these parasites occur in Africa and are zoonoses because the usual hosts for the cysticerci are cattle and swine respectively, from which humans become infected with the adult tapeworm. The term 'taeniosis' has been avoided as a current review of parasitological nomenclature (Kassai *et al.* 1989) recommends that this should be applied to infection with either the adult or the larval stages.

Scientific meetings on the problems associated with taeniasis and cysticercosis have been organized in many parts of the world over the last few years. The venues include San Miguel de Allende, Mexico (Flisser, Willms *et al.* 1982), Ceske Budejovic, Czechoslovakia (Prokopic 1983), Liverpool, United Kingdom (Craig *et al.* 1984) and Antwerp, Belgium (Geerts *et al.* 1987). There was also a series of meetings held in Nairobi, Warsaw, Prague and Geneva under the auspices of the World Health Organization at which guidelines for the surveillance, prevention and control of taeniasis and cysticercosis were drawn up (Gemmell *et al.* 1983). In addition, the European Economic Community has a working group on cestode zoonoses, including taeniasis/cysticercosis, which meets regularly to monitor the prevalence of infection with these parasites within the Community. These meetings reflect the continuing awareness in both the scientific community and the United Nations agencies of the public health importance of human taeniasis and cysticercosis and of the economic importance of cysticercosis in cattle and pigs.

The purpose of this review is to outline the history and biology of these parasites, their current medical/veterinary and economic importance and to discuss in more detail the prospects for improving control by introducing new methods of treatment or animal management and by adopting a rational approach to the development of diagnostic assays and immunoprophylaxis.

History and biology

Tapeworms were first documented as parasites of man in 1500 BC when the Ebers papyrus described worms that were probably species of *Taenia*,

54

but it is likely that the existence of such large and obvious parasites was known well before that time (Hoeppli 1956). Parasites of the family Taeniidae have two mammalian hosts in their life cycle (Fig. 3.1). The strobilated adult tapeworms inhabit the small intestine of the final host (Fig. 3.2). *T. solium* and *T. saginata* occur in Africa, the adult tapeworms of both these species being parasitic in man and causing human taeniasis. The larval cysticerci of these species (Fig. 3.3) were also known to man in ancient times and Neito (1982) reviewed the history of their discovery. However, it was only in 1760 that Pallas recognized the association of the larval cyst and the adult tapeworm (Foster 1965) and it was not until the following century that the life cycles of the various taeniid species were really understood (Rausch 1975, Muller 1975).

The morphological characteristics and life cycles of these parasites are well described in the literature (Soulsby 1982). As with most taeniids *T. solium* bears a double row of hooks on its rostellum (Fig. 3.4a), whereas *T. saginata* does not have a rostellum or hooks (Fig. 3.4b). The gravid terminal segments of the adult tapeworms are either passed out in the faeces or actively migrate out of the anus. The gravid segments of *T. saginata* and *T. solium* contain up to 80000 and 40000 eggs, respectively. The eggs released from these segments form the only free-living stage in the life cycle and are immediately infective for the

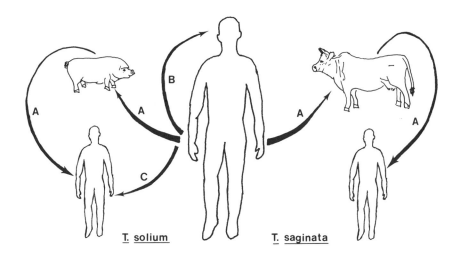

Figure 3.1 Schematic life cycles of *Taenia solium* and *T. saginata*. The adult tapeworm of both species is located in the small intestine of the human host. When *T. solium* eggs are ingested by pigs or *T. saginata* eggs are ingested by cattle the eggs hatch and the larvae migrate to the tissues where they develop into cysticerci. Man becomes infected with either tapeworm when he consumes meat containing viable cysticerci (Route A). However, *T. solium* cysticerci may develop in humans (Routes B and C) including those already infected with the tapeworm.

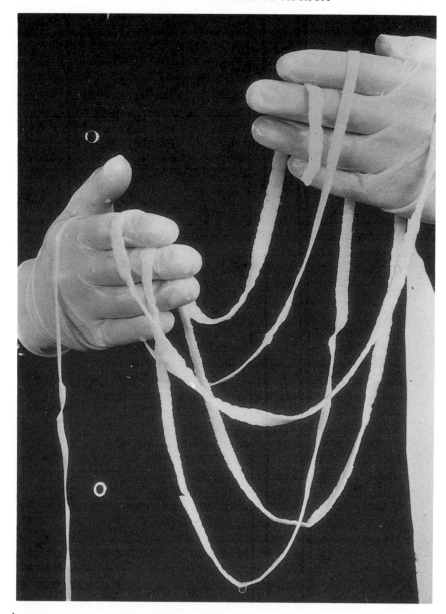

Figure 3.2 A *T. saginata* tapeworm, which can be 3–10 metres in length. A *T. solium* tapworm looks very similar but is usually only 2–7 metres long. (Photograph courtesy of Dr W. Crewe, Liverpool School of Tropical Medicine.)

intermediate host. After the eggs have been ingested by a suitable host, the six-hooked hexacanth larvae or oncospheres (Fig. 3.5) hatch and activate in the small intestine before migrating to the development site via the lymphatics and blood stream. The biology of the oncospheres was reviewed by Lethbridge (1980).

Cysticercosis is the term used to describe infection of the mammalian intermediate host by those taeniid cestodes, including both *T. solium* and *T. saginata*, whose larvae are called cysticerci. They take the form of unilocular cysts each containing a single scolex. The development and biology of cysticerci in the mammalian host was reviewed by Slais (1970). Although the oncospheres of *T. saginata* and *T. solium* require at least 12 weeks to develop to fully infective cysticerci, the characteristic bladder form may be identified within 2–4 weeks. The most common intermediate host for *T. solium* is the domestic pig. However, the cysticerci of this cestode can also infect a number of other mammalian hosts, including dogs and, unfortunately, man. Such infections in man can cause severe symptoms and even fatalities. In contrast, the cysticerci of *T. saginata* are more restricted in their host range and are usually only found in domestic cattle, both *Bos indicus* and *B. taurus*. Mature cysticerci of *T. saginata* and *T. solium* are up to 10 and 20 mm in diameter, respectively.

The characteristic site for the cysticerci of both species is in the skeletal, cardiac, diaphragmatic or lingual musculature but some cysts may develop in the liver and in other viscera. There is reported to be a strain of *T. saginata* in the Republic of Sudan, and in Somalia, for which the predilection site is the liver of cattle (El-Sadik Abdullah 1979, Cornaglia & Lo Schiavo 1985). In humans the cysticerci of *T. solium* may occur in the brain (Fig. 3.6) or in the eye (Zenteno-Alanis 1982). It is this characteristic which is responsible for the major risk to human health caused by this parasite.

Humans become infected with the tapeworm when they consume meat containing viable cysticerci.

Epidemiology

The variations in the epidemiological patterns of taeniasis/cysticercosis throughout Africa are a reflection of the numbers and distribution of the human, cattle and pig populations and also of the different management methods used with these domestic animals. Pigs are not commonly kept in Islamic countries including most of northern Africa. In African countries where pigs are kept they tend to be held in much smaller numbers than cattle. In East Africa, because of the regulations governing the control of African swine fever, pigs tend to be maintained in tick-free commercial piggeries. Commercial piggeries are also found in west, central and southern Africa but in these areas village-reared pigs are common. Such pigs are left to forage around human dwelling places, where they have access to human detritus. This practice, combined with the omnivorous and coprophagous habits of pigs probably accounts for

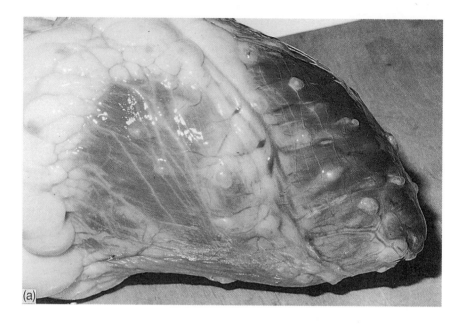

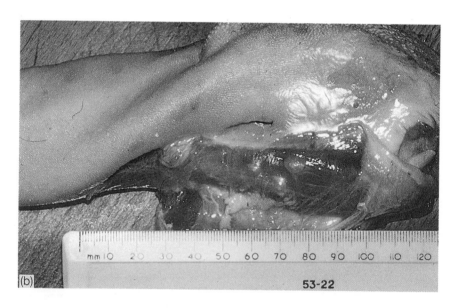

Figure 3.3 *T. saginata* cysticerci in the heart and tongue of a calf (a and b) and *T. solium* cysticerci in the heart and tongue of a pig (c and d). (Photographs courtesy of Elsevier Publications, Cambridge.)

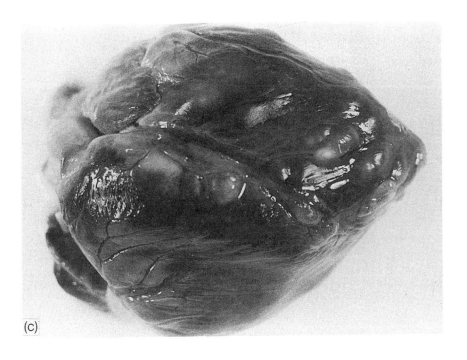

(c)

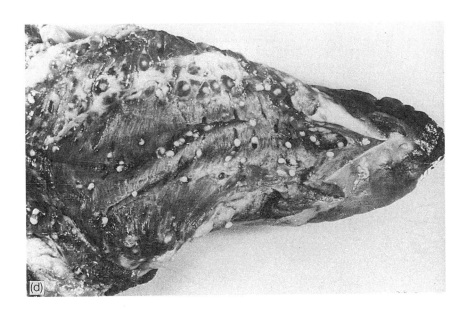

(d)

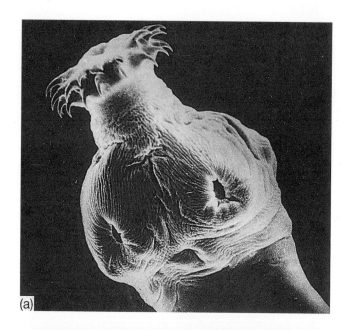

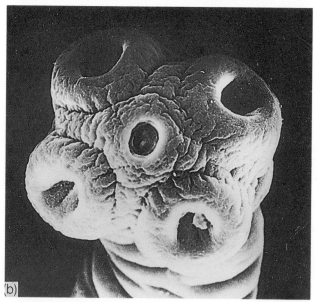

Figure 3.4 A scanning electron micrograph (SEM) of a *T. saginata* scolex which is characterized by a lack of hooks and is usually about 2 mm in diameter (b) – and an SEM of the scolex of *T. solium* with its hooks. The *T. solium* scolex is usually about 1 mm in diameter (a). (Photograph courtesy of Dr A. Flisser, Institute of Biomedical Investigations, Mexico.)

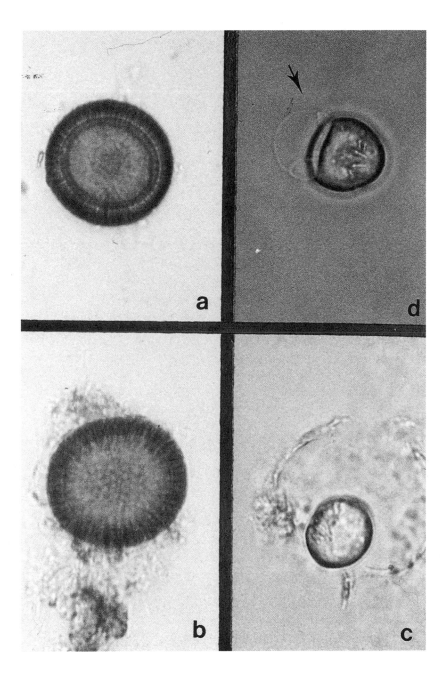

Figure 3.5 A *T. saginata* egg (a) which is 35–40 µm in diameter and indistinguishable from a *T. solium* egg. The egg shell disrupts in the stomach of the calf (b) releasing the oncosphere in its oncospheral membrane (c). The oncosphere which is about 20 µm in diameter is activated in the intestine and burrows through the intestinal wall. Note the material released from the excretory glands (d).

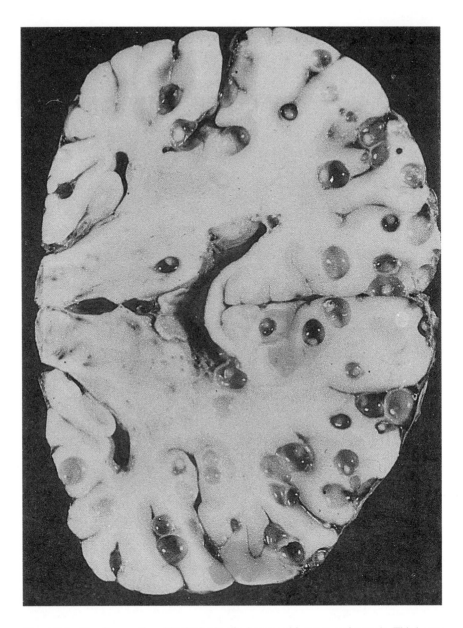

Figure 3.6 *T. solium* cysticerci in the brain of a human with neurocysticercosis. This is one of the characteristics of *T. solium* which makes this parasite of medical importance. (Photograph courtesy of Elsevier Publications, Cambridge.)

the fact they are often found to have massive burdens of *T. solium* cysticerci.

Cattle in Africa are most numerous away from the tsetse fly belt and the deserts but they extend northward into sub-Saharan Africa and along the north coast. Ethiopia, Somalia and the area along the Nile are particularly heavily populated. Cattle extend down eastern Africa in a patchy distribution, avoiding areas of tsetse fly and then become more prevalent again in southern Africa (FAO, WHO, OIE Animal Health Yearbook 1988). Cattle being herbivorous are usually taken to graze in areas away from direct contact with human dwellings and their associated detritus, although they may still be in close contact with their herdsmen.

Consequently, although both cattle and the tapeworm and the cysticerci of *T. saginata* are more common and more widespread in Africa than pigs and *T. solium*, the intensity of infection with the former is usually less (Gemmell 1986). Indeed *T. solium* is virtually unknown in the Islamic countries of northern Africa. Merle (1958) reported that *T. solium* was less widespread in Africa than *T. saginata* but was prevalent in Madagascar, Cameroon, parts of Zaire and South Africa. This may be an underestimate since pigs with heavy infections of cysticerci were common on the Bauchi Plateau of Nigeria in the 1960s (M.M.H.S. personal observation) and Nelson *et al.* (1965) cited cases reported from Kenya, Zimbabwe, Rwanda, Burundi and Zaire. There are few reports since then, although Pandey and Mbemba (1976) confirmed that *T. solium* occurs in pigs of Zaire, while Zoli *et al.* (1987) cited cases of infected pigs in Chad, Nigeria and Cameroon. The disease in man was first recorded in West Africa on the Ivory Coast (Bowesman 1952) while Proctor *et al.* (1966) identified spinal cysticercosis as a major cause of paraplegia in man in Ghana.

In Africa cattle often graze in close proximity to non-domesticated antelopes and like them can be prey for wild carnivores if not closely guarded. The question was raised as to whether wild animals, and in particular wild ruminants, play a role in the transmission of *T. saginata*. The definitive study on this was that by Nelson *et al.* (1965), who examined the carcasses of over 2000 wild animals, including 72 ruminants. They found cysticerci in only 25 animals, mostly ruminants, but only one wildebeest (*Connochaetes taurinus*) contained cysticerci with a hookless scolex, characteristic of *T. saginata*. These authors also cited another finding of a few such cysts in a wild bushbuck (*Tragelaphus scriptus*) and some infections in captive animals. Their overall conclusion, however, was that 'wild animals are of little importance as intermediate hosts of *T. saginata* in Kenya'. They reached a similar conclusion for *T. solium* and indeed considered that some of the reports of the cysts of this parasite being found in man, and even in pigs, in Africa may have resulted from incorrect identification of the cysticerci of other taeniod cestodes.

These conclusions can probably be applied to Africa in general. Indeed, the role of wildlife and their taeniod infections may even be beneficial, for it has recently been suggested that the self-limiting infections that result when cattle ingest the eggs of taeniod parasites of other ruminants (e.g. *Taenia crocutae*) may serve to stimulate or enhance the resistance of cattle to *T. saginata* infection (Harrison *et al.* 1985).

The opportunity in Africa for contact between pigs and wild animals is very much more limited than that for cattle, since the pigs tend to be housed or kept in villages.

Following a series of epidemiological studies, Gemmell (1986) concluded that the difficulty in eradicating the large tapeworms such as *T. saginata* and to a lesser extent *T. solium* may lie in an interaction between their high biotic potential and the effective immunity acquired by their intermediate hosts. Thus the balance between parasitic survival and attrition, on the one hand, and host immunity and fitness on the other, are the principal components in a complex system of interactions. For example in areas of high prevalence of *T. saginata*, such as Africa, young cattle rapidly develop resistance to reinfection as a result of ingesting eggs from pasture (Urquhart 1961, Froyd 1964a) and will retain that resistance because of the continual challenge. A result of this acquired resistance is that the numbers of cysticerci in the cattle when they are slaughtered are lower than would otherwise be expected from the high infection pressure. If the numbers of eggs on the pasture drops, perhaps as a result of the application of control measures, young cattle may fail to acquire this resistance and remain susceptible to sporadic challenge. This is the situation which occurs in areas of low prevalence such as Britain, where cattle involved in localized outbreaks may become heavily infected (Sewell & Harrison 1978).

Naive cattle of all ages are susceptible to primary infection with *T. saginata* (Vegors & Lucker 1971), however, they rapidly acquire a strong resistance to reinfection and will usually begin to destroy those cysts which become established following the primary infection within 10–12 months (Penfold 1937). Thus an explanation is needed for the fact that adult cattle of five or more years of age are commonly found to be carrying viable *T. saginata* cysticerci at slaughter in Kenya (Urquhart 1961, Froyd 1964a, 1964b). One possibility is that these cysticerci are those which have survived since their hosts were first infected very early in life, as it has been shown experimentally that cysticerci acquired during the neonatal period are able to survive in an otherwise resistant host (Gallie & Sewell 1972). Although *T. solium* has a high biotic potential, it appears to be more susceptible to control by conventional means. This may be partly due to the relatively lower egg production of adult *T. solium* as compared with adult *T. saginata* and to the ease with which management practices for pigs can be modified.

Zoonotic importance

Taeniasis causes various symptoms and signs which probably depend very much on the psychological and physical characteristics of the host: some patients lose their appetite and thus lose weight, others gain weight; some tolerate the infection while others do not. Statistical cluster analysis has been very useful in analysing the symptoms (Pawlowski 1983). Taeniasis is known to cause pathomorphological changes in the jejunal mucosa and also functional disorders; reversible achlorhydria or lowered gastric secretion being found in 70% of patients (Pawlowski 1986). As a result of the reduced gastric acidity, bacterial or viral gastroenteritis is common, especially in those who are permanently exposed to various intestinal pathogens (Cook 1985).

Despite the relatively mild symptoms caused by the adult tapeworm, the infection is aesthetically unacceptable. In addition, infection with adult *T. solium* is associated with the risk of an acute infection with the cysticerci. Infected persons require treatment while the rest of the population require protection from infection and the assurance that their beef and pork do not contain viable cysticerci of either parasite.

In humans the symptoms of cysticercosis are also generally of minor importance, in the absence of nervous system involvement, being limited principally to muscular aches and pains. Neurocysticercosis is however a serious condition, often debilitating and sometimes fatal. The most frequent feature is epilepsy but almost any focal neurological abnormality may occur (Shanley & Jordan 1980).

Locally acquired human cysticercosis continues to be a major cause of morbidity in parts of Africa. It has been found to be the commonest identifiable cause of epilepsy in black South Africans, accounting for 32% of the epileptic patients in this population (Powell *et al.* 1966). However, the prevalence of human cysticercosis in most of Africa is not as well recorded (Zoli *et al.* 1987) as it is in other areas of the world such as Mexico (Flisser *et al.* 1976, Flisser *et al.* 1980), India (Chopra *et al.* 1981), Northern China (Jing & Wang 1985) and West Irian (Gajdusek 1978).

A recent epidemiological survey carried out in northern Togo also points to a link between cysticercosis and epilepsy in that area (Dumas *et al.* 1988) with 38% of epileptic patients also found to have cysticercosis.

It has been commonly assumed that, apart from its zoonotic importance, cysticercosis *per se* is of little importance in domestic animals, only causing occasional acute disease following rare massive infections or leading to aesthetically unacceptable infections in pet dogs (Botha 1980). However, there is some evidence which suggests that these parasites may not be totally innocuous. This is because, in common with many other parasites, such infections appear to be associated with general debilitation and some degree of suppression of their host's immune response (Nichol & Sewell 1984, Laclette *et al.* 1986, Wes Leid *et al.*

1986). If this suppression results in the hosts being less able to control concurrent diseases or to respond adequately to vaccines, these parasites may be of greater general importance than has been previously believed. Clearly such considerations may also apply to human cysticercosis patients.

Meat inspection

Although most animals slaughtered in Africa under village conditions are not subject to any form of meat inspection by qualified personnel, infected meat may be noted and rejected as unfit for consumption as was the case with carcasses of *T. solium* infected pigs (measly pork) in Europe as early as the thirteenth century, well before the association with tapeworm infections was recognized (Viljoen 1937, quoting Von Ostertag).

The public health importance of larval taeniids in meat has led to the practice of condemnation or treatment of infected organs or whole carcasses observed during meat inspection in centralized modern slaughter houses in an attempt to prevent completion of the parasitic life cycle. Research into the control of *T. saginata* taeniasis/cysticercosis was stimulated in Africa largely as a result of the desire by the beef producing nations to develop an export trade in beef to Europe and elsewhere. The presence of cysticerci in the meat would be a serious obstacle to meeting the import regulations of the recipient countries.

Studies were made by veterinary surgeons supervising meat inspection in abattoirs, who were primarily concerned with finding the most reliable routines for detecting the cysts in naturally infected carcasses by direct knife-and-eye inspection (Viljoen 1937, Mann & Mann 1947) and also ways of handling such carcasses, usually by prolonged freezing or boiling, so as to eliminate the possibility of them transmitting the parasite to man. At the same time the data obtained in abattoirs was used to assess the prevalence of cysticercosis in cattle and pigs in the catchment areas.

Meat inspection still relies exclusively on visual examination of the intact and cut surfaces of the carcass in the slaughter house by meat inspectors who follow officially laid-down procedures. These vary from country to country and between the two parasitic infections. In the case of *T. saginata* the organs examined usually include the heart, diaphragm, tongue and cheek muscles, these being on the whole the less valuable parts of the carcass. Several of these are also the sites at which the largest concentration of metacestodes are found in experimentally infected animals (Table 3.1) and may therefore be the optimal sites to examine in areas where the cattle have usually acquired the infection relatively recently. However under African conditions, where the infections are most commonly long-standing, the heart in particular is usually less heavily infected at slaughter. It was also shown some years ago that it is

Table 3.1 The distribution of *Taenia saginata* cysticerci in various organs of 97 experimentally infected cattle (the standard error was < 1.0% in each case) (G. J. Gallie & M. M. H. Sewell, unpublished).

Organ	Proportion of total burden (%)	Average number of cysts per 10 kg
Heart	21.7	160
Diaphragm	6.9	100
Head	5.5	80
Oesophagus	0.5	20
Forelimbs	14.4	9
Tongue	1.3	9
Lungs	2.0	8
Hind limbs	20.8	5
Neck	7.0	6
Trunk	19.0	3
Elsewhere	< 0.5	< 1

advantageous to include one or more cuts into the shoulder muscles of African cattle (Mann & Mann 1947).

Although light infections with *T. saginata* are common, only the more heavily infected carcasses can be reliably detected by these traditional methods of knife-and-eye inspection. Hence it is recognized that existing procedures fail to detect many of the infected carcasses (Mann & Mann 1947, Walthers and Koseke 1980). Accordingly meat inspection cannot be relied upon to adequately control these parasites, although it may prevent the aesthetically unacceptable heavily infected carcasses from appearing in the market place. Unfortunately, it is precisely these lightly infected carcasses, which usually escape detection during meat inspection, that are also most readily overlooked by consumers, so that the parasites in them are more likely to be eaten by humans and so infect those with a preference for raw or lightly cooked meat.

In the case of *T. solium* in pigs generalized infections are more common and the cysticerci are larger so they are more easily detected at meat inspection. This may be another reason why this parasite has proved easier to control by conventional means.

Public health measures

The consequences to man of infection with the cysticerci of *T. solium* together with the economic losses cased by the disease in man and animals have led to many national efforts to control or eradicate tapeworm diseases (Abduladze 1970, Oliver 1974, Schwabe 1984).

In most developed countries a combination of meat inspection and other factors, such as general improvements in sanitation and the availability of effective drug treatment for humans infected with tapeworms, have succeeded in almost or completely eradicating *T. solium* but not *T. saginata*. Such measures have not so far been consistently applied in most of the less developed countries. At present only a few such countries even have definite plans for controlling these parasites (Pawlowski 1980, Flisser *et al.* 1980, Engelbrecht *et al.* 1985). The large-scale use of praziquantel has proved useful in the control of *T. solium* taeniasis (Pawlowski 1986b). It must therefore be expected that taeniasis/cysticercosis will continue to be a problem in much of Africa for the foreseeable future.

Economic impact

Attempts to reduce the prevalence of *T. solium* and *T. saginata* in humans and their cysticerci in pigs and cattle may have a considerable impact on the economics of the meat production industries. This will be particularly important where export industries are involved, since most importing countries have stringent regulations designed to prevent the importation of infected meat.

The costs can be broken down into those involved in treating human taeniasis (Pawlowski 1986) and more importantly human cysticercosis (Velasco-Suarez *et al.* 1982); those arising when pig (Acetavedo-Hernandey 1982) and cattle carcasses are treated or condemned; and the costs involved in the inspection procedures themselves. In most countries there are no accurate data on any of these factors, especially in relation to *T. solium*. Grindle (1978) estimated that bovine cysticercosis was costing the economies of both Kenya and Botswana about one million UK pounds each per annum. In the former country this mainly arose from the loss of value that results in small abattoirs from boiling the meat to kill the cysts. In Botswana, a lower prevalence in fewer animals caused similar losses because of the reduced export potential. The losses will also vary with the prevalence, which fell for *T. saginata* in Kenya between 1975 and 1980 (Cheruyiot 1983), but rose again during the drought of 1984/85.

Diagnosis

Taeniasis

It is particularly desirable to identify and treat *T. solium* infections in man because of the risk of either the host or his contacts developing *T. solium* cysticercosis. It is important, therefore, to be able to discriminate between *T. saginata* and *T. solium* tapeworm infection in humans. The morphological characteristics of the gravid proglottides such as their size and the number of uterine branches are usually used as the criteria for such identification. However, there can be considerable difficulty in distinguishing the two species because such features may overlap (Proctor 1972) and accordingly many clinical reports refer only to '*Taenia* spp.'

The two species can be differentiated biochemically by the electrophoretic profiles of their proteins (Bursey *et al.* 1980) or by their isoenzyme patterns (Le Riche & Sewell 1977, 1978). Unfortunately the samples required for use in these procedures are labile and must be either processed immediately or carefully frozen, preferably in liquid nitrogen. Fixatives such as alcohol or formalin cannot be used, so samples cannot easily be transported from one reference laboratory to another.

The application of species specific DNA probes to the identification of these tapeworm proglottides has immediate utility precisely because DNA is very stable and samples can be stored in alcohol for transportation, prior to DNA extraction and analysis. In the clinical situation it is important to have DNA probes which will positively identify each individual sample. This is now feasible with the production of three DNA probes, one specific for *T. saginata*, one reactive with both *T. saginata* and *T. solium* (Harrison, Delgado *et al.* 1987, 1988) and a third specific for *T. solium* (Rishi & McManus 1987). We may also expect DNA probes to make a major impact in epidemiological studies as specific reagents for the identification of discrete strains or geographical isolates of the parasite.

Cysticercosis

The most urgent requirement in the diagnosis of taeniid cysticercal infections is a serological test for the presence of viable parasites. This could be used to provide an immediate estimate of the prevalence of these infections in humans, cattle or pigs and, as a rational indicator for clinical action for *T. solium* neurocysticercosis in man. The progress made in this direction will be reviewed later in this section, but a more general review with some clinical perspectives will be presented first.

In contrast to adult tapeworm infections, where the proglottides or eggs may be found in the faeces, the identification of hosts harbouring cysticerci by direct parasitological examination poses problems. This is because the cysticerci are relatively small and widely disseminated in the tissues of the host. In humans *T. solium* cysticerci can sometimes be identified by biopsy of subcutaneous nodules, while in pigs it is often possible to observe them under the tongue, especially in relatively heavy infections. Cysticercosis is rarely diagnosed during life in cattle. The presence of dead calcified cysticerci can be detected by X-rays, but while this may be of some value in assisting the diagnosis of human cysticercosis, it is of little practical value in domestic animals. These reservations have stimulated a search for alternative methods of parasite detection.

Computerized tomography (CT) scanning has provided a major advance in the reliable diagnosis of neurocysticercosis in humans (Byrd *et al.* 1985). However, the procedure is expensive and rarely available in rural areas where *T. solium* persists, nor is it practical to screen large numbers of people by this technique.

An alternative diagnostic procedure is serodiagnosis, which allows the rapid screening of patients with neurological symptoms. Infected domestic animals might also be identified by *ante mortem* serological screening and treated to kill the cysticerci, excluded from feedlot systems or subjected to close *post mortem* examination followed by treatment of the infected carcasses by freezing or cold storage.

All of these parasitological or clinical procedures require specially trained personnel and have the added disadvantage that each individual test is processed serially. Modern serology, through the widespread adoption of the enzyme-linked immunosorbent assay (ELISA) on the other hand, is rapid, cheap, may be automated and has the major logistic advantage that the samples are processed simultaneously with consequent decrease in turnover time.

Requirements for a serodiagnostic assay include levels of sensitivity, specificity and stability and ease of performance, especially when used under tropical conditions. It is particularly important when diagnosing human cysticercosis that each infected patient is accurately identified. The criteria could perhaps be less stringent for herd animals, such as cattle, which can be treated on a group basis. Diagnostic assays for larval cestode infections should be directed towards the detection of those hosts harbouring viable larvae. They should also provide an indication of the severity and progress of the infection, thus identifying those hosts which are suitable candidates for drug or surgical treatment.

It has long been known that taeniids and other helminth parasites contain many cross reactive antigenic epitopes (Biguet *et al.* 1965, Capron *et al.* 1968). Accordingly it is not surprising that the previous assays, which were based on the use of crude parasite extracts for antibody detection experienced limited success. The problems included lack of specificity due to crossreactions with other helminths (Gottstein *et al.* 1986) and lack of sensitivity due to poor signal to background ratios (Flisser & Larralde 1986). Hence there have been many reports of unacceptably high numbers of false negative and/or false positive results. The problem is particularly serious in the diagnosis of helminth infections in domestic animals as they tend to be exposed to a large range of parasites (Geerts *et al.* 1977). Also, although not generally the case with *T. solium* infection in pigs, *T. saginata* tends to be over-dispersed in the cattle population with the majority of animals harbouring only a small number of cysticerci (Yong *et al.* 1984) so that they do not develop a strong serological response to infection.

The detection of specific antibodies in cerebro-spinal fluid is still the main diagnostic tool for suspected cases of neurocysticercosis in man (Estrada & Kuhn 1985, Mohammad *et al.* 1985, Rosas *et al.* 1986). Although false positive and false negative reactions may again cause interpretation difficulties, serodiagnosis of *T. solium* cysticercosis is undoubtedly of epidemiological value (Larralde *et al.* 1986). Attempts to improve the performance of serological tests by partial purification of antigen extracts have been limited (Costa *et al.* 1982, Nascimento & Mayrink 1984, Espinoza *et al.* 1982). More recently assays based on cyst fluid have yielded improved results (Larralde *et al.* 1986).

The reason why assays of an acceptable degree of sensitivity and specificity for the detection of *T. saginata* and *T. solium* infection have not been developed is primarily a failure to identify sufficiently specific antigenic epitopes (Harrison & Parkhouse 1985). In addition there is evidence of strain differences occurring in *T. saginata* (Woulters *et al.* 1987). Such differences may be related to the antigenic make-up of these parasites and could be important factors in the design of diagnostic assays. More detailed work on this aspect is still required and further evidence for such variation could be sought through the use of DNA probes and restriction enzyme polymorphism. A detailed study at the molecular level of parasite antigens is a prerequisite for the logical design of serodiagnostic assays (Parkhouse *et al.* 1987). This is particularly the case when dealing with complex metazoan parasites such as the taeniids (Harrison & Parkhouse 1985). The application of modern analytical techniques such as Western blotting (Harrison, Parkhouse *et al.* 1984, Grogl *et al.* 1985, Larralde *et al.* 1986) and the direct or biosynthetic radiolabelling combined with SDS–PAGE analysis should allow the identification of those parasite products which are of diagnostic potential.

A monoclonal antibody ELISA capture assay has recently been developed which detects surface and excreted products of cysticerci in the serum of *T. saginata* infected bovines and, due to a convenient cross reaction, the assay also detected similar excretions/secretions of *T. solium* cysticerci in the serum and in the cerebrospinal fluid of humans with confirmed *T. solium* cysticercosis (Harrison *et al.* 1989). This latter finding was confirmed by further studies (Correa *et al.* 1989). Limited additional studies also suggest a use for this monoclonal-antibody-based capture assay in the diagnosis of *T. solium* cysticercosis of pigs (Rodrigues del Rosal *et al.* 1989).

Once a specific serodiagnostic system has been identified, the sensitivity of modern ELISA-based technology, notwithstanding the possibility of further refinements, offers the necessary sensitivity and versatility for field use. Effective serodiagnosis will undoubtedly lead to an increased number of pigs or cattle identified as infected with *T. solium* or

T. saginata since the direct knife-and-eye methods for detecting cysticerci are known to be inaccurate (Mann & Mann 1947, Walthers & Koske 1980, Heath *et al.* 1985). The costs involved in maintaining a full-scale control programme may then be considerable and would meet with resistance from farmers or the meat trade unless a suitable compensation scheme was initiated.

Treatment

Treatment of infection with the adult tapeworms of *T. saginata* or *T. solium* is relatively easy with the appropriate dose of suitable chemotherapeutic drugs. The drugs of choice are niclosamide, introduced in 1960 (Pearson & Hewlett 1985), and praziquantel (Davis & Wegner 1979, Groll 1980) which was introduced in the late 1970s. Andrews *et al.* (1983) reviewed the action and activity of praziquantel and the drug has been used in wide-scale treatment programmes for the control of *T. solium* tapeworm infections in humans (Pawlowski 1987).

Until the beginning of the 1980s there was no effective treatment for cysticercosis in man other than surgical intervention to excise cysticerci or to relieve hydrocephalus in neurocysticercosis. Praziquantel is not only an effective treatment for adult taeniids but will destroy viable cerebral cysts of *T. solium* (Groll 1982, Sotelo *et al.* 1984) . Adverse effects are fortunately few, and inflammatory reactions due to dying cysticerci are treated with corticosteroids.

Chemotherapy has only been used in the experimental treatment of cysticercosis in pigs and cattle. Albendazole, which appeared to offer promise as a dual-purpose drug already licensed for use in many countries as a nematocide in cattle and pigs, is not fully effective against the cysticerci of *T. saginata* (Geerts & Kumar 1981, Stevenson *et al.* 1981, Kassai *et al.* 1984). Another benzimadazole derivative, fenbendazole, is effective against the cysticerci of *T. solium* (Tellez-Giron, *et al.* 1981) in pigs and high doses of fenbendazole are effective against immature and mature cysticerci of *T. saginata* (Bessonov *et al.* 1984). These results require confirmation.

Praziquantel kills *T. saginata* cysticerci in cattle at dose rates of 100 mg/kg and above but there is some controversy as to its efficacy against immature cysticerci (Thomas & Gonnert 1978, Gallie & Sewell 1978, Bessonov *et al.* 1980). Unfortunately praziquantel is at present an expensive drug and it is not economic to use it in cattle, although this may not continue to be the case as patents run out and/or alternative sources or analogues become available, such as the Chinese pyquiton (Xu *et al.* 1982). Another new Chinese drug Mienang No 5 also kills cysticerci (Jing & Wang 1985).

Even when all the cysts have been killed by a suitable drug, any calcified residues derived from those cysts that were already dead or dying at the time of treatment remain detectable in the carcass for months (Harrison, Gallie *et al.* 1984). Meat inspectors would have to be fully informed that drug treatment has taken place and appreciate that such residues, although possibly unacceptable aesthetically, are quite innocuous as regards the transmission of the parasite to man. This is of particular importance while meat inspection is relied on as the main means of post-mortem diagnosis and control. It would clearly be preferable in the long term to seek means of preventing infection in the first place.

Immunoprophylaxis

Because of the importance of acquired immunity in regulating the intensity of infection with cysticerci of *T. saginata*, Gemmell (1986) concluded that a vaccine is needed to replace this natural immunity until the parasite is eradicated and thus reduce the risk of cysticercosis storms. Harrison, Parkhouse *et al.* (1984) considered that this would apply even if effective chemotherapy was introduced as an additional control measure for *T. saginata* infection in cattle.

The development of vaccines against *T. saginata* cysticercosis in cattle and *T. solium* cysticercosis in pigs may be feasible as there is good evidence that taeniid larval infections can be prevented by immune mechanisms. Resistance to secondary infection can be produced either by primary infection with the parasite (Gallie & Sewell 1972, 1974, 1981) or by immunization using homologous or heterologous parasite extracts (Gallie & Sewell 1976, Lloyd & Soulsby 1976, Lloyd 1979). The oncosphere, or material secreted or shed by the oncosphere, is thought to be one potent source of immunogens (Rickard & Brumley 1981, Lloyd 1984). Extracts of cysticerci have also been used to vaccinate pigs against *T. solium* infection (Molinari *et al.* 1983b), as have extracts of *T. saginata* proglottid to immunize cattle against *T. saginata* infection (Gallie & Sewell 1976). However, until recently, little work had been done on the molecular characterization of the antigens involved (Harrison & Parkhouse 1985).

A considerable amount of evidence points to the importance of humoral immunity in resistance to the early stages of taeniid parasites. Protection by passive transfer of immune sera from animals infected with *T. saginata* and *T. taeniaeformis* has indicated that antibodies play a role in immunity and that it is the invasive larvae or oncospheres which are vulnerable to such attack (Lloyd & Soulsby 1976, Soulsby & Lloyd 1982).

Although the invasive oncosphere is considered the prime 'target' of a protective immune response, components of the cysticercal stage also constitute possible 'targets' and cysticercal extracts have been used to vaccinate animals (Harrison & Joshua 1987, Lightowlers & Rickard

1988). Studies on *T. solium* have suggested that resistance against the older cysticerci may have a cellular component in particular involving eosinophils (Molinari *et al.* 1983a). Evidence from the model system *T. taeniaeformis* in mice also points to a cellular involvement in the killing of larval cestodes (Letonja & Hammberberg 1987a, 1987b).

There would be clear advantages in having a vaccine which was directed against both the early invasive and later developmental stages of the parasites. If any invasive parasites survived the first line of immune defence they could then be killed at a later stage in development. Up till now most work has concentrated on either invasive oncospheres or on the fully developed metacestodes. Intermediate developmental forms now require study (Bogh *et al.* 1986, Heath & Harrison 1986). Using monoclonal antibodies raised against *T. saginata* oncospheres, Harrison & Parkhouse (1986) have shown that some components may be shared between oncospheres and metacestodes. This also appears to be the case with *T. taeniaeformis*, although other metacestode components may be unique (Gibbens *et al.* 1986). Studies on the surface, excretory/secretory, somatic and cyst fluid components of *T. saginata* indicate that an antibody response to certain components of the cysticerceri may also be associated with resistance to infection (Parkhouse & Harrison 1987, Joshua *et al.* 1988a, 1988b). In addition investigations into possible antigenic variation between different strains of these parasites should be initiated to ascertain if they have the same antigenic and genomic characteristics and to ensure that vaccines are directed towards uniformly available antigens.

Progress is being made on the application of genetic engineering to the search for vaccines against taeniids. For instance such an approach with *T. taeniaeformis* has led to immunization experiments in mice (Bowtell *et al.* 1984, Lightowlers *et al.* 1986). Similar studies on *T. ovis*, a sheep/dog taeniid which is closely related to *T. saginata* and *T. solium*, have resulted in the cloning of genes which express parasite proteins shared between the invasive oncosphere and the adult tapeworms of this species (Fisher & Howell 1986, Howell & Hargreaves 1988), although the exact nature of some of these proteins has still to be determined. Recently, however, Johnson *et al.* (1989) produced a recombinant peptide, from a *T. ovis* oncosphere cDNA library, with a molecular weight of approximately 45 kDa, which gave a 94% protection in sheep against ovine cysticercosis. Although there are many practical difficulties to be overcome in the production of vaccines by genetic engineering (Simpson *et al.* 1986), extending the application of these techniques to parasites such as *T. solium* and *T. saginata* could do much to overcome the problems encountered when trying to produce vaccines against a parasite, where the main source of parasitic material is man or cattle. The vaccination of cattle and pigs in endemic areas must be a long-term aim for control of *T. saginata* and *T. solium* infection.

Conclusions

Conventional approaches that can be taken for the control of cysticercosis in man and domestic animals and taeniasis in man include improved standards of human hygiene, public health education and a more realistic approach to management practices and to the efficacy of meat inspection. It would be rational to adopt the attitude that when any carcasses in a group are infected then all the animals from that group are presumed to be infected and treated accordingly by prolonged cold-storage or boiling.

These approaches may well eliminate *T. solium* but only reduce the prevalence of *T. saginata*. Additional control methods such as immuno-prophylaxis and wide-scale application of chemotherapy are needed to augment conventional methods, speed up control programmes and possibly eliminate *T. saginata* from the cattle population.

Improvements in the accuracy of detection of infected animals or humans by the development of highly specific and sensitive monoclonal antibody-based assays which detect the characteristic products of metabolism of viable cysts in the circulation would be very advantageous. This would allow the identification of human patients, young animals or groups of animals which require treatment. The detection of those slaughter-weight animals which present a genuine hazard to human health and the exclusion of animals which only contain dead cysts could be a major factor in rendering such serological screening acceptable to owners, butchers and public health authorities.

Within the last decade the possibility of effective chemotherapy against cysticercosis has become a reality. Praziquantel is still too expensive for routine use in animals although this situation may well change. There is a continuing need to develop commercially viable larvicidal cestocides suitable for treating both humans and animals. Indeed, in view of the appearance of resistance to previously highly effective anthelmintics by other helminths it would be preferable not to place total reliance on any one drug. However, the instigation of a control programme involving the use of chemotherapy would probably need to be undertaken by government since such a programme could not be properly co-ordinated if it were to rely solely on the initiative and action of individuals.

It will be desirable, especially in hyperendemic areas, that cysticidal drugs should mainly be used in young animals to reduce treatment costs and to allow time for complete resorption of dead cysts before slaughter. They should also be used in conjunction with an effective and cost-effective vaccination procedure so as to prevent reinfection. Such vaccines should preferably be non-living and might with advantage be incorporated in with other vaccine or vaccines used in the same youngstock. Future studies should be aimed at devising such cost-effective and practical diagnostic and immunoprophylactic methods by the application of modern immunochemical and bioengineering techniques.

References

Abduladze, K. I. 1970. *Essentials of cestodology*. Taeniasis of animals and man and diseases caused by them. K. I. Shrjabin (ed.), Vol. 4, US Department of Agriculture.

Acetavedo-Hernandey, A. 1982. Economic impact of porcine cysticercosis. In *Cysticercosis: present state of knowledge and perspectives*. A. Flisser, K. Willms, J. P. Laclette, C. Larralde, *et al.* (eds), 63–8. New York: Academic Press.

Andrews, P., H. Thomas, R. Pohlke & J. Seubert 1983. Praziquantel. *Med Res Rev.* **3**, 147–200.

Bessonov, A. V., N. S. Arkipova, S. P. Ubiraer, A. N. Uspenskii *et al.* 1984. Treatment of cattle infected with *Cysticercus bovis*. *Trudy Vsesoyuznogo Institute Gel'mintologii im KO Skrjabnina*, **27**, 13–7.

Bessonov, A. V., A. N. Uspenskii, Yu. B. Komarov, Sh. A. Azimov, *et al.* 1980. Treatment of *Cysticercus bovis* infection in calves. *Vet. Moscow*, **3**, 40–2.

Biguet, J., F. Rose, A. Capron & P. Tran Van Ky 1965. Contribution de l'analyse immunoélectrophoretique à la connaissance des antigènes verminaux. Incidence pratiques sur leur standardisation, leur purification et le diagnostic d'helminthosis par immunoélectrophorèse. *Rev. Immunol. Therap. Microb.* **29**, 5–23.

Bogh, H. O., M. D. Rickard & M. W. Lightowlers 1986. Stage specific immunity following vaccination against *Taenia taeniaeformis* infections in mice. In *Immunology and molecular biology of cestode infections*. Coopers Animal Health Symposium, Melbourne, p. A5.

Botha, W. S. 1980. Cerebral cysticercosis in a dog. *J. South Afr. Vet. Assoc.* **51**, 127.

Bowesman, C. 1952. Cysticercosis in West Africa. *Ann. Trop. Med. Hyg.* **46**, 101–2.

Bowtell, D. D. L., R. B. Saint, M. D. Rickard & G. F. Mitchell 1984. Expression of *Taenia taeniaeformis* antigens in *Escherichia coli*. *Mol. Biochem. Parasitol.* **13**, 173–85.

Bursey, C. D., J. A. McKenzie & M. B. H. Burt 1980. Polyacrylamide gel electrophoresis in the identification of *Taenia* (Cestoda) by total protein. *Int. J. Parasitol.* **10**, 167–74.

Byrd, S. E., J. Daryabagy, R. Thompson, J. Zant *et al.* 1985. The computed tomographic spectrum of cerebral cysticercosis. *J. Nat. Med. Assoc.* USA **77**, 553–60.

Capron, A., J. Biguet, A. Vernes & D. Afchain 1968. Structures antigènique des helminthes. Aspects immunologiques des relations hôte–parasite. *Pathol. Biol.* **16**, 121–8.

Cheruyiot, H. K. 1983. Bovine helminth parasites of economic importance. Abattoir survey in Kenya 1979–80. *Bull. Anim. Hlth Prod. Afr.* **31**, 367–75.

Chopra, J. S. U. Kaur, & R. C. Mahajan 1981. Cysticercosis and epilepsy: a clinical and serological study. *Trans. R. Soc. Trop. Med. Hyg.* **75**, 518–20.

Cook, G. C. 1985. Infective gastroenteritis and its relationship to reduced gastric acidity. *Scand. J. Gastro-enterol.* **20**, Suppl. III, 17–23.

Cornaglia, E. & A. Lo Schiavo 1985. Massive hepatic cysticercosis in the zebu. *Bolletino Scientifico della Facolta di Zootecnia e Veterinaria Università Nationale Somalia*, **5**, 101–6.

Correa, D., M. A. Sandoval, L. J. S. Harrison, M. E. Parkhouse *et al.* 1989. Human neurocysticercosis: comparison of monoclonal and polyclonal based EIA capture assays for the detection of parasitic products in cerebrospinal fluid. *Trans. R. Soc. Trop. Med. Hyg.* **83**, 814–6.

Costa, J. M., A. W. Ferreira, M. M. Makino & M. E. Camargo. 1982. Spinal fluid immunoenzymatic assay (ELISA) for neurocysticercosis. *Rev. Inst. Med. Trop. São Paulo*, **24**, 337–41.

Craig, P. S., W. Crewe & R. R. Owen (eds) 1984. Trends in research on cestode infections. *Ann. Trop. Med. Parasitol.* **78**, 181–263.

Davis, A. & D. H. G. Wagner 1979. Multicentre trials of praziquantel in human schistosomiasis: design and technique. *Bull. Wld Hlth Org.* **57**, 767–71.

Dumas, M., E. Grunitzky, M. Deniau, F. Dabis *et al.* 1988. Epidemiologic study of neuro-cysticercosis in north of Togo (West Africa). XIIth International Congress on Tropical Medicine and Malaria; C–Now Satellite Symposium on human neurocysticercosis, Rotterdam.

El–Sadik, A. 1979. Distribution of bovine cysticercosis in Sudan from an abattoir survey in Khartoum (Sudan). *S. Nauch. Tru. Mosk. Vet. Akad.* **108**, 120–2.

Engelbrecht, H., U. A. Mentzel & H. Hudemann 1985. Grundlagen und Kritesien eines Komplexen Programms zur Bekämpfung von *Taenia saginata* im Kreis Wittstock (Bez. Putsuum/DDR). *Zeit. Gesamte Hyg. Grenzgeh.* **31**, 234–6.

Espinoza, B., A. Flisser, A. Plancarte, A. & C. Larralde 1982. Immunodiagnosis of human cysticercosis: ELISA and immuno–electrophoresis. In *Cysticercosis: present state of knowledge and perspectives*. A. Flisser, K. Willms, J. P. Laclette, C. Larralde *et al.* (eds), 163–170. Suppl., 235–6. New York: Academic Press.

Estrada J. J. & R. E. Kuhn 1985. Immunochemical detection of antigens of larval *Taenia solium* and anti–larval antibodies in the cerebrospinal fluid of patients with neuro-cysticercosis. *J. Neurol. Sci.* **71**, 39–48.

FAO, WHO, OIE. Animal Health Yearbook 1988. FAO, Rome.

Fisher, M. & M. J. Howell 1986. Isolation, cloning and characteristics of different genes in *Taenia ovis*. In *Immunobiology and molecular biology of cestode infections*. Coopers Animal Health Symposium, Melbourne, A7.

Flisser, A. & C. Larralde 1986. Cysticercosis. In *Immunodiagnosis of parasitic diseases*. K. W. Walls & P. M. Schantz (eds), 109–61. New York: Academic Press.

Flisser, A., I. Bulnes, M.L. Dias & R. Luna 1976. Estudio sero–epidemiologico de la cysticercosis humana en poblaciones predominantemente indigenas y rurales del estado de Chiapas. *Arch. Invest. Medica* **7**, 107–13.

Flisser, A., E. Woodhouse & C. Larralde 1980. Human cysticercosis: antigens, antibodies and non–responders. *Clin. Exp. Immunol.* **39**, 27–37.

Flisser, A., K. Willms, J. P. Laclette, C. Larralde *et al.* (eds) 1982. *Cysticercosis: present state of knowledge and perspectives*. New York: Academic Press.

Flisser. A., L. Rivera, J. Trueba, B. Espinoa *et al.* 1982. Immunology of human neuro-cysticercosis. In *Cysticercosis: present state of knowledge and perspectives*. A. Flisser, K. Willms, J. P. Laclette, C. Larralde *et al.* (eds), 549–63. New York: Academic Press.

Foster, W. P. 1965. *A history of parasitology*. Edinburgh and London: E. & S. Longstone.

Froyd, G. 1964a. The artificial oral infection of cattle with *Taenia saginata* eggs. *Res. Vet. Sci.* **5**, 434–40.

Froyd, G. 1964b. The longevity of *Cysticercus bovis* in bovine tissues. *Brit. Vet. J.* **120**, 205–11.

Gajdusek, D. C. 1978. Introduction of *Taenia solium* into West New Guinea with a note on an epidemic of burns from cysticercus epilepsy in the Ekari people of the Wissel Lakes area. *Papua New Guinea Med. J.* **21**, 329–42.

Gallie, G. J. & M. M. H. Sewell 1972. The survival of *Cysticercus bovis* in resistant calves. *Vet. Rec.* **91**, 481–2.

Gallie, G. J. & M. M. H. Sewell 1974. The serological response of three–month old calves to infection with *Taenia saginata* (*Cysticercus bovis*) and their resistance to reinfection. *Trop. Anim. Hlth Prod.* **6**, 163–71.

Gallie, G. J. & M. M. H. Sewell 1976. Experimental immunisation of six–month old calves against infection with the cysticercus stage of *Taenia saginata*. *Trop. Anim. Hlth Prod.* **8**, 233–42.

Gallie, G. J. & M. M. H. Sewell 1978. The efficacy of praziquantel against the cysticerci of *Taenia saginata* in calves. *Trop. Anim. Hlth Prod.* **10**, 36–8.

Gallie, G. J. & M. M. H. Sewell 1981. Inoculation of calves and adult cattle with oncospheres of *Taenia saginata* and their resistance to reinfection. *Trop. Anim. Hlth Prod.* **13,** 147–54.

Geerts, S. & V. Kumar 1981. The effect of albendazole on *Taenia saginata* cysticerci. *Vet. Rec.* **109,** 217.

Geerts, S., V. Kumar & J. Brandt (eds) 1987. *Helminth zoonoses with particular reference to the tropics.* Dordrecht: Martinus Nijhoff.

Geerts, S., V. Kumar & J. Vercruysse 1977. *In vivo* diagnosis of bovine cysticercosis. *Vet. Bull.* **47,** 653–64.

Gemmell, M. A. 1986. A critical approach to the concepts of control and eradication of echinococcosis/hydatidosis and taeniasis/cysticercosis. *Proc. Sixth Int. Cong. Parasitology, Parasitology – Quo vadit?* M. J. Howell (ed.), 469–72. Australian Academy of Science.

Gemmell, M. A., Z. Matyas, Z. Pawlowski & E. J. L. Soulsby (eds) 1983. *Guidelines on surveillance, prevention and control of taeniasis/cysticercosis.* WHO, VPH 83/49.

Gibbens, J. C., L. J. S. Harrison & R. M. E. Parkhouse 1986. Immuno-globulin class response to *Taenia taeniaeformis* in susceptible and resistant mice. *Parasite. Immunol.* **8,** 491–502.

Gottstein, B., V. C. W. Tsang & P. M. Schantz 1986. Demonstration of species-specific and cross-reactive components of *Taenia solium* metacestode antigens. *Am. J. Trop. Med. Hyg.* **35,** 308–13.

Grindle, R. J. 1978. Economic losses resulting from bovine cysticercosis with special reference to Botswana and Kenya. *Trop. Anim. Hlth Prod.* **10,** 127–40.

Grogl, M., J. J. Estrada, G. MacDonald & R. E. Kuhn 1985. Antigen antibody analysis in neurocysticercosis. *J. Parasitol.* **71,** 433–42.

Groll, E. W. 1980. Praziquantel for cestode infections of man. *Acta Tropica* **37,** 293–6.

Groll, E. W. 1982. Chemotherapy in human cysticercosis with praziquantel. In *Cysticercosis: present state of knowledge and perspectives.* A. Flisser, K. Willms, J. P. Laclette, C. Larralde *et al.* (eds), 207–18. New York: Academic Press.

Harrison, L. J. S. & G. W. P. Joshua 1987. Immunoprophylaxis of *Taenia saginata* cysticercosis. In *Helminth zoonoses with particular reference to the tropics.* Geerts, S., V. Kumar & J. Brandt (eds), 81–4. Dordrecht: Martinus Nijhoff.

Harrison, L. J. S. & R. M. E. Parkhouse 1985. Antigens of taeniid cestodes in protection, diagnosis and escape. *Curr. Top. Microbiol. Immunol.* **120,** 159–68.

Harrison, L. J. S. & R. M. E. Parkhouse 1986. Passive protection against *Taenia saginata* by a mouse monoclonal antibody reactive with the surface of the invasive oncosphere. *Parasite. Immunol.* **8,** 319–32.

Harrison, L. J. S., J. Delgado & R. M. E. Parkhouse 1987. Differentiation of *Taenia saginata* and *Taenia solium* by the use of cloned DNA fragments. *Trans. R. Soc. Trop. Med. Hyg.* **82,** 174.

Harrison, L. J. S., J. Delgado & R. M. E. Parkhouse 1988. Differential diagnosis of *Taenia saginata* and *Taenia solium* with DNA probes. *Parasitology* **100,** 459–61.

Harrison, L. J. S., G. J. Gallie & M. M. H. Sewell 1984. Absorption of cysticerci in cattle after treatment of *Taenia saginata* cysticercosis with praziquantel. *Res. Vet. Sci.* **38,** 378–80.

Harrison, L. J. S., G. W. P. Joshua, S. H. Wright & R. M. E. Parkhouse 1989. Specific detection of circulating surface/secreted glycoproteins of viable cysticerci in *Taenia saginata* cysticercosis. *Para. Imm.* **11,** 351–70.

Harrison, L.J.S., E.K.M. Muchemi & M. M. H. Sewell 1985. Attempted infection of calves with *Taenia crocutae* cysticerci and their subsequent serological response. *Res. Vet. Sci.* **38,** 383–5.

Harrison, L. J. S., R. M. E. Parkhouse & M. M. H. Sewell 1984. Variation in 'target' antigens between appropriate and inappropriate hosts of *Taenia saginata* metacestodes. *Parasitology* **88,** 659–63.

Heath, D. D. & G. B. L. Harrison 1986. Protective antigen sources from *Taenia taeniaeformis* larvae and from larval cestodes in sheep. In *Immunobiology and molecular biology of cestode infections*. Coopers Animal Health Symposium, Melbourne, A6.

Heath, D. D., S. B. Lawrence & H. Twaalhoven 1985. *Taenia ovis* cysts in lamb meat: the relationship between the numbers of cysts observed at meat inspection and the numbers of cysts found by fine slicing of tissue. *New Zealand Vet. J.* **33**, 152–4.

Hoeppli, R. 1956. The knowledge of parasites and parasitic infections from ancient times to the 17th century. *Exp. Parasitol.* **5**, 398–419.

Howell, M. J. & J. J. Hargreaves 1988. Cloning and expression of *Taenia ovis* antigens in *Escherichia coli. Mol. Biochem. Parasitol.* **28**, 21–30.

Jing, J. S. & P. Y. Wang 1985. Advance of studies on cysticercosis of man and pigs at home and abroad in recent years. *Chin. J. Vet. Sci. Technol. (Zhonyo Shouyi Keji)*, **6**, 33–6.

Joshua, G. W. P., L. J. S. Harrison & M. M. H. Sewell 1988a. Surface and excreted antigens of *Taenia saginata*. Proceedings of 'Parasitic helminths: genes membranes and antigens'. *Trans. R. Soc. Trop. Med. Hyg.* **82**, 188–9.

Joshua, G. W. P., L. J. S. Harrison & M. M. H. Sewell 1988b. Excreted/secreted products of developing *Taenia saginata* metacestodes. *Parasitology* **97**, 477–87.

Kassai, T., M. Cordero del Campillo, J. Euzeby, S. Gaafar, Th. Hiepe & C. A. Himonas 1989. Standardized nomenclature of animal parasite diseases (SNOAPAD). *Vet. Para.* **29**, 299–326.

Kassai, T., C. Takats, P,. Rendl & E. Fok 1984. Effect of albendazole and praziquantel on *Taenia saginata* cysticerci. *Helminthol.* **21**, 295–302.

Laclette, J. P., L. Arcos and K. Willms 1986. Complement depletion by the vesicular fluid from metacestodes of *Taenia solium*. Coopers Animal Health Symposium, Melbourne C4.

Larralde, C., J. P. Laclette, R. Bojalil, M. L. Diaz *et al.* 1986. Towards a reliable serology of *Taenia solium* cysticercosis with vesicular fluid antigens. In *Immunobiology and molecular biology of cestode infections*. Coopers Animal Health Symposium, Melbourne, E4.

Larralde, C., J. P. Laclette, C. S. Owen, I. Madrazo, *et al.* 1986. Reliable serology of *Taenia solium* cysticercosis with antigens from cyst vesicular fluid: ELISA and haemagglutination tests. *Am. J. Trop. Med. Hyg.* **35**, 965–73.

Le Riche, P. D. & M. M. H. Sewell 1977. Differentiation of *Taenia saginata* and *Taenia solium* by enzyme electrophoresis. *Trans. R. Soc. Trop. Med. Hyg.* **71**, 237–8.

Le Riche, P. D. & M. M. H. Sewell 1978. Differentiation of taeniid cestodes by enzyme electrophoresis. *Int. J. Parasitol.* **8**, 479–83.

Lethbridge, R. C. 1980. The biology of the oncosphere of cyclo–phyllidean cestodes. *Helminthol. Abstr.* **49**, 59–72.

Letonja, T. & C. Hammerberg 1987a. *Taenia taeniaeformis*: early inflammatory response around developing metacestodes in the liver of resistant and susceptible mice. I. Identification of leucocyte response with monoclonal antibodies. *J. Parasitol.* **73**, 962–70.

Letonja, T. & C. Hammerberg 1987b. *Taenia taeniaeformis*: early inflammatory response around developing metacestodes in the liver of resistant and susceptible mice. II. Histochemistry and cyto–chemistry. *J. Parasitol.* **73**, 971–9.

Lightowlers, M. W. & M. D. Rickard 1988. Excretory–secretory products of helminth parasites: effects of host immune responses. *Parasitology* **96**, S123–S66.

Lightowlers, M. W., R. D. Honey, M. D. Rickard & G. F. Mitchell 1986. Development of a model vaccine against cysticercosis. *Immunobiology and molecular biology of cestode infections*. Coopers Animal Health Symposium, Melbourne, A6.

Lloyd, S. 1979. Homologous and heterologous immunisation against metacestodes of *Taenia saginata* and *Taenia taeniaeformis* in cattle and in mice. *Z. ParasitenKde.* **60**, 87–96.

Lloyd, S. 1984. Passive and active immunisation against cysticercosis. In *Agriculture: Some important parasitic infections in bovines considered from economic and social points of view.* J. Euzeby & J. Gevrey (eds), 187–8. ECSC–EEC–EAEC Brussels and Luxembourg Office for publication of the European Communities.

Lloyd, S. & E. J. L. Soulsby 1976. Passive transfer of immunity to neonatal calves against metacestodes of *Taenia saginata. Vet. Parasitol.* **2,** 355–62.

Mann, I. & E. Mann 1947. The distribution of measles (*Cysticercus bovis*) in African bovine carcasses. *Vet. J.* **103,** 239–51.

Merle, A. 1958. Les cysticercoses communes à l'homme et aux animaux. *Bull. Off. Int. Epizoot.* **49,** 483–500.

Mohammad, I. N., D. C. Heiner, B. L. Miller, M. A. Goldberg *et al.* 1985. Enzyme–linked immunosorbent assay for the diagnosis of cerebral cysticercosis. *J. Clin. Microbiol.* **20,** 775–9.

Molinari, J. L., R. Meza & P. Tato 1983a. *Taenia solium*: cellular reactions in the larvae (*Cysticercus cellulosae*) in naturally parasitized, immunized hogs. *Exp. Parasitol.* **56,** 327–38.

Molinari, J. L., R. Meza, B. Suarez, S. Palacias *et al.* 1983b. *Taenia solium* immunity in hogs to the cysticercus. *Exp. Parasitol.* **55,** 340–57.

Muller, R. 1975. *'Worms and disease'. A manual of medical helminthology,* London: William Heinemann Medical Books.

Nascimento, E. & W. Mayrink 1984. Avaliacão de antigenos de cysticercus cellulosae no immunodiagnostico da cysticercosae humana pela hemaglutinacão indirecta. *Rev. Inst. Med. São Paulo* **26,** 289–94.

Neito, D. 1982. Historical notes on cysticercosis. In *Cysticercosis: present state of knowledge and perspectives.* A. Flisser, K. Willms, J.P. Laclette, C. Larralde *et al.* (eds), 1–7. New York: Academic Press.

Nelson, G. S., F. R. N. Pester & R. Rickman 1965. The significance of wild animals in the transmission of cestodes of medical importance in Kenya. *Trans. R. Soc. Trop. Med. Hyg.* **59,** 507–24.

Nichol, C. P. & M. M. H. Sewell 1984. Immunosuppression by larval cestodes of *Babesia microti* infections. *Ann. Trop. Med. Parasitol.* **78,** 228–33.

Oliver, L. J. 1974. The economics of human parasitic infections. *Z. ParasitenKde.* **45,** 197–210.

Pandey, V. S. & Z. Mbemba 1976. Cysticercosis of pigs in the Republic of Zaire and its relation to human taeniasis. *Ann. Soc. Med. Trop.* **51,** 43–6.

Parkhouse, R. M. E. & L. J. S. Harrison 1987. Cyst fluid and surface associated glycoprotein antigens of *Taenia* sp. metacestodes. *Parasite. Immunol.* **9,** 263–8.

Parkhouse, R. M. E., N. Almond, Z. Cabrera & W. Harnett 1987. Nematode antigens in protection diagnosis and pathology. *Vet. Immunol. Immunopathol.* **17.**

Pawlowski, Z.S. 1980. Epidemiologica tasemcncy i wagrmygey *Taenia saginata.* [Studies on epidemiology of *T. saginata* taeniasis and cysticercosis] *Wiad. Parazytol.* **26,** 539–52.

Pawlowski, Z. S. 1983. Clinical expression of *Taenia saginata* infection in man. In *Proceedings of the first international symposium of human taeniasis and cattle cysticercosis.* 138–44. I. Prokopic (ed.), Prague: Academia Praha.

Pawlowski, Z. S. 1986. Intestinal helminthiasis and human health: Recent advances and future needs. *Proc. Sixth Int. Cong. Parasitol. Parasitology – Quo vadit?* M. J. Howell (ed.), 159–67 Australian Academy of Science.

Pawlowski, Z. S. 1987. Large–scale use of chemotherapy of taeniasis as a control measure for *Taenia solium* infections. In *Helminth zoonoses with particular reference to the tropics.* S. Geerts, V. Kumar & J. Brandt, (eds), 100–5. Dordrecht: Martinus Nijhoff.

Pearson, R. D. & E. L. Hewlett 1985. Niclosamide therapy for tapeworm infection. *Ann. Intern. Med.* **102**, 550–1.

Penfold, H. B. 1937. The life history of *Cysticercus bovis* in the tissues of the ox. *Med. J. Australia* **1**, 579–83.

Powell, S. Y., E. M. Proctor, A. J. Wilmot & N. Macleod 1966. Cysticercosis and epilepsy in Africans: A clinical and neurological study. *Ann. Trop. Med. Parasitol.* **60**, 152–8.

Proctor, E. M. 1972. Identification of tapeworms. *S. Afr. Med. J.* **46**, 234–8.

Proctor, E. M., S. J. Powell & R. Elsdon-Dew 1966. The serological diagnosis of cysticercosis. *Ann. Trop. Med. Parasitol.* **60**, 146.

Prokopic, J. (ed.) 1983. First international symposium of human taeniasis and cattle cysticercosis. Ceste Budejovice, Cczechoslovakia, 20–24 Sept. 1982. Prague: Academia.

Rausch R. L. 1975. Taenidae. In *Diseases transmitted from animals to man*. W. T. Hubbert, W. F. McCullock & P. R. Schnurrnberger (eds), 678–707. 6th edn. Illinois: Thomas Sprinfield.

Rickard, M. D. & J. L. Brumley 1981. Immunization of calves against *Taenia saginata* infection using antigens collected by *in vitro* incubation of *T. saginata* oncospheres or ultrasonic disintegration of *T. saginata* and *T. hydatigena* oncospheres. *Res. Vet. Sci.* **30**, 99–103.

Rishi, A. K. & D. P. McManus 1987. DNA probes which unambiguously distinguish *Taenia solium* from *T. saginata*. *Lancet* **ii**, 1275.

Rodriguez–del Rosal, D. Correa & A. Flisser 1989. Swine cysticercosis: detection of parasite products in serum. *Vet. Rec.* **124**, 488.

Rosas, N., J. Sotelo & D. Nieto 1986. ELISA in the diagnosis of neurocysticercosis. *Arch. Neurol.* **43**, 353–6.

Schwabe, C. W. 1984. *Veterinary medicine and human health*. 3rd edn. Baltimore and London: Williams & Wilkins.

Sewell, M. M. H. & L. J. S. Harrison 1978. Bovine cysticercosis. *Vet. Rec.* **102**, 223.

Shanley, J. D. & M. C. Jordan 1980. Clinical aspects of CNS cysticercosis. *Arch. Intern. Med.* **140**, 1309–13.

Simpson, A. J. G., T. Walker & R. Terry 1986. An introduction to recombinant DNA technology. *Parasitology* **91**, S7–S14.

Slais, J. 1970. *The morphology and pathogenicity of the bladder worms Cysticercus cellulosae and Cysticercus bovis*. Prague: Academia.

Sotelo, J., F. Escobedo, J. Rodriguez–Carajal, B. Torres and F. Rubio–Donnadieu. 1984. Therapy of parenchymal brain cysticercosis with praziquantel. *N. Eng. J. Med.* **310**, 1001–7.

Soulsby, E. J. L. 1982. *Helminths, arthropods and protozoa of domesticated animals*. 7th edn. London: Baillière and Tindall.

Soulsby, E. J. L. & S. Lloyd 1982. Passive immunization in cysticercosis: characterization of the antibodies concerned. In *Cysticercosis: present state of knowledge and current perspectives*. A. Flisser, K. Willms, J.P. Laclette, C. Larralde *et al.* (eds), 539–48. New York: Academic Press.

Stevenson, P., P. W. Holmes & J. M. Muturi 1981. Effect of albendazole on *Taenia saginata* cysticerci in naturally infected cattle. *Vet. Rec.* **109**, 82.

Tellez-Giron, E., L. Dufour & M. Montante 1981. Effect of flubendazole on *Cysticercus cellulosae* in pigs. *Am. J. Trop. Med. Hyg.* **30**, 135–8.

Thomas H. & R. Gonnert 1978. The efficacy of praziquantel against experimental cysticercosis and hydatidosis. *Z. ParasitenKde.* **55**, 165–79.

Urquhart, G. M. 1961. Epizootiological and experimental studies on bovine cysticercosis in East Africa. *J. Parasitol.* **47**, 857–69.

Vegors, H. H. & I. T. Lucker (1971). Age and susceptibility of cattle to initial infection with *Cysticercus bovis*. *Proc. Helminthol Soc. Washington,* **38,** 122–7.

Velasco–Suarez, M., Bravo–Becherelle & F. Quirasco 1982. Human cysticercosis: Medical and social implications and economic impact. In *Cysticercosis: present state of knowledge and current perspectives*. A. Flisser, K. Willms, J. P. Laclette, C. Larralde, *et al.* (eds), 45–72. New York: Academic Press.

Viljoen, N. F. 1937. Cysticercosis in swine and bovines with special reference to South African conditions. *Ond. J. Vet. Sci. Anim. Indust.* **9,** 412–570.

Walthers, M. & J. K. Koske 1980. *Taenia saginata* cysticercosis: a comparison of routine meat inspection and carcass dissection results in calves. *Vet. Rec.* **106,** 401–2.

Wes Leid, R., C. M. Suquet, R. F. Grant, L. Tanigoshi *et al.* 1986. Taeniastatin, a cestode proteinase inhibitor with broad host regulatory activity. In *Immunobiology and molecular biology of cestode infections*. Coopers Animal Health Symposium, Melbourne, D1.

Woulters, G., J. Brandt & S. Geertz 1987. Observations on possible strain differences in *Taenia saginata*. In *Helminth zoonoses with particular reference to the tropics*. S. Geerts, V. Kumar & J. Brandt, (eds), 76–80 Dordrecht: Martinus Nijhoff.

Xu, Z. B., M. P. Chan, M. L. Hung & W. N. Chung. 1982. Efficacy of pyquiton (praziquantel) in treatment of human cysticercosis. *Beijing Med. J.* **4,** 326–8.

Yong, W. K., D. D. Heath & F. Van Knapen 1984. Comparison of cestode antigens in an enzyme–linked immunosorbent assay for the diagnosis of *Echinococcus granulosus, Taenia hydatigena* and *T. ovis* infections in sheep. *Res. Vet. Sci.* **36,** 24–31.

Zenteno-Alanis, G. H. 1982. A classification of human cysticercosis In *Cysticercosis: present state of knowledge and current perspectives*. A. Flisser, K. Willms, J.P. Laclette, C. Larralde *et al.* (eds), 197–226. New York: Academic Press.

Zoli, A., S. Geerts & T. Vervoort 1987. An important focus of porcine and human cysticercosis in West Cameroon. In *Helminth zoonoses with particular reference to the tropics*. S. Geerts, V. Kumar & J. Brandt, (eds), 85–91, Dordrecht: Martinus Nijhoff.

4 Trichinella in Africa and the *nelsoni* affair

Introduction

It used to be that the epidemiology of trichinosis, as generally understood, was fairly simple. There was one species of *Trichinella*, namely *T. spiralis*. As a tool for biological research it was very popular, infecting laboratory animals easily and generally behaving in a laudable, that is to say predictable, manner. As a public health problem it was a matter of man and his pigs. If man would behave himself in a laudable manner, that is to say if he would be more punctilious in the management of his kitchen and his piggery, the disease should be readily controlled. *Trichinella* larvae occurred, to be sure, in various other animals, but that was of zoological rather than medical interest. The parasite had a wide geographical distribution; but apart from a few small outbreaks on the Mediterranean border of north Africa, the African continent was thought to be free of trichinosis as a human disease.

All that has changed. The situation is now vastly more complicated – and the man whom we honour in this volume must be held largely responsible! The genus *Trichinella* now boasts a species called *nelsoni*. Its status as a species has been challenged (a process watched by Professor Nelson with stoicism and amusement) but whether it retains that rank or is demoted to a sub-species, the name *nelsoni* will always be with us. Indeed there are now four declared species of *Trichinella*, at least two of them being widely accepted. Now there are innumerable isolates, with various biological characteristics. Now the sylvatic cycle of transmission is an integral part of the trichinosis picture, not only maintaining the parasite independently of man, but also providing a direct source of infection for man. Now trichinosis is known to be enzootic in many parts of Africa and several outbreaks of human trichinosis have been reported south of the Sahara. In view of the role played by Professor Nelson in these developments, it is fitting that African trichinosis should be reviewed in this book. The problem from the outset has been that nobody knows as much about the subject as Nelson himself. He, being as aware as anyone of the dilemma thus presented, has graciously encouraged me to proceed with the task; and I do so in the comfort of having his own writings as a secure base.

North Africa

Egypt occupies a position of special distinction in relation to trichinosis. It is the most recently 'discovered' endemic area – yet it has given us the oldest known incident of *Trichinella* infection in man or animal anywhere in the world! The latter episode involved human infection, but needs to be divorced from the general discussion because the infection occurred many centuries before the discovery of the parasite. The case is now old but the patient was not; he was a teenage boy named Makht, believed to have been a weaver who lived near the River Nile about 12 000 BC. In modern times his mummified body was found to contain an object that was identified as a *Trichinella* cyst on the basis of its appearance in histological section (Millet *et al.* 1980).

The existence of trichinosis in Egypt was first reported in 1879. This time the infection was not in man but in a hippopotamus – actually an animal that must have been something of a 'white elephant' because it had been presented by a khedive to a city in France where it was continuously sickly and where, despite the tender loving care of the local zookeepers, it died after four months (Heckel 1879). *Trichinella* cysts were found at necropsy and since the animal had been given a non-meat diet in France, it is highly probable that it had become infected in Egypt.

Perhaps it was the hippopotamus report that prompted Ostertag to list Egypt as an endemic area more than half a century ago. Ostertag's treatise on meat inspection (Ostertag 1919) contained a little noticed but extremely comprehensive review of trichinosis. In it he cited not only Egypt but also Algiers and East Africa as areas where trichinosis occurs. In view of the later developments in those lands (see below) one must wonder where Ostertag got the information that so enabled him to anticipate the later findings. On the other hand, Ostertag mentioned in the same sentence the occurrence of trichinosis in Australia – suggesting either that his information was not so well founded after all, or that the present consensus about the absence of trichinosis in Australia rests upon correspondingly shaky ground.

The recently held view that trichinosis was absent from Egypt was not based merely on a lack of reported clinical cases. El Afifi & El Sawah (1962) stated that the Egyptian authorities had a 'general belief' in the absence of the infection. The position taken by the authorities was evidently more modest than that of Wahby (1943) who had worked as a veterinary inspector at the Cairo abattoir from 1939 to 1943, and who concluded that 'there is definitely no *Trichinella spiralis* in Egyptian pigs'. Yet the authorities must have been tempted to become more authoritarian as a result of a new systematic attempt to find the parasite in pigs slaughtered at the Cairo abattoir. El Afifi and El Sawah (1962) examined some 48 000 squashed slivers of muscle from the diaphragms of 1000 pigs, and used the digestion method to examine larger (20 g) samples from 100 of them. They found not a single larva.

El Afifi and El Sawah, however, were wisely aware that no amount of not finding something proves that it is not there. They cautioned against complacency and urged continuous surveillance. For a while their caution seemed excessive. Selim and Youssef (1968) examined 777 wild animals, and although nematode larvae were found in the musculature of a few of them, the authors concluded on morphological grounds that they were not *Trichinella*. Rifaat *et al.* (1969) examined 1097 wild animals and found no larvae. In both of these studies, however, the animals examined were almost all small rodents (rats and mice) and it is now well established that such animals are not good indicators of the endemicity of *Trichinella* in wildlife populations. Indeed in less than a decade after El Afifi and El Sawah issued their warning, *Trichinella* had been found in Egypt. Prompted by a 1975 outbreak of a human disease that was thought to be trichinosis (though never confirmed) new attempts were made to find the parasite in Egyptian animals. Using trichinoscopy, Sedik and his colleagues reported finding it in pigs slaughtered at the Cairo abattoir (cited by Morsy *et al.* 1980 and by Siam *et al.* 1979). Tadros and Iskander (1977) reported finding it in 4 of 50 native pigs slaughtered at the Cairo abattoir, though the report itself is unclear in its parasitological aspect. Further examination of porcine flesh at the same slaughterhouse showed that *Trichinella* was indeed present in abundance (Siam *et al.* 1979; El Nawawi 1981). Where Sedik found a prevalence rate of about 1%, which would have to be considered extremely high, Siam *et al.* found a prevalence of 3% and El-Nawawi found a prevalence of 4.53% – which would have to be considered quite phenomenal. These values can hardly be considered suspect on grounds of statistical insufficiency. El-Nawawi examined no less than 40670 swine by direct microscopy in the period from December 1975 to December 1976. Siam *et al.* examined only 100 samples of fresh pork, but they did it not by trichinoscopy, but by digestion of ten-gram samples of mixed muscle (including diaphragm and masseter tissue) and inoculation of isolated larvae into rats for confirmation purposes. Poor management practices, especially the feeding of uncooked swill to swine, were thought to be responsible for the high rate of infection (El-Nawawi 1981). Similar, or even higher, prevalence rates were formerly reported for swill-fed pigs in the USA (Zimmermann 1974). Pigs were not the only animals found to be infected. Dogs and rodents, previously given a clean bill of health, were also shown to be infected. In Alexandria, an examination of animals in the vicinity of piggeries and the abattoir revealed infection in 21 of 222 dogs, 6 of 79 climbing rats (*Rattus rattus alexandrinus*); 45 of 258 burrowing rats (*R. norvegicus*); 6 of 64 mice and 0 of 67 cats (Barakat *et al.* 1982). In a serological study of animals in the Suez Canal zone, *Trichinella* antibodies were detected in 13% of the highly carnivorous

R. norvegicus and 1.4% of the more herbivorous *R. rattus* (Morsy *et al.* 1980). Finally in 1978, the first confirmed human case was reported in Egypt (Morcos *et al.* 1978). The patient had recently eaten underdone pork; she had many of the signs and symptoms of trichinosis, including periorbital oedema; she had a high eosinophilia and a positive intradermal test; she apparently responded to thiabendazole treatment; and despite the absence of biopsy evidence, she undoubtedly had trichinosis. More recently still, meat from Egypt was held responsible for a small but fairly severe outbreak of human trichinosis in Germany (Bomer *et al.* 1985). In that instance, a traveller had illicitly brought from Cairo a piece of dried meat sold as camel meat. This delicacy was undoubtedly responsible for the outbreak of trichinosis, but it was not possible to determine whether the meat actually came from a camel. If confirmed, this Egyptian episode would not only have provided new evidence of trichinosis in that country but also the first report of *Trichinella* in camels.

Egypt is not the only country on the Mediterranean border of Africa in which trichinosis has made an appearance. In 1945 about a dozen (European) individuals in Algeria fell ill with what was diagnosed as trichinosis on the basis of clinical findings but with no parasitological confirmation (Gerard 1946). These patients undoubtedly got their illness locally, and if indeed it was trichinosis they would represent the first diagnosed autochonous cases on the African continent. The source of infection appeared to be domestic pigs (though one of the patients was a Muslim). In 1962 11 additional cases were diagnosed in Algeria and this time five were confirmed by biopsy (Lanoire *et al.* 1963). It was thought probable that the infection had been acquired from pigs. In the same year trichinosis was diagnosed in a young French soldier lately repatriated from Algeria. His illness became apparent four days after reaching France, and was attributed to eating the meat of a wild boar killed in Algeria. A series of biopsies revealed an intensive and recent *Trichinella* infection (Verdaguer 1963).

Africa south of the Sahara

The word 'trichina' was by no means absent from medical reports in sub-Sahara Africa even before the discovery of the first parasitologically proven case in 1959. Hundreds of cases of 'trichina' were recorded in reports of the Kenya Medical Department in the years 1927–48, but it is thought that in all cases the intended diagnosis was *Trichuris* (Nelson & Mukundi 1962). The case reported from Sierra Leone (Burrows 1910) must similarly be a case of mistaken identity. The breezy conversational tone of Burrows' report (so sadly missing from the typical terse and efficient communication of our day) is matched only by its parasitological ineptness; the cysts in which '*Trichinella*' specimens resided were said to be like dried peas, and were found in the liver and on the mesentery and peritoneum. Across the continent, beyond the eastern and southern borders of the Sahara, the occurrence of *Trichinella* was again doubtful. Corsi (1939) reviewed the health statistics of Italian colonial military operations in 1935 and 1936 and cited a single case of trichinosis in an Italian soldier of the Eritrean battalion. The basis of diagnosis was not given and it is by no means certain that the soldier acquired his illness in the Somalia or Eritrea areas. In commenting on this case Ricci (1940) suggested that trichinosis would be very uncommon in that region because of the spread of Islam and the scarcity of pigs. Caronia (1937) gave a brief account of a worker (presumably Italian) who had become ill in East Africa and was removed to Rome where *Trichinella* cysts were found in his muscles. It is quite likely that he acquired his illness in Africa but again the evidence is inconclusive. A single case of a disease 'indistinguishable from trichinosis' was reported from South Africa (Levy *et al.* 1960) but infection with *Trichinella* was not confirmed parasitologically or immunologically. At the time of the 1959 outbreak in Kenya, Nelson and his colleagues were able to gather many expert testaments to the absence of trichinosis from Africa south of the Sahara (Forrester *et al.* 1961). These included statements that the infection had not been recorded in domestic or wild animals in that immense region in general, and that it was specifically unknown in the Union of South Africa, Swaziland, Bechuanaland (Botswana), Southern Rhodesia (Zimbabwe), Northern Rhodesia (Zambia), Madagascar, Tanganyika (Tanzania), Uganda, Kenya, Sudan, Nigeria, Angola, Ghana, Togo, Senegal, Ivory Coast and Dahomey (Benin). In the case of Kenya, the official position was backed up by the failure to find the parasite in the diaphragmatic muscle of 10 000 domestic pigs (unpublished data, cited by Forrester *et al.* 1961).

In June 1959 a rural hospital near Mt Kenya admitted 11 Kikuyu patients (a young man, three teenage boys and seven younger boys) suffering from a variety of complaints, but all having muscle pain and eosinphilia and most having trismus and oedema of face or limbs (Forrester *et al.* 1961; Nelson and Forrester 1962). Some of the cases were initially diagnosed as typhoid fever, and it will be recalled that this was also the diagnosis in the first case in which *Trichinella* was shown to be capable of causing severe disease in man – the well known case of a young woman who died in Dresden in 1860 (Zenker 1860). A suspicion of trichinosis was entertained because of these shared signs and symptoms and because, as Nelson (1970) so aptly stated: 'Fortunately, the patients were seen by a recently appointed physician with no preconceived ideas as to what diseases might occur in Africa'. In August 1959, muscle biopsies were taken and sent to Nairobi where the diagnosis of trichinosis was confirmed. (The date was inadvertently given as August 1960 in a later report.) This, then, was the episode that led in subsequent years to the numerous contributions of Nelson and his colleagues to our understanding of trichinosis – contributions that included valuable and well-controlled experiments, original epidemiological observations, and examples of scientific writing from which we can learn that the dullness that afflicts so much of our scientific prose is a preventable malady. The reports in question are: Forrester *et al.* 1961, Nelson, Rickman *et al.* 1961, Nelson & Forrester 1962, Nelson & Mukundi 1962, Nelson & Mukundi 1963, Nelson, Guggisberg *et al.* 1963, Forrester 1964, Nelson & Blackie 1964, Nelson, Blackie *et al.* 1966, Nelson 1968, Nelson 1970, Nelson 1972, Nelson 1982, Nelson 1983.

The Kenyan outbreak of 1959 was special for a number of reasons. It was the first confirmed outbreak south of the Sahara. There had apparently been no 'intestinal phase', the patients being admitted to hospital with evidence of muscle involvement arising about two weeks after ingestion of infected meat. The parasite was of wildlife origin: the meat in question was the flesh of a bushpig (*Potamochoerus porcus*) which had been 'toasted' over an open fire in the forest where the bushpig had been killed. The disease was atypical also in that most of the patients had polyglandular enlargement. None had 'splinter' haemorrhages under the fingernails. In some cases the disease was severe: pain was intense and four patients were essentially unable to walk or eat because of their muscle contracture and 'lockjaw'. One of them, a boy aged seven years, died before the diagnosis was confirmed. The intensity of infection was extraordinarily high, the four most severely affected patients having a mean of 2715 larvae per gram of thigh muscle. In the fatal case, the tongue, jaw, thigh and diaphragm muscles had 5190, 4560, 4060 and 2090 larvae per gram, respectively. The remarkable thing was not that one patient died, but that all the others survived and responded to treatment. Never before had such heavy *Trichinella* infections been recorded in human beings. More than a year after recovery from illness and discharge from hospital, some of the former patients were found to still be harbouring large numbers of living larvae. These individuals appeared healthy, and one was engaged in arduous daily labour despite an infection that would ordinarily be expected to be lethal (1555 larvae per gram). The acute disease responded well to prednisone therapy (some patients also received diethylcarbamazine or dithiazanine, without clearcut benefit) and the Kenyan outbreak was thus not unusual in its response to corticosteroid therapy. Despite the severity of the disease in some individuals, the huge numbers of larvae that were associated with merely moderate or mild disease in other individuals suggested that the indigenous people of the region were unusually tolerant of *Trichinella*, or that the strain of *Trichinella* responsible for this outbreak was atypical in its pathogenicity. The latter eventually proved to be the case, and the repercussions of that discovery are still being felt (see below).

After the 1959 outbreak, Nelson suggested that the paucity of reports of trichinosis in Africa might reflect a lack of diagnosis rather than a lack of the parasite (Nelson & Forrester 1962). Further evidence of the potential for human infection in Africa soon appeared, with two cases diagnosed in 1961 in farm labourers in the Rift Valley (near Nakuru, about 100 miles west of the site of the original outbreak); one case in the same year in a labourer at Mt Londiani (near Lake Victoria); and at least 30 cases in 1963, with at least two deaths, among labourers at a farm on the edge of Tinderet Forest, south of the Rift Valley (Forrester 1964, Nelson & Forrester 1962, Nelson 1970). The Tinderet Forest outbreak was severe, with three deaths, and differed from the Mt Kenya outbreak in that the early phase of the disease was characterized by abdominal pain and diarrhoea. In all of these outbreaks, infection was attributed to the meat of a wild pig, probably the bush pig, *Potamochoerus porcus* (but perhaps the giant forest hog, *Hylochoerus meinertzhageni*, in the case of the Tinderet Forest outbreak). Within a few years the disease was reported on the other side of Africa, with nine cases reported in Europeans living in Senegal (Gretillat & Vassiliades 1967). The West African cases were attributed to consumption of the flesh of warthog (*Phacochoerus aethiopicus*). Since then, human trichinosis has been reported twice more in Kenya (Hutcheon & Pamba 1972, Kaminsky & Zimmerman 1977) and once in Tanzania (Bura & Willett 1977). The source of the infection was not established in the Kenyan cases. In the Tanzanian outbreak, it was almost certainly a local wild animal of the pig family (called a 'wild pig' by the authors in their text, but described in the summary as the warthog, *P. aethiopicus*). The Tanzanian cases were similar to the original Kenyan cases in that the intensity of infection was extremely high, but different in that the classic early phase of the disease, with diarrhoea and abdominal pain, was evident. Two of the 11 patients (members of the Iraqw tribe from the highlands of north central Tanzania) died, but one survived an infection intensity of 5560 larvae per gram and another survived with 6530 larvae per gram – the highest rate ever recorded in man. Muscular contractions persisted in these patients after recovery from acute illness, and the 10-year-old girl who had 5560 larvae per gram was walking tiptoe because of contraction in the ankle region more than a half year after her hospitalization (Bura & Willett 1977).

Epidemiology

On the Mediterranean border of Africa the epidemiology of trichinosis is likely to simulate that of Europe, and the high incidence of the parasite found at an abattoir in Cairo (see above) suggests that the domestic pig, here as elsewhere, serves as the focus of a classical 'domestic cycle'. The scarcity of reported human cases in North Africa probably reflects the dietary habits of the human populations (including religious and cultural proscriptions) rather than any lack of source material.

South of the Sahara trichinosis appears to be a zoonosis that is essentially sylvatic. Human infections have arisen from bushpig and warthog rather than from domestic pig. Extension of swine raising into new areas and changes in swine husbandry may increase the significance of the domestic cycle (Nelson *et al.* 1962, Bura & Willett 1977) but at present neither the domestic pig nor synanthropic rats appear to be involved. Domestic dogs have become infected after eating the flesh of bushpig; but although dogs are eaten by some African tribes, their significance in *Trichinella* transmission is uncertain (Nelson 1970).

More than 50 species of wildlife in Africa have been examined for *Trichinella*. Those found infected were: lion, *Panthera leo*; serval, *Felis serval*; leopard, *P. pardus*; bushpig, *P. porcus*; spotted hyena, *Crocuta crocuta* (Sachs & Taylor 1966); side-striped jackal, *Canis adustus*; striped hyena, *Hyaena hyaena* (Nelson, Guggisberg *et al.* 1963); warthog, *P. aethiopicus*; golden jackal, *C. aureus* (Gretillat 1970 a, 1970b); silver-backed jackal, *C. mesomelas*; and multimamate rat, *Mastomys natalensis* (Young & Kruger 1967). A reference to infection in a cheetah, *Acinonyx jubatus* (Nelson & Forrester 1962) was later found to be in error, the animal in question being a serval (Nelson, Guggisberg *et al.* 1963).

As in other parts of the world, rodents in Africa have rarely been found infected with *Trichinella*, and their significance in transmitting the infection by serving as prey to carnivores is now thought to be minimal. In Kenya more than 2000 rodents, of 19 species, were examined but none was found infected; and the lack of infection in rodents is strongly borne out by the failure of investigators to find larvae in small carnivores, such as the genet and mongoose, that feed on the rodents (Nelson 1983). More and more, the important element in trichinosis transmission appears to be carrion. Carnivores and omnivorous scavengers feed on the carcasses of other carnivores and scavengers, and become infected in doing so. The animals in Africa whose carcasses provide the infective *Trichinella* larvae for the infection of other animals should not be considered with those listed above as known susceptible species. Virtually all mammals are thought to be susceptible to *Trichinella* (about 150 species have been reported susceptible; Campbell 1983) though not all are equally susceptible to all parasite strains. Even herbivores are susceptible, and their slight role in transmission is a reflection of their dietary habits not their innate insusceptibility. Thus trichinosis in Africa might be regarded as a zoonosis sustained by carnivores and scavengers in general and propagated directly and simply by the eating of uncooked flesh (Fig. 4.1).

In practice, of course, certain species of wild animals will be more important than others in the transmission of trichinosis. For example, the habits of the hyena would be conducive to transmission in the wild and

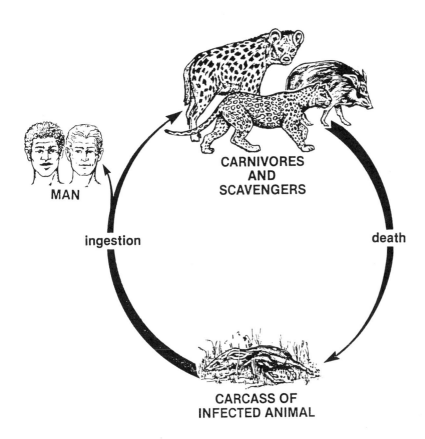

MAN

CARNIVORES
AND
SCAVENGERS

ingestion

death

CARCASS OF
INFECTED ANIMAL

Figure 4.1 Transmission cycle of *Trichinella* in tropical Africa. The animals shown (hyena, leopard, bushpig) are merely representative of the many wild carnivores and scavengers that may play a role in transmission. Wild suids such as the bushpig are the principal source of infection for man. (Reproduced with permission from Campbell 1989.)

indeed prevalence is known to be high in this animal (Nelson, Guggisberg *et al.* 1963, Young & Kruger 1968). The habits of man make it much more likely that he will become infected by eating animals of the pig family (*Suidae*) than by eating the flesh of large carnivores. African people rarely eat the flesh of wild carnivores or scavengers such as the hyena (Nelson 1970, Nelson, Guggisberg *et al.* 1963). Thus in West Africa, where prevalence appears to be low in warthogs and high in golden jackals (Gretillat 1970a, 1970b) human infection was attributed to the former (Gretillat & Vassiliades 1967); and the East African outbreak of 1959 was caused by consumption of bushpig rather than the more commonly infected hyena. Consumption of the flesh of wild pigs is itself rare in much of Africa, and that is undoubtedly one of the reasons for the rarity of outbreaks of human trichinosis.

It is apparent from the foregoing that although man is a celebrated omnivore, his dietary habits (and thus his exposure to food-borne disease) are dictated by cultural as well as biological forces. The practice of cooking flesh is everywhere a major factor in limiting man's exposure to trichinosis. The reason for the low consumption of wild pig meat in the savannah lands of West Africa is that Muslim law, which forbids pork consumption, prevails in that region (Nelson 1982). In many areas of Africa, however, the bushpig is hunted for food, and because it destroys crops (Bura & Willett 1977, Nelson 1982). Domestic dogs, which can become infested with *Trichinella*, are eaten by certain peoples in West Africa and the Sudan (Nelson 1983). The Maasai and other Nilo-Hamitic tribes may eat the flesh of lion and leopard in a ritual context, even though the meat of wild carnivores is rarely eaten in Africa (Nelson 1982). The Kikuyu people of Kenya have traditionally eschewed the flesh of wild animals; but the Mau Mau uprising of the 1950s caused an erosion of traditional values, and the Kikuyu youths involved in the important 1959 outbreak of trichinosis admitted eating the 'toasted' meat of a bushpig while camping on the slopes of Mt Kenya during an illegal hunting expedition (Forrester *et al.* 1961, Nelson 1972).

The cultural determinants of infection are not necessarily of a culinary nature. Burial rites are deeply entrenched; but bodies sometimes are not. According to Huntingford (1953) several Nilo-Hamitic groups of equatorial East Africa (the Nandi, Dorobo, Keyo and Suk peoples) have traditionally disposed of the bodies of their dead by taking them outside the village and leaving them on the ground for the hyenas to eat. Such practices could, at least in theory, provide an exception to the rule that man is everywhere a 'dead end' host for *Trichinella*. According to Huntingford, individuals who were particularly distinguished by virtue of youth, age or wealth were generally disposed of in ways that would seem more conventional to Western observers; but the remainder of the population could provide a more or less continuous source of infection for feral animals. Forrester (1964) noted that Dorobo people inhabited each of four areas in Kenya where trichinosis had occurred, and suggested the existence of a man–pig–man cycle of infection involving forest-dwelling tribes and various wild suid species.

The *nelsoni* affair

The remarkable Kenyan outbreak of June 1959 suggested that the parasite itself might be remarkable. Biopsy material from the patients was therefore fed to laboratory albino rats (Forrester *et al.* 1961). Very few adult worms were recovered from the gut of the rats; and very few larvae, some of them dead or calcified, were recovered from diaphragm tissue (0–20 larvae per gram resulting from inocula from about 200–400 larvae). At the same time an English laboratory strain of *Trichinella*, administered in the form of infected rat muscle, was highly infective to rats (2000–4000 larvae/gram from inocula of 200–400 larvae).

Soon other studies confirmed the distinctiveness of the Kenyan *Trichinella*. The establishment of laboratory isolates for experimental purposes was itself a problem, because infection generally failed to 'take' in common laboratory animals. Larvae recovered from a serval cat were passed through a mongoose and through monkeys and baboons before enough larvae could be collected for comparative infectivity trials. It was then shown that laboratory rats and wild rats and domestic pigs were extremely resistant in comparison to an English strain of cat origin (Nelson & Mukundi 1963). Three strains of rat, including a 'hooded' variety, differed slightly in resistance to infection, but all were highly refractory in contrast to their marked susceptibility to the English strain of the parasite (Nelson *et al.* 1966). It was also shown that the Kenyan strain shared its low degrees of infectivity for rats with an Alaskan strain but did not share it with Polish strains (Nelson, Blackie *et al.* 1966). But the most remarkable finding, and one which took a long time to become accepted was that both the Kenya and Alaska strains were virtually non-infective to domestic pigs. The same authors showed that Kenya-strain infections in rats, with their modest numbers of adult worms in the gut and minimal dissemination of progeny into the musculature, could be used to induce a considerable degree of immunity against challenge with the English strain. Meanwhile, in Poland, Professor and Mrs Kozar carried out experiments showing that their laboratory strains of mice and rats, too, had a relatively low degree of susceptibility for the Kenyan strain, though the refractoriness was not as marked as in the experiments of Nelson and his colleagues. They also found that the survival of mice following exposure to very large numbers of Kenyan larvae was greater after the strain had been passed through mice for one generation than when the strain was first tested in mice (Kozar & Kozar 1965). Although this evidence of altered infectivity was of a preliminary nature, it was consonant with a report that the infectivity of sylvatic isolates of *Trichinella* in Poland increased rapidly upon consecutive passages in mice and guinea pigs (Zimoroi, cited by Kozar & Kozar 1965). In the same series of experiments, the Kozars also showed that an isolate of *Trichinella* from man in Poland was intermediate between the Kenyan strain and the Polish laboratory strain in infectivity for mice. These authors pointed out that in areas where 'the circle of hosts is closed for a prolonged time, as for instance in East Africa' this ability to adapt to new hosts could lead to the development of a regional 'strain' of the parasite. When *Trichinella* was discovered in the western side of the African continent, the parasite proved highly infective for cats but had very little infectivity for mice and rats (Gretillat & Vassiliades 1967) further supporting the concept that African *Trichinella* was different from what had been regarded as 'normal'. Thus the 1960s, in 'trichinellology' as in other aspects of life, were years in which established 'truths' and conventional wisdom were severely shaken.

Infectivity was not the only biological characteristic in which the Kenyan *Trichinella* differed from others. Boev & Sokolova (1981) reported that the new variety had (as might befit a denizen of the tropics) an increased tolerance to heat.

The differences in infectivity and pathogenicity between the Kenyan and the 'ordinary' strain of *Trichinella* were of evident medical and epidemiological significance; but they did not arouse the adversarial instincts of investigators until Britov and Boev (1972) elevated the new *Trichinella* to the status of a species: *Trichinella nelsoni*. The essence of species distinction between similar organisms is the existence of irreconcilable differences between mates. The differences are genetically controlled, and are in practice signalled by the absence of offspring when male and female representatives of the different populations are given an opportunity to mate. Boev and his colleagues have adduced such differences in erecting the new species *T. nelsoni* (originally described on the basis of Nelson's Kenyan isolate) as well as *T. nativa* (Britov & Boev 1972).

Many other investigators have hesitated to adopt the species classification of Britov and Boev. One reason is the inevitable resistance to change. As Shaikenov and Boev (1983) put it, after a century and a half during which *Trichinella* was regarded as a single species 'some investigators couldn't get used to the reality of other opinions'. This resistance was fairly easily overcome in the case of *T. pseudospiralis*, Garkavi 1972, where several subtle differences in morphology were abetted by a striking difference in host response (lack of encapsulation by transformed host muscle fibres). Another reason, also pointed out by Shaikenov & Boev (1983) is that morphological differences, which have been the traditional basis of species erection for a very long time, are virtually impossible to discern in the case of *nelsoni* and *nativa* populations. It was indeed on this basis that Lichtenfels *et al.* (1983) argued that 'subspecific designations provide a more accurate reflection of our knowledge of the nematodes that cause trichinosis than would designation of species'.

Boev and his colleagues, on the other hand, championed the validity of ecological and zoogeographical factors, especially when dealing with sibling species, and further pointed out that ultrastructural and biochemical differences may show differences in parasite composition that are not apparent by more traditional morphological studies. Such factors, as well as immunological studies, have recently been brought to bear on the question of *Trichinella* isolates. Even without calling upon the use of isoenzyme or antigenic determinants, however, Boev and his colleagues presented a scheme for 'primary diagnosis' of the sibling species. Setting aside cross-breeding experiments (which are difficult and which are undertaken by very few investigators) *T. pseudospiralis* can be distinguished by lack of encapsulation as already mentioned; *T. nativa* can be distinguished by its ability to tolerate temperatures of −12 to −17°C for at least six months. The remaining two species, *T. nelsoni* and *T. spiralis*, can be distinguished from each other on the basis of various factors including the relatively lower infectivity of the former for laboratory rodents and pigs, and relatively stronger cellular reaction that the former induces in pig muscle, and the relative infrequency of gastrointestinal symptoms induced by the former in human infections (Boev & Sokolova 1981). Biological characteristics subsequently found to differ among populations of *Trichinella* include the behaviour of larvae in response to cold or heat (Shaikenov 1983) and response to chemotherapy of the host animal (Chadee *et al.* 1983, Ozeretskovskaya *et al.* cited by Shaikenov *et al.* 1983). For a broader consideration of infraspecific variation in *Trichinella*, see Boev & Sokolova 1981, Dick 1983, and Flockhart 1986.

Presumably every structural and biochemical feature has a spectrum of tolerable variation, and the total number of potential differences between isolates of *Trichinella* must be virtually infinite. The greater the separation in time and place between isolates, the greater those physical and chemical differences are likely to be. The closer the isolates are in time and place, the harder one will have to look to find the subtle differences that will even then exist. Where populations are reproductively isolated, species designation usually poses no problem. Where interbreeding occurs, but differences are marked and perhaps even of practical significance, then the matter is not so simple. *T. nelsoni* is no longer the Kenyan strain, or even the tropical strain. It is an organism described as a species and reported from places as far south as South Africa, as far north as Sweden and Estonia, as far west as Portugal and as far east as the Kazakh Soviet Socialist Republic in central Asia (Shaikenov & Boev 1983). These authors point out that *T. nelsoni* is not adapted to prolonged periods of cold, and that the northern boundary of its habitat is approximately coincident with the isotherm line separating temperatures of −8 to 0°C from those of −16 to −8°C in the month of January. In addition to the places mentioned above, the *nelsoni* designation has been given to isolates from Bulgaria, Czechoslovakia, Italy, Spain, Switzerland and Iran. Not surprisingly, these isolates, though sexually compatible, are not identical. The early studies of Nelson and his colleagues suggested that the Kenyan strain was not, as had generally been assumed, a serious threat to man as a potential participant in the 'domestic cycle' of trichinosis (though it could, and had, infected man directly from the 'sylvatic cycle'). However, South African and Italian isolates designated *nelsoni* were moderately infective for domestic pigs (Young & Kruger 1968, Pozio *et al.* 1985). Pigs from Bilorussia were found infected with *T. nelsoni* after being fed on scraps of wild animals; and Ozeretskovskaya and her colleagues reported at least 84 cases of trichinosis (thought to be due to *T. nelsoni* infection) caused by 'natural strains of *Trichinella* passaged through pigs' (cited by Shaikenov & Boev 1983). The Kenyan strain, when serially passaged in mice, appeared to acquire (as mentioned above) a higher degree of infectivity for such animals (Kozar & Kozar 1965), just as an Indian feline strain (now thought to have probably been *T. nelsoni*) had readily become adapted to rats (Schad *et al.* 1967). An Italian isolate, later designated *T. nelsoni*, was obtained from the horse but was infective to many host species including the mouse and rat (Pampiglione *et al.* 1978). It was capable of developing to the adult stage in the kestrel (Cancrini *et al.* 1985). Thus even the characteristic that first alerted Nelson and his fellow workers in Kenya to the existence of a new strain of *Trichinella* is subject to modification under conditions of passage and varies in degree among the geographical isolates subsequently designated *T. nelsoni*.

To refer to an organism as *Trichinella nelsoni* is merely to conclude that the process of evolution and speciation has gone further than is suggested by the term *Trichinella spiralis nelsoni*. This is not the place to argue the merits of one form of address over the other, nor indeed do I have the expertise to do so. I share the scientific conservatism that engenders a reluctance to accept new species designations as long as there are reasonable doubts about them. As Flockhart (1986) points out, we need labels for our isolates both for interpreting our current research and for making use of future findings. Designation of the latitude and longitude of the site of origin of isolates is a useful adjunct to taxonomic nomenclature (Dick 1983) and its use should be especially encouraged in situations, like the present one, that enjoy a high level of taxonomic controversy. But names are irresistibly handy; and if we can agree on the connotations and denotations of *nelsoni*, it will not matter so much whether we use the word as a specific or subspecific designation.

L'envoi

We know a lot more about the epidemiology of trichinosis now than we did in 1959 when a bunch of Kikuyu youths, flouting the mores of their elders, brought on themselves a terrible sickness and gave Nelson and his colleagues of the Kenya Medical Department something to think about. We know a lot about host species, about the biology and immunology of infections, and about the factors governing prevalence in particular places. Only in the past few years have we applied the techniques of molecular biology to the study of infraspecific variation (Dame *et al.* 1987). Only in the past few years have we shown that trichinosis in a given domestic pig herd might be maintained, not by swill feeding or by ingestion of farm rats nor yet by ingestion of wildlife carrion, but by cannibalism within the herd (Hanbury *et al.* 1986). Only in the past few years have we learned that birds may have a species of *Trichinella* especially adapted to such hosts, and that the species in question lacks the feature that we have regarded as the most distinctive of all *Trichinella* features – the ability to induce in host muscle the formation of a thick collagen-rich capsule or 'nurse cell'. Only in the past few years . . . but the list could go on and on!

Most of all, perhaps, we have learned how much we don't know. The summing-up offered by Nelson (1970) two decades ago was prophetic: 'An inescapable conclusion from these observations concerning geographic strains is that it is no longer possible to generalise regarding the epidemiology and epizoology of trichinosis on the basis of observations made only in the more developed countries of the world. Moreover, the idea that the rat and the pig are hosts *par excellence* of *T. spiralis* must be abandoned. Further studies may show that there are many geographic races of *T. spiralis*, each adapted for transmission within a particular complex of meat-eating animals, and that the variation in the infectivity of a particular strain are as important in determining the epidemiology of trichinosis as are the eating habits of the hosts'. We are still learning that trichinosis, while a global parasitic infection, is also a mosaic of regional adaptations.

Afterword

Since this chapter was written, the complexity of intraspecific variation in *Trichinella* has become more evident, and measures have been taken to conserve and catalogue the isolates used by various investigators (see Campbell 1989 and Pozio *et al.* 1989). *T. nelsoni* is one of seven major types, and more than 100 isolates, cryopreserved by Dr Pozio and his colleagues at the Istituto Superiore di Sanità in Rome. The known geographic distribution of *Trichinella* has become significantly expanded with the discovery of the parasite in Australia. Ostertag's treatise of 1919 (see section on North Africa above) now seems more remarkable than ever.

References

Barakat, R. M., M. R. El-Sawy, M. K. Selim & A. Rashwan 1982. Trichinosis in some carnivores and rodents from Alexandria, Egypt. *J. Egypt. Soc. Parasitol.* **12**, 445–51.

Boev, S. N. & L. A. Sokolova 1981. Primary diagnostics of sibling species of *Trichinellae*. *Wiad. Parazyt.* **27**, 483–7.

Bomer, W., H. Kaiser, W. Mannweiller, H. Mergerian *et al.* 1985. An outbreak of trichinellosis in northern Germany caused by imported air-dried meat from Egypt. In *Trichinellosis*, C. W. Kim (ed.). 314–20. Albany, NY: State University of New York.

Britov, V. A. & S. N. Boev 1972. Taxonomic rank of various strains of *Trichinella* and their circulation in nature. *Vestn. Akad. Nauk. SSSR*, **28**, 27–32.

Bura, M. W. T. & W. C. Willett 1977. An outbreak of trichinosis in Tanzania. *E. Afr. Med. J.* **54**, 185–93.

Burrows, D. 1910. A case of trichiniasis in a native of Sierra Leone. *J. Trop. Med. Hyg.* **13**, 102–3.

Campbell, W. C. 1983. Epidemiology. I. Modes of transmission. In *Trichinella and trichinosis*, W. C. Campbell (ed.). 425–44. New York: Plenum Press.

Campbell, W. C. 1989. Trichinosis revisited – another look at modes of transmission. *Parasit. Today*, **4**, 83–6.

Cancrini, G., G. Canestri-Trotti, R. Constantini, F. Franceschini *et al.* 1985. Recent research on trichinellosis in the fox and other animals in Italy. In *Trichinellosis*, C. W. Kim (ed.) 263–7. Albany, NY: State University of New York.

Caronia, G. 1937. Sulla terapia della trichinosi. *Arch. Italiano Sci. Med. Coloniale*, **18**, 283.

Chadee, K., T. A. Dick & G. M. Faubert 1983. Sensitivity of *Trichinella* sp. isolates to thiabendazole. *Can. J. Zool.* **61**, 139–46.

Corsi, A. 1939. Dalla relazione medico-statistica sulle condizioni sanitarie delle forze armate nelle colonie negli anni 1935 e 1936. *Giorn. Med. Milit.* **87**, 339–66.

Dame, J. B., K. D. Murrel, D. E. Worley & G. A. Schad 1987. Genetic evidence for the presence of synanthropic *Trichinella spiralis* in sylvatic hosts. *Exp. Parasitol.* **64**, 195–203.

Dick, T. A. 1983. Species and intraspectific variation. In *Trichinella and Trichinosis*, W. C. Campbell (ed.) 31–73. New York: Plenum Press.

El-Afifi, A. & H. M. El Sawah 1962. Trichinosis in the United Arab Republic. *J. Arab. Vet. Med. Assoc.* **22**, 341–4.

El-Nawawi, F. A. 1981. Swine trichinellosis in Egypt. *Arch. Lebensmittel hyg.* **32**, 156–8.

Flockhart, H. A. 1986. *Trichinella* speciation. *Parasit. Today* **2**, 1–3.

Forrester, A. T. T. 1964. Human trichinellosis in Kenya. In *Proceedings 1st International Conference Parasitology, Rome*, A. Corradetti (ed.) 669–71. Oxford: Pergamon.

Forrester, A. T. T., G. S. Nelson & G. Sander 1961. The first record of an outbreak of trichinosis in Africa south of the Sahara. *Trans. R. Soc. Trop. Med. Hyg.* **55**, 503–13.

Gérard, R. 1946. Petite épidémie de trichinose en Algérie. *Rev. Med. Nav.* **1**, 355–62.

Gretillat, S. 1970a. Epidemiology of trichinosis of wild animals in West Africa – warthog receptivity of the West-African strain of *Trichinella spiralis*. *J. Parasitol.* (Supplement, Part 1) **56**, 124.

Gretillat, S. 1970b. Epidémiologie de la trichinose sauvage au Sénégal. *Wiad. Parazyt.* **16**, 109–10.

Gretillat, S. & G. Vassiliades 1967. Présence de *Trichinella spiralis* (Owen, 1835) chez les carnivores et suides sauvages de la région du delta du fleuve Sénégal. *C. R. Acad. Sci.* **264**, 1297–300.

Hanbury, R. D., P. B. Doby, H. O. Miller & K. D. Murrel 1986. Trichinosis in a herd of swine: Cannibalism as a major mode of transmission. *J. Am. Vet. Med. Assoc.* **188**, 1155–9.

Heckel, S. 1879. Trichinosis in the hippopotamus. *Am. J. Micr.* **4**, 183–4.

Huntingford, G. W. B. 1953. *The Southern Nilo-Hamites*, East Central Africa, Part VIII *Ethnographic Survey of Africa*, D. Forde (ed.) London: International African Institute.

Hutcheon, R. A. & H. O. Pamba 1972. Report of a family outbreak of trichinosis in Kajiado District – Kenya. *E. Afr. Med. J.* **49**, 663–6.

Kaminsky, R. G. & R. R. Zimmermann 1977. *Trichinella spiralis*: Incidental finding. *E. Afr. Med. J.* **54**, 643–6.

Kozar, Z. & M. Kozar 1965. A comparison of the infectivity and pathogenicity of *Trichinella spiralis* strains from Poland and Kenya. *J. Helminthol.* **39**, 19–34.

Lanoire, Duchier & Freiz 1963. Onze observations de trichinose en Algérie. *Gazette des Hôpitaux* **135**, 305–10.

Levy, L., R. Morris & G. McLeish 1960. A parasitic infestation indistinguishable from trichinosis – a new disease? *S. Afr. Med. J.* **34**, 1069–72.

Lichtenfels, J. R., K. D. Murrel & P. A. Pilitt 1983. Comparison of three subspecies of *Trichinella spiralis* by scanning electron microscopy. *J. Parasitol* **69**, 1131–40.

Millet, N. B., G. D. Hart, T. A. Reyman, M.R. Zimmermann *et al.* 1980. ROM I: Mummification for the common people. In *Mummies, disease and ancient cultures*, A. Cockburn & E. Cockburn (eds), 71–84. Cambridge: Cambridge University Press.

Morcos, W. M., E. G. Mikhail & M. M. Youssef 1978. The first diagnosed case of trichinosis in Egypt. *J. Egypt. Soc. Parasitol.* **8**, 121–9.

Morsy, T. A., S. A. Michael & S. F. A. El-Seoud 1980. Trichinosis antibodies sought in rodents collected at Suez Canal Zone, A.R.E. *J. Egypt. Soc. Parasitol.* **10**, 259–65.

Nelson, G. S. 1968. The transmission of *Trichinella spiralis* from wild animals to man and domestic animals. In *Some diseases of animals communicable to man in Britain*, 77–89. Oxford: Pergamon Press.

Nelson, G. S. 1970. Trichinosis in Africa. In *Trichinosis in man and animals*, S. E. Gould (ed.) 473–92. Springfield, Ill: Charles C. Thomas.

Nelson, G. S. 1972. Human behaviour in the transmission of parasitic diseases. In *Behavioural aspects of parasite transmission*, E. U. Canning & C. A. Wright (eds), 109–22. London: Zool. J. Linnean Soc.

Nelson, G. S. 1982. Carrion-feeding cannibalistic carnivores and human disease in Africa with special reference to trichinosis and hydatid disease in Kenya. *Symp. Zool. Soc. London.* **50**, 181–98.

Nelson, G. S. 1983. Wild animals as reservoir hosts of parasitic diseases of man in Kenya. In *Tropical parasitoses and parasitic zoonoses*, J. D. Dunsmore (ed.) 59–72. Tenth Meeting of the World Association for the Advancement of Veterinary Parasitology, Australia.

Nelson, G. S. & E. J. Blackie 1964. An Alaskan strain of *Trichinella spiralis* of low infectivity to laboratory rats. *Trans. R. Soc. Trop. Med. Hyg.* **58**, (Abstracts of Laboratory Meeting) 6–7.

Nelson, G. S. & A. T. T. Forrester 1962. Trichinosis in Kenya. *Wiad. Parazyt.* **8**, 17–28.

Nelson, G. S. & J. Mukundi 1962. The distribution of *Trichinella spiralis* larvae in the muscles of primates. *Wiad. Parazyt.* **8**, 629–32.

Nelson, G. S. & J. Mukundi 1963. A strain of *Trichinella spiralis* from Kenya of low infectivity to rats and domestic pigs. *J. Helminthol.* **37**, 329–38.

Nelson, G. S., E. J. Blackie & J. Mukundi 1966. Comparative studies on geographical strains of *Trichinella spiralis*. *Trans. R. Soc. Trop. Med. Hyg.* **60**, 471–80.

Nelson, G. S., C. W. A. Guggisberg & J. Mukundi 1963. Animal hosts of *Trichinella spiralis* in East Africa. *Ann. Trop. Med. Parasitol.* **57**, 332–46.

Nelson, G. S., R. Rickman & F. R. N. Pester 1961. Feral trichinosis in Africa. *Trans. R. Soc. Trop Med. Hyg.* **55**, 514–7.

Ostertag, R. 1919. *Handbook of meat inspection.* 472. Chicago: American Veterinary Publishing.

Pampiglione, S., R. Baldelli, C. Corsini, S. Mari *et al.* 1978. Infezione sperimentale del cavallo con larve di trichina. *Parassitologia* **20**, 183–93.

Pozio, E., M. Gramiccia, Al. Mantovani, O. Massi *et al.* 1985. Distribution of *Trichinella nelsoni* in muscles of experimentally infected swine. In *Trichinellosis*, C. W. Kim (ed.) 246–50. Albany, NY: State University of New York.

Pozio, E., G. La Rosa & P. Rossi 1989. *Trichinella* reference centre, *Parasit. Today* **5,** 169–70.

Ricci, M. 1940. Elmintologia umana dell'Africa Orientale. *Riv. Biol. Colon.* **3,** 241–95.

Rifaat, M. A., A. H. Mahdi & M. S. Arafa 1969. Further evidence that Egypt is trichina-free. *J. Egypt. Pub. Hlth Assoc.* **44,** 193–6.

Sachs, R. & A. S. Taylor 1966. Trichinosis in a spotted hyena (*Crocuta crocuta*) of the Serengeti. *Vet. Rec.* **78,** 704.

Schad, G. A., S. Nundy, A. B. Chowdhury & A. K. Bandyopadhyay 1967. *Trichinella spiralis* in India. II. Characteristics of a strain isolated from a civet rat in Calcutta. *Trans. R. Soc. Trop. Med. Hyg.* **61,** 249–58.

Selim, M. K. & L. B. Youssef 1968. Survey on trichinosis affecting some carnivores and rodents in Egypt. *Vet. Med. J.* (Cairo University) **15,** 235–41.

Shaikenov, B. 1983. Biokinetic differences of *Trichinella* species. *Wiad. Parazytol.* **29,** 587–93.

Shaikenov, B. & S. N. Boev 1983. Distribution of *Trichinella* species in the Old World. *Wiad. Parazyt.* **29,** 595–608.

Siam, M. A., S. A. Michael & N. H. Ghoneim 1979. Studies on the isolation of the infective stages of *Trichinella spiralis* and *Toxoplasma gondii* from fresh and processed pork in Egypt. *J. Egypt. Pub. Hlth Assoc.* **54,** 163–70.

Tadros, G. & A. R. Iskander 1977. Trichinosis among swine in Egypt. *Zool. Soc. Egypt Bull.* **27,** 104–10.

Verdaguer, S. C. Hiltenbrand, J. Dangoumau & J. P. Vaillant 1963. A propos d'un cas de trichinose humaine. *J. Med. Bordeaux* **140,** 1687–94.

Wahby, M. M. 1943. Gleanings of an Egyptian abattoir. *Vet. J.* **99,** 189–90.

Young, E. & S. P. Kruger 1967. *Trichinella spiralis* (Owen, 1935) Railliet, 1895, infestation of wild carnivores and rodents in South Africa. *J. South Afr. Vet. Med. Assoc.* **38,** 441–3.

Zenker, F. A. 1860. Ueber die Trichinen-krankheit des Menschen. *Virchow's Arch. Pathol. Anat.* **18,** 561–71.

Zimmermann, W. J. 1974. The current status of trichinellosis in the United States. In *Trichinellosis* C. W. Kim (ed.) 603–9. New York: Intext Educational Publishers.

5 The African hookworm problem: an overview

Importance and distribution

In a recent review on the importance of parasitic diseases, Davis (1983) has pointed out that 'the population of developing countries suffer from a multitude of disadvantages: poverty, malnutrition, low income, high birthrates, illiteracy and parasitic disease'. Africa typifies such a developing continent and the hookworms comprise a significant component of the parasitic diseases.

Davis then went on to emphasize that 'epidemiologically, many of these parasitic infections are advancing, not receding . . . despite the technological revolution of the 20th century'.

Outlining the complexity of the consequences of parasitic diseases, Davis continued:

> What is perhaps not so well appreciated is the impact of these diseases on the economy of developing countries by the production of high morbidity rates, low working days, decline in output at both individual and community level, and nonproductive allocation of sparse financial resources for curative treatment, which is frequently nullified because the prevailing environmental and socioeconomic background conditions lead to reinfection and perpetuation of epidemiologic transmission cycles.

Further, there is an increasing belief that chronic parasitic infections in school-age children can and do affect cognitive performance, thus minimizing the limited schooling available in these Third World countries (Kvalsvig 1988, Halloran *et al.* 1989).

Hookworm infection exemplifies this parasitic situation and Africa characterizes the struggling Third World economy.

Stoll (1949) in his classic review of 'this wormy world', estimated the world prevalence of hookworm to be some 457 million cases. Later he revised his estimate to about 600 million cases (Stoll 1962) and the latest calculations put the figure at between 800 million and 1500 million cases with about 55000 to 60000 deaths a year resulting from hookworm disease (WHO 1984, Schofield 1985, Bruyning 1985).

Hookworm infections are widely reported from most parts of Africa but the prevalences reported vary depending on the area, the population and the techniques used. Despite this, the fact remains that hookworm infection continues to be a major problem over much of Africa (Van Oye 1961).

Recorded prevalence figures for hookworm in Africa when given by country often oversimplify a complex picture resulting from variation, even within national boundaries, due to variable ecological and other relevant factors. Overall records of prevalence in Africa vary widely from zero through to over 80% (Bruyning 1985), with the higher levels of infection being found in the lower, warmer and moister areas and the lower levels of infection being found in the higher, cooler and drier parts of the continent. In broad terms, high prevalences and high loads are usually encountered in the tropical regions of Central Africa, with the problem decreasing as one moves to the south. The whole situation was put in perspective by Van Oye (1961) who stated: 'It is impossible for us to interpret the official statistical data at their true value' and furthermore, he went on: 'To arrive at an estimate of the importance of ancylostomiasis throughout any given country, presents a difficulty sometimes insoluble'.

The complexity of the hookworm situation, even in individual countries can be illustrated from studies in Zimbabwe. Thus Goldsmid (1968b) recorded the prevalence of hookworm at a large urban hospital as 5.7%. Detailed rural studies, however, revealed prevalences of zero to 63% in the country (Goldsmid 1972a). Further, although high levels of endemic infection were found to exist in parts of the country, especially the Zambezi Valley (60%) and the Lowveld (63%), many of the hookworm infections detected were imported (Goldsmid 1972a, Goldsmid et al. 1974). In fact, in the former study, a series of 301 people were investigated for hookworm, of whom 26 (8.6%) were diagnosed as being infected. Of these subjects, 17 (5.6%) came from Malaŵi and 2 (0.7%) were from Mozambique, even though of the total of 301 people investigated, only 33 (11%) originated from Malaŵi and 9 (3%) from Mozambique. In other words, of the total of 33 people from Malaŵi, 51.5% were infected and of the nine from Mozambique 22.2% were infected. In contrast, of the Zimbabweans included in the study, only 2.7% (7/259) were found to carry hookworms. Thus overall, 73% of the hookworm cases were of foreign origin. It was further found that with extended residence in Zimbabwe, the percentage of cases infected with hookworm fell as did the loads when adjudged by Stoll egg counts (Fig. 5.1).

The conclusions from these studies were therefore that while some parts of Zimbabwe, and namely the warmer, moister and lower areas, were suitable for significant hookworm transmission, much of the country was not ideal and the hookworm problem in these parts was essentially an introduced one due largely to the importation of the infections with migrant labour from neighbouring territories of high prevalence. The decrease in both load and prevalence with time of residence would seem to be due to interrupted transmission resulting from the adverse

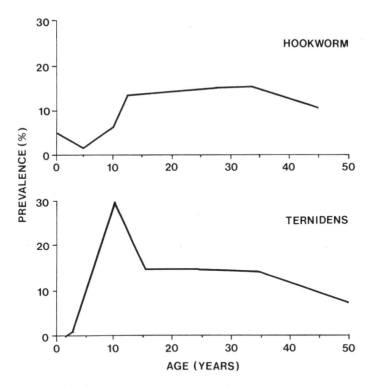

Figure 5.1 Prevalence of hookworm and *Ternidens deminutus* in Zimbabweans.

ecological conditions found over much of Zimbabwe, together with the fact that some of the workers at least, would have gained employment in seweraged urban areas and would additionally have taken to wearing shoes in keeping with their newly acquired affluence.

Urbanization *per se* does not automatically provide protection from parasitic infections and can, in fact, result in a deteriorating situation due to overcrowding. In fact Davis (1983) has stated that 'urbanization with rural populations attracted to towns in search of work, inevitably results in an outstripping of the available sanitary facilities'. The interrelating effects of ecological factors, including urbanization, on parasitic infections including hookworm, have been discussed in general terms by Croll (1983), Schad *et al.* (1983) and Mwosu (1983) but the overall complexity of urbanization and sanitation has been highlighted by studies such as that of Feacham *et al.* (1983) who concluded from a comparative study covering Botswana, Zambia and Ghana, that their 'findings suggest that the provision of superior water and sanitation facilities to small clusters of houses, or to houses scattered through an area, may not protect those families from infection if the over-all level of faecal contamination of the environment is high'.

In the smaller villages of Nigeria, helminth burdens of the population have been shown to be high and the studies of Mwosu (1983) in this area are probably representative of much of tropical Africa. Referring to helminth infections, he recorded the triad of hookworm, *Ascaris lumbricoides* and *Trichuris trichiura* to be common and concluded that the helminth infection levels for the villages studied were high 'paralleling those found in villages in similar biogeographical zones of the country more than a decade ago'.

Even in large cities, parasitic infection rates may remain high. Nnochiri (1968) in a detailed analysis of parasitic disease and urbanization in Nigeria, postulated that there were fundamental differences in the effects of urbanization in different regions of Africa. He suggested that these basic differences could be seen in the large cities of Southern Africa when compared to those of West Africa. He commented that the urban areas of West Africa often lacked even the basic sanitary advantages one might expect in a large urban centre. He then went on to state that 'the lack of services and facilities is one of the most important features differentiating most Nigerian urban areas from their Western counterparts', and he ranked towns and cities in what he termed 'those countries in Africa in

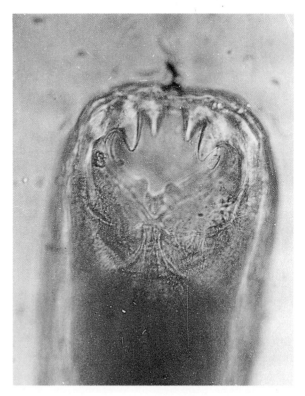

Figure 5.2 Photomicrograph of anterior end of *Ancylostoma duodenale* showing characteristic teeth.

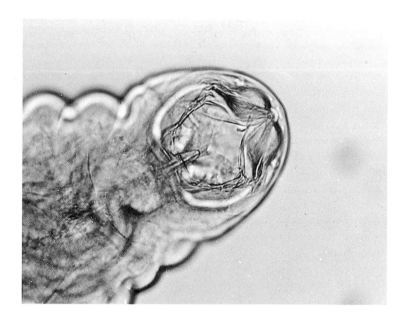

Figure 5.3 Photomicrograph of anterior end of *Necator americanus* showing cutting plates.

which substantial European settlement had occurred over the years', as being more comparable to Western developed cities in this regard. He specifically mentioned Johannesburg in South Africa and Salisbury (Harare) in Rhodesia (Zimbabwe) as representative of this group with a long history of European influence. It is therefore interesting to compare the 5.7% hookworm prevalence and decreasing prevalence and load associated with residence in Harare as recorded by Goldsmid (1968b, 1972a) with the prevalence of 37% to 60% recorded in Nigeria by Nnochiri (1968) in similar urban groups.

Of the common species of hookworm infecting humans, *Ancylostoma duodenale* (Fig. 5.2) tends to be more sub-tropical in distribution (Miller 1979) and the ratio between this species and *Necator americanus* (Fig. 5.3) changes (within certain limits) with altitude – the latter species increasing as altitude increases (Pawlowski 1986). In Africa, *A. duodenale* is generally considered to be the dominant species in the coastal areas of northern Africa, while *N. americanus* comprises the dominant species in central and southern Africa (Muller 1975, Beaver *et al.* 1984, Bruyning 1985). However, these species probably coexist over much of their range today, perhaps due to human migrational trends (Miller 1979, Ukoli 1984).

Publications from Egypt still seem to reflect the presence of *A. duodenale* only (Farid *et al.* 1966, Farid *et al.* 1973) but '*N. americanus* seems to dominate in central Africa, either alone or with *A. duodenale* being present also, but at a lower level'.

In Tanzania, although *N. americanus* was the main species or the only species recorded by Sturrock (1966), Forsyth (1970) and Kilala (1971). *A. duodenale* was stated by Meakins *et al.* (1981) to be the commonest species in the north of the country.

Gelfand and Warburton (1967) quote work from Uganda referring to *N. americanus* as the only species present, but Hutton and Somers (1961) imply the presence of *A. duodenale* as well, stating that they found 'predominantly *N. americanus*'.

The situation in Kenya seems to be similar, with Rees *et al.* (1974) commenting that 'in general *Necator americanus* is ubiquitous' and that *A. duodenale* is confined to the coast.

Among the pygmies of Zaire *N. americanus* was recorded as probably the only species occurring by Pampiglione *et al.* (1979).

In West Africa, Topley (1968), in The Gambia, recorded only *N. americanus* and Nnochiri (1968) stated that this species was the one 'nearly always' found in West Africa.

Farther south, in the Kalahari Bushmen, Heinz (1961) found *N. americanus* to be the only intestinal helminth detected in his survey and Fleisher (1985), also in Botswana, again recorded this species as the only hookworm found, believing that it survived here being better able to withstand the dry climatic conditions – an interesting view in the light of the comment by Pawlowski (1986) that *Necator*, as compared to *Ancylostoma*, 'requires more moisture'.

A similar situation was found to prevail in Namibia, where in Owamboland, *N. americanus* was the only hookworm recorded by Kyronseppa and Goldsmid (1978) although their findings are at variance with those of Gildenhuys (1970) who had reported the species occurring there as *A. duodenale*. In a more recent study of the Kavongo territory of Namibia, Evans & Joubert (1989) reported the prevalence of hookworm but did not elaborate on the species involved.

In Zimbabwe, Gelfand and Warburton (1967) cited *N. americanus* and *A. duodenale* to be present in equal numbers but the studies of Goldsmid (1965, 1968, 1972a), Macdonald and Goldsmid (1973), Goldsmid *et al.* (1974), and McCabe and Goldsmid (1976) showed that although both species were found at significant levels, *N. americanus* was undoubtedly the commoner species – a situation that seems true for Mozambique as well (de Azevedo 1964).

Buckley (1946) in a detailed helminthological survey of Zambia, reported the finding of *N. americanus* only, but *A. duodenale* has recently been reported there for the first time (Hira & Patel 1984), suggesting that it might be a recent import, either from Mozambique or Zimbabwe, where it has long been known to occur (Blackie 1932).

This suggests the interesting hypothesis that *A. duodenale* has spread inland from the coast with migrant African labour. It has long been traditional for people from Malaŵi, Zambia and Mozambique to seek and obtain work in South Africa and Zimbabwe. It would seem not inconceivable that *A. duodenale* could have been carried inland to Zimbabwe from Mozambique and become established there in the early days of Rhodesia (or perhaps even further back in history), where it now coexists with *N. americanus*, which probably moved into southern Africa from the north with the original human migrational movements of the African tribes. Subsequently, returning Zambian workers might have carried *A. duodenale* back with them when they returned home, with the result that it has become established there.

Another possibility for the spread of *A. duodenale* into Zambia, of course, could be the recent influx of refugees into that country as a result of the present troubles in Mozambique.

Such a hypothesis for a continuing spread of *A. duodenale* through southern Africa, perhaps from the Mozambique coast into Zimbabwe and then directly or indirectly into Zambia, seems rather contradictory to the view expressed by Pawlowski (1986) that in the last few decades the *N. americanus/A. duodenale* ratio has changed in favor of *Necator* due to the fact that this species is better adapted to spread and survive in changing ecological and social conditions. *A. duodenale* may, in reality, prove to be far more adaptable than suggested by Pawlowski, having been able to adapt, among other things, to an oral route of infection (Migasena & Gilles 1987) and even, perhaps, being able to infect humans via paratenic carriage in meat such as chicken, lamb, pork or beef when it is consumed undercooked (Schad *et al.* 1984).

It is interesting that the spread of *A. duodenale* has also been suggested in Kenya by Rees *et al.* (1974) who feel that *N. americanus* was probably the original Kenyan species and that *A. duodenale* may have been introduced to a coastal focus from Asia or the Middle East – and this may well have also been the origin of the species in Mozambique, which has a long history of contact with the East.

It would indeed be fascinating if studies on biochemical strain variations could be attempted on the African hookworms, both *A. duodenale* and *N. americanus*, as reviewed for parasitic helminths in general by Bryant and Flockhart (1986), in order to show the degree of relatedness between the hookworms from the various countries and regions of Africa to allow us to trace the past (and perhaps present) spread of these species through the continent.

The difficulty in critically assessing the true accurate distributions of *A. duodenale* and *N. americanus* in Africa may reflect the tendency for diagnostic and non-parasitologically trained research investigators to diagnose hookworm infections on the basis of egg morphology only and without recourse to a differential study by larval culture (Goldsmid 1967, 1982), autopsy studies or post-anthelmintic collection of adult worms to confirm the species involved. This oversimplification of the true complexity of the diagnosis of such infections, especially as it relates to Africa, is summed up by the statement of Migasena and Gilles (1987) that: 'the diagnosis of hookworm infection is straightforward; the characteristic eggs are found in the faeces.' This simplistic approach has bedevilled attempts to unravel the complex situation in Africa.

Sometimes such a generalized approach is taken even further, with a species name being unjustifiably tagged to cases in which only eggs are used as the diagnostic criterion (e.g. McCullough & Friis-Hansen 1961, who described the hookworm eggs they recovered from patients in Zambia as *A. duodenale* as discussed by Goldsmid 1968b).

In point of fact, scientifically planned studies of hookworm species distribution should utilize autopsy, post-anthelmintic and larval culture techniques in combination, as, for example, if culture methods alone are used, it might result in under-detection of *A. duodenale* as, when *N. americanus* and *A. duodenale* occur in mixed infections, egg production of the *A. duodenale* can be suppressed (Pawlowski 1986).

Other hookworm-like parasites

The generalized approach by reporting only 'hookworm' in surveys is an extreme oversimplification of the true situation, especially in Africa, where the whole 'hookworm' picture is complicated by a range of strongylid and rhabditid species (Lavine 1968, Dunn 1978) of intestinal nematodes, including *A. duodenale*, *N. americanus*, *Oesophagostomum* spp., *Ternidens deminutus*, *Trichostrongylus* spp., and *Strongyloides fuelleborni*, infecting humans – and all of which have hookworm-like eggs (Goldsmid 1967, 1968a, 1972b).

The point is exemplified in Zimbabwe where, although *T. deminutus* was first described to be a relatively common human parasite over fifty years ago (Sandground 1929, 1931, Blackie 1932) it was not again reported from humans there until Goldsmid (1967, 1968b), once more drew attention to its continuing presence. In the intervening years, all diagnostic laboratories and surveys in the country had lumped them together as hookworm or *N. americanus* in their reports, ignoring such reviews as that of Amberson and Schwarz (1952) and Nelson (1965) questioning what had happened to *Ternidens* in the intervening years. In fact, Nelson (1965) specifically and pointedly remarked in his review of the parasitic helminths of baboons with particular reference to species transmissible to humans, that there was 'a curious absence of recent records' of human *Ternidens* infections at that time.

The similarity of the eggs of all these intestinal nematodes can provide diagnostic problems. Thus the eggs of *N. americanus* and *A. duodenale*, and for all practical purposes of *Oesophagostomum* spp., are indistinguishable. The eggs of *T. deminutus* in most cases can be separated on their larger size by measurement, while eggs of *Trichostrongylus* can be distinguished as they differ from the previous species in shape, tapering at one or both ends. In fresh stool specimens, the eggs of *S. fuelleborni* can be identified by their small size and by the fact that they can be seen to contain a fully developed and motile rhabditiform larva when passed from the host (Fig. 5.6). Once stool specimens have been passed by the host, the larva rapidly hatches and is then indistinguishable from the rhabditiform larva of *S. stercoralis* (Goldsmid 1968a, 1971a, 1972b, 1982).

A complete species identification (or at least a generic identification) of *Oesophagostomum* or *Trichostrongylus* is possible using the Harada Mori Cultivation or similar larval rearing technique to allow the rearing of characteristic filariform larvae of all the parasitic species in these groups of nematode as well as allowing a more accurate separation of such coprophilic species as *Rhabditis* and *Rhabditella* (Goldsmid 1967, 1971a, 1982). Such larval rearing techniques, although very sensitive, especially in strongyloidiasis, are rarely used in routine laboratories and are not even often used in surveys, being time and space consuming and taking 8–10 days for completion in most cases (Rowbottom *et al.* in press).

Ternidens

This whole issue complicates the African hookworm problem and led Nelson (1972) to comment: 'I also suspect that many of the so-called hookworm infections recorded in the Bushmen, Hadza and Pygmies are in fact *Ternidens deminutus*, the 'false hookworm' of monkeys with eggs difficult to distinguish from *N. americanus* and *A. duodenale*'.

T. deminutus is, in fact, an intriguing enigma in itself. It is closely related to *Oesophagostomum*, which it morphologically resembles (Figs. 5.4 & 5.5) (Goldsmid & Lyons 1973, Goldsmid 1971a, 1982).

The distribution of *T. deminutus* as we know it is wide but discontinuous and patchy. While infection of an extensive range of Old World monkey species has been recorded from Africa, Asia, Southeast Asia and certain Pacific islands, infections from humans have only been described from central and southern Africa, the Comoros Islands and Mauritius (Goldsmid 1971a, 1974a, 1982).

Most reports of human infection with *T. deminutus* give prevalence as being generally low and leading to the belief that the species is of minor significance (Cook 1986). However, studies in Zimbabwe have indicated that in this country at least, the species is widespread in both monkeys and humans, and in some African communities the human prevalence may reach 87% (Goldsmid 1982).

The discontinuous human distribution of *T. deminutus* in Africa is not, however, simply a reflection of misdiagnosis for hookworm, although that does comprise part of the problem. Thus the species has been found in

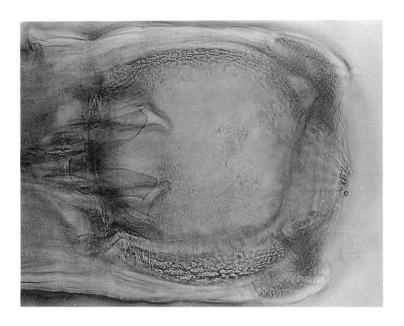

Figure 5.4 Photomicrograph of anterior end of *Ternidens deminutus* showing anterior corona radiata and deep-set teeth.

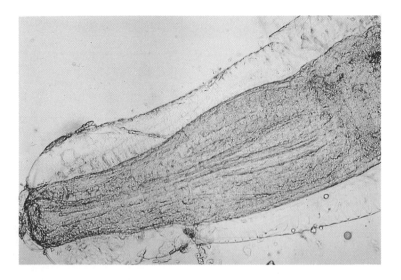

Figure 5.5 Photomicrograph of anterior end of *Oesophagostomum* sp., showing absence of teeth.

humans in Zaire by Van den Berghe (1934) and it was first recorded in Tanzania by Kilala (1971) when differentiated on egg size/volume as outlined by Goldsmid (1968a) (Fig. 5.6). The species has not been detected in studies from Namibia (Kyronseppa & Goldsmid 1978) or Zambia (Hira & Patel 1984), although Blackie (1932) did record one case in this latter country. In these studies, *T. deminutus* was specifically looked for, and Goldsmid (1972a), investigating the problem in Zimbabwe, found that in 127 cases of *Ternidens* he investigated, 124 were from Zimbabwe, three were from Malaŵi (but had all lived in Zimbabwe for in excess of 10 years) and none were from Zambia, Mozambique or South

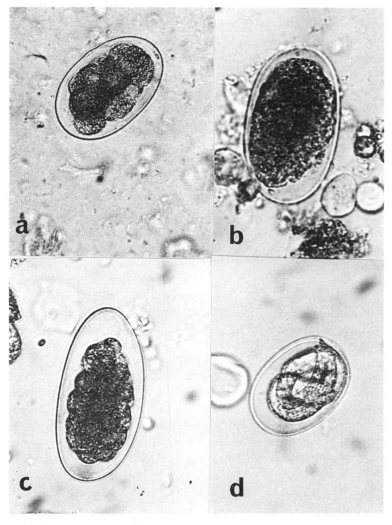

Figure 5.6 Eggs of hookworm (a); *T. deminutus* (b); *Trichostrongylus* (c) and *S. fuelleborni* (d).

Africa, even though *T. deminutus* had been diagnosed from humans in the latter two countries in earlier studies (Sandground 1931, Amberson & Schwarz 1952). Although detecting no cases in his survey in Mozambique, de Azevedo (1964) did comment that he 'presumed that this parasite still exists', while Anthony and McAdam (1972) have diagnosed a case in Uganda, but from a surgical biopsy, not from eggs. In Kenya, Rees *et al.* (1974) have remarked that although *T. deminutus* has not been reported from humans in that country, it is common in monkeys and baboons there and it is thus 'possible that human infections occur in Kenya'.

Trichostrongylus

Compared to the prevalence of hookworm infections, records of human infection with *Trichostrongylus* are rare. The trichostrongyloids infect a wide range of wild and domestic animal species, but of the group, only one subfamily, the Trichostrongylinae, has been recorded from humans (Durette-Desset 1985) and high prevalences in humans tend to be recorded in areas where there is a very close association between humans and sheep (Bruyning 1985). In Africa records of the genus *Trichostrongylus* in humans are sporadic, even within individual countries, being recorded in some investigations but not in others by the same investigators. Thus it has been recorded in Zimbabwe by Blackie (1932), Goldsmid (1968b) and Macdonald and Goldsmid (1973) as well as by Pampiglione *et al.* (1979) in Zaire. However, many carefully conducted studies have failed to detect this helminth in humans and the overall conclusion would seem to be that the genus only sporadically infects humans and as such is of little clinical significance, a view also held by Cook (1986). However, experience with *Trichostrongylus* suggests that although human cases will occur sporadically wherever cattle and sheep are kept, as suggested by Muller (1987) and Bruyning (1985), it remains a relatively rare parasite of humans over most of Africa. However, if looked for, the odd case will be found. Its rarity was well illustrated in the survey by Feacham *et al.* (1983) who, in a comparative study covering Ghana, Zambia and Botswana, detected one case only – in Ghana. It is of interest to note that some surveys in Africa have recorded an extremely high prevalence of human infection with *Trichostrongylus*. Thus reports from Egypt have noted prevalence of 11–60% in some populations (Tongson & Eduardo 1982) and Ethiopia is also reported to have a high prevalence of human infection (Bruyning 1985).

Oesophagostomum

The genus *Oesophagostomum* is also widespread in Africa as a parasite of domestic stock (Ukoli 1984) and non-human primates (Goldsmid 1974a,

Goldsmid & Rogers 1978) and cases of human infection have been recorded from Uganda (Elmes & McAdam 1954, Lothe 1958, Anthony & McAdam 1972) and Zimbabwe (Gordon *et al.* 1969) and the northern Togo–Ghana border (Gigase *et al.* 1978).

In most instances of human oesophagostomiasis, diagnosis has been made after surgical biopsy of a helminthoma containing the worms. With the genus being so common in animals in Africa, it would seem surprising if it is not a fairly widespread parasite of rural people – albeit rather rare. However, details of transmission in human communities are not known. The fact that the eggs are indistinguishable from those of the hookworms makes its routine diagnosis impracticable without resorting to larval culture methods.

In a recent review of the nodular worms of genus *Oesophagostomum* and their reclassification into several subgenera, Stewart & Gasbarre (1989) have pointed out that this species from primates fits into two subgenera, *Conowberia* (which contains *O. stephanostomum*) and *Ihlea* (containing *O. bifurcum*). They feel it is unclear whether humans or primates are the natural hosts and *Conowberia*, in their opinion, seems to have arisen in Africa and then dispersed to Southeast Asia and probably even re-entered Africa and thence to South America – in the latter case purely in humans as there are no other known suitable hosts for *O. stephanostomum* in the New World. They then imply that hookworm from slaves would have included cases of *Oesophagostomum* and this in turn would seem to suggest that human oesophagostomiasis must be much commoner than presently available prevalence figures would suggest, again partly due to confusion with the hookworm species when diagnosing rests on egg recovery alone.

Strongyloides

S. stercoralis is found over the whole of Africa where hookworms are encountered. Reported figures of its occurrence, however, probably underestimate its prevalence, particularly if based upon single stool specimens, due to recovery difficulties resulting from irregular shedding of larvae from the host, a factor of great significance in chronic infection (Rowbottom *et al.* in press). Cavalho (1988) feels that the longer life and 'greater resistance' of hookworm larvae compared to those of *S. stercoralis* might explain the greater frequency of hookworm infections compared to those caused by *Strongyloides*. However, difficulty of diagnosis in strongyloidiasis, a point of special note by Genta (1987), may also explain this fact. Thus in chronic strongyloidiasis, Pelletier (1984) reported that up to five hours of microscopy could be required to diagnose 90% of cases (Rowbottom *et al.* in press)

S. fuelleborni is a common monkey parasite in Africa (Goldsmid 1947a,

1982, Goldsmid & Rogers 1978) and as a parasite of humans it has been recorded as far south as Zimbabwe (Goldsmid 1968b) and Namibia (Kyronseppa & Goldsmid 1978) but it tends to be commoner in the tropical areas of the continent. Thus it has been found to be a significant human parasite over most of the countries of central, central east and west and west Africa (Buckley 1946, Pampiglione & Ricciardi 1971, 1972, Hira & Patel 1977; 1980). Hira & Patel (1980) again ascribe the scarcity of reports of human infection with *S. fuelleborni* in Africa to a lack of awareness of its existence and a consequent misidentification of its eggs.

S. fuelleborni is relatively easy to differentiate in the laboratory from the strongyle species and from *S. stercoralis* if an awareness is present and fresh or formalin-fixed fresh stool specimens are examined.

Zoonotic aspects

A. duodenale and *N. americanus* are essentially parasites of humans only and infect by skin penetration, although *A. duodenale* can also infect by ingestion of the filariform larva (Pawlowski 1986) and may even be able to make use of paratenic hosts to extend its period of survival prior to achieving infection of the human host when meat of the paratenic host is eaten undercooked (Schad *et al.* 1984).

S. stercoralis too, is essentially a parasite of humans, although it is possible that animals such as dogs may also harbour this species and thus serve as a reservoir for human infection, especially in tropical regions such as Nigeria, where Ugochukwu & Ejimadu (1985) recorded prevalences of 15–25% in various dog breeds. It differs from the hookworms in its ability to multiply in a free-living cycle, although there seems to be some strain variability here (Galliard 1967) and in being able to autoinfect and so persist for periods in excess of 40 years or so in infected humans (Rowbottom *et al.* in press).

Oesophagostomum and *Trichostrongylus* are undoubtedly zoonotic helminthic infections with humans usually becoming infected by ingestion of the infective larvae with vegetable foods contaminated by infected animal droppings (Tongson & Eduardo 1982, Marcus 1982), although the latter worms may at times infect by skin penetration (Muller 1975).

S. fuelleborni also infects by skin penetration and has always in the past been considered to be zoonotic, with humans becoming infected from monkeys. However, in a detailed study of the species from Africa, Pampiglione and Ricciardi (1971) concluded that '*Strongyloides fuelleborni* is a parasite of man not just accidental as was believed until now, but adapted to the human species and present in a vast zone which includes the rain forest belt of Cameroon and Central African Republic between 2° and 4°N as well as several forest zones in the mountains of Caffa

(Ethiopia) and in Rhodesia (Zimbabwe) and Zambia'.

The epidemiology of the 'false hookworm' *T. deminutus* has not to date been uncontroversally elucidated.

The enigma of its distribution in monkeys and humans has suggested that two species of this genus might exist – one in humans and limited to central and southern Africa and one in non-human primates and having a much wider distribution through the Old World tropics (Goldsmid & Lyons 1973, Goldsmid 1971a, 1974a, 1982).

This latter hypothesis had some tentative support from the fact that Yamaguti (1954) had described a second species, *T. simiae*, from monkeys in the Celebes, based upon morphological criteria.

However, studies by Goldsmid and Lyons (1973) including scanning electron microscopy, on specimens obtained from humans and baboons in Zimbabwe, failed to detect any morphological differences which could justify considering the helminths as belonging to different species.

Differences were noted in the worms of baboon or human origin, notably in size and colour, with the human worms being darker in colour and larger in size (10.62 ± 0.72 mm) than those from the baboons, which were lighter in colour and smaller in size (8.28 ± 0.20 mm), as discussed by Goldsmid and Lyons (1973), and Goldsmid (1974a, 1982) (Fig. 5.7).

These differences were further studied to determine their significance in relation to a possible species difference. In terms of the difference in size, it seemed more probable that the baboon specimens were smaller

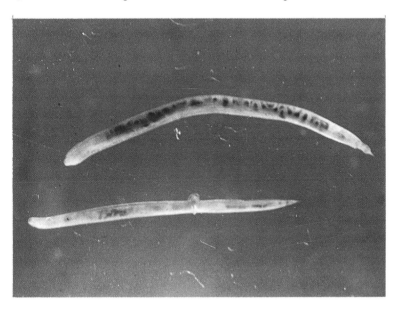

Figure 5.7 *Ternidens deminutus* from human (top) and baboon (below) to show differences in size.

due to stunting, resulting from the heavier load (249 ± 44.5 worms) and consequent crowding effect of the helminths (both *Ternidens* and *Oesophagostomum*) carried in the baboon's large intestine when compared to the almost invariably low loads (19.5 ± 4.9 worms) carried by humans.

The cause of such stunting could also, perhaps, be immunological, with longstanding and more intense infections in baboons being reflected in the adult worm being smaller.

It must at this point be emphasized that this is hypothesis only and it is interesting that in recent reviews such as that of Anderson (1986) on the population dynamics and epidemiology of intestinal nematode infections, although the effects of worm burden on fecundity are considered, no mention is made on the effects of burden (including that of mixed species) on individual worm size or mean size of the worm population.

The mode of infection of the host with *T. deminutus* also remains to be elucidated. The sluggish nature of the third stage free-living larva of *Ternidens* in the soil has long suggested a passive mode of infection such as larval ingestion as occurs in *Oesophagostomum*, when compared to the very active third stage infective larvae of such skin-penetrating species as *N. americanus* and *S. stercoralis* (Sandground 1931, Blackie 1932, Goldsmid 1971b) and attempts to infect known susceptible hosts via the skin have invariably failed (Amberson & Schwarz 1952, Goldsmid 1971b).

In the literature can be found references to infection with *T. deminutus* by ingestion of the third stage, free-living (and presumably infective) larvae. These are based upon a report of successful oral infection by Sandground (1929) and overlook his latter report on the experiment in which he states, in a footnote, that this earlier successful experiment subsequently turned out to be with *Trichostrongylus instabilis* and not *T. deminutus* at all (Sandground 1931).

Subsequent efforts to infect baboons by feeding them with third stage *Ternidens* larvae also proved unsuccessful (Goldsmid 1970) but it should be noted that these captured baboons had an unknown history of previous infection and were thus of unknown immune status. The mode of infection with this species thus remains to be established with certainty.

In a series of experiments performed in an attempt to shed some light on the mode of infection, investigations were carried out on the exsheathing behaviour of the third stage, free-living larvae of *T. deminutus* (Rogers & Goldsmid unpublished).

In these experiments, it was found that exsheathing of larvae reared by such methods as the Harada Mori Test Tube Cultivation Technique (Goldsmid 1967) could be achieved by incubating larvae in solutions of 0.05% or higher of sodium hypochlorite – a process long known to stimulate exsheathing in nematode larvae (Lapage 1935).

The actual process of exsheathing in *Ternidens* is similar to that seen in

N. americanus and involves the formation, at the anterior end of the sheath around the larva, of a swelling, resulting in the formation of a 'cap', the shedding of which results in the release of the larva (Goldsmid 1982, Rogers *et al.* unpublished).

However, further experiments with *T. deminutus* larvae to stimulate exsheathing by simulating human intestinal environmental conditions by exposure of third stage larvae to human intestinal secretions (bile, stomach and small intestine extracts obtained at autopsy) all failed (Rogers & Goldsmid unpublished).

Further, exposure of these larvae to varying pH and CO_2 concentrations at 4°C, 22°C and 37°C gave the interesting observation that larval activity was greatest in most cases at 22°C rather than at 4°C or 37°C for any given pH or CO_2 concentration (Rogers *et al.* unpublished). This leaves us with the as yet unproven postulate that infection of the primate host occurs not by direct larval ingestion as happens in the related genera *Oesophagostomum* and *Chabertia*, but perhaps involves the incorporation in the life cycle of some form of intermediate host – a suggestion first put forward by Amberson and Schwarz (1952) and subsequently amplified by Goldsmid (1971a, 1971b, 1974a, 1982).

Such a hypothesis could explain some of the epidemiological enigmas of *Ternidens* infections. It could explain the lack of success to date of experimentally infecting either baboons or humans – animals known to be highly susceptible to infection – by the final free-living larval stage of the parasite, as well as the failure to achieve exsheathing or further larval development by exposure of these larvae to intestinal secretions and simulated physiological conditions found in the primate intestine.

Of course, even if ingestion by such an intermediate host was the source of infection, it would require ecological contact between humans and the monkeys, together with a shared dietary requirement for utilization of the intermediate host. These facts, taken together, suggested that the role of intermediate host, if such exists, could conceivably be filled by termites (Goldsmid 1971a, 1974a, 1982) or some other edible insect.

In Africa, such ecological contact does exist between baboons (which continually raid village crops and would thus contaminate the ground with infected droppings) and humans. Termites could then become infected by the free-living third stage *Ternidens* larvae in the soil – and termites comprise a basic component of the diet of both the people and baboons. It is indeed intriguing to speculate that fungus-growing termites, such as *Macrotermes natalensis* (which has a wide distribution through the warmer parts of Africa, according to Skaife 1953), might be rearing *Ternidens* larvae in their fungus gardens due to the utilization of infected baboon faeces, and then becoming themselves infected from these gardens or from the fungus grown there.

Clinical and immunological aspects

The work of Foy and Kondi (1960), Foy and Nelson (1963) and Gilles *et al.* (1964) all contributed to an understanding and acceptance of the concept of hookworm infection and hookworm disease.

The symptoms of infection with hookworm may begin early, with the development of a hookworm dermatitis ('dew itch') at the site of larval penetration of the skin, to be followed sometimes, especially in cases of heavy infection, with a pneumonitis associated with a Loeffler's syndrome (Miller 1979). The clinical effects of the adult hookworms on their host is a complex interrelationship involving worm species, worm load, host diet and iron reserves (and even iron overload) as well as such factors as sex of the host as discussed by Goldsmid (1965) and host immune response (Ogilvie *et al.* 1978). The role of the hookworms in causing anaemia (with such secondary sequelae as progressive insufficiency); hypoalbuminemia and oedema; mesenteric lymphadenitis; diarrhoea and duodenitis (but not malabsorption) as well as possibly contributing to impaired overall physical and mental development has been discussed in detail by Miller (1965, 1979), Woodruff (1965), Corachan *et al.* (1981), Cook (1986), Pawlowski (1986), Migasena & Gilles (1987) and Nesheim (1987).

All in all, the hookworm species can be accepted as parasites of major significance and as a primary cause, or important contributing factor, to human disease (especially when coexisting with other parasites, as is usually the case in Africa, where multiple species infections are often the rule rather than the exception) or with such dietary deficiencies as kwashiorkor (Allen & Dean 1965) or in pregnancy.

Of the other intestinal nematode species, infection with *S. stercoralis* varies from asymptomatic to symptomatic and presenting with severe abdominal involvement including profuse diarrhoea and malabsorption (WHO 1980, Cook 1986) or dysentery (Genta 1987). It is the ability of *S. stercoralis* to persist for years in chronic form and to autoinfect the host and involve virtually every tissue of the body (Genta 1987) which poses a new and significant threat in Africa today. Few studies have been carried out on chronic strongyloidiasis in Africa, but with the widespread occurrence of Acquired Immunodeficiency Syndrome (AIDS) on that continent, and particularly in the tropical areas of central Africa (Biggar 1986) the threat of disseminated strongyloidiasis becomes one of major concern and in need of urgent investigation.

Strongyloidial hyperinfection syndrome, with massive and often fatal autoinfection of immunologically compromised hosts has been described in Hodgkin's disease, various forms of leukaemia, lymphosarcoma and in organ transplant recipients (Pelletier 1984, Weinkel *et al.* 1988). It has also been suggested that the syndrome might follow infection with HIV but there is little evidence as yet to support this hypothesis (Genta, 1987, Lockwood and Weber 1989) and *Strongyloides* has been removed from the CDC classification of AIDS while a recent review of opportunistic infections in AIDS in the tropics by Piot and Mann (1988) makes no mention of it at all.

The immune response in *Strongyloides* infection is certainly complex, involving immune mechanisms at the intestinal, pulmonary and cutaneous levels depending, at least in part, on whether infection is primary, secondary or multiple (Butterworth 1984).

Genta too, has demonstrated that in chronic strongyloidiasis there is not only a depression of T-cell activity, but also an involvement of factors in the patient's serum which inhibit parasite-specific cellular responses *in vitro* (Genta et al. 1983, Genta 1984).

Pawlowski (1986) feels that the threat of life-threatening strongyloidiasis in cases of AIDS might not become a major clinical problem as 'the reproduction of *S. stercoralis* in the human intestine and the clinical expression of strongyloidiasis are much influenced by local gut immunological mechanisms . . . but probably not those that are impaired in AIDS patients' a stance also taken by Petithory and Derouin (1987) who have pointed out that disseminated strongyloidiasis has not emerged as a major consequence of HIV infection in central African countries such as Zaire or the Central African Republic.

It is interesting that Weinkel et al. (1988) found that while eosinophilia was common in the normal host infected with *Strongyloides*, it was frequently absent in patients with hyperinfection syndrome. By contrast, Oliver et al. (1989) found it to be an important feature in their diagnostic index for the diagnosis of strongyloidiasis in a non-endemic area.

Studies by Pampiglione and Ricciardi (1972a) on the clinical effects of human infection with *S. fuelleborni* have suggested that symptomatology due to this species is milder than with *S. stercoralis* but again more studies under field conditions are needed to supplement these observations made on otherwise healthy human volunteers. In this regard, it might be worth noting the severe effects of the *S. fuelleborni*-like species in Papua New Guinea which is associated with the Swollen Belly Syndrome of young children (Vince et al. 1979, Ashford et al. 1979, Ashford & Barnish 1987, Barnish & Ashford 1989).

Trichostrongylus being a relatively uncommon intestinal parasite of humans in most areas and usually being recorded in light infections, is probably of little practical clinical significance (Muller 1975, Cook 1986), but presumably in heavy loads it could be associated with blood loss from the host as recorded for the related species *Haemonchus contortus* in domesticated animals by Baker and Douglas (1966).

Oesophagostomum on the other hand, although not commonly reported, may be more common than at present believed. Despite this, these worms are clinically important in individual cases due to the fact that their invasion of the gut wall results in the development of helminthomas which often require surgical intervention. This role of helminths, and particularly *Oesophagostomum*, in causing helminthomas has been considered in depth by Anthony and McAdam (1972). It is worth noting that *A. duodenale* may also be associated with helminthoma formation and that *T. deminutus* (a species closely related to *Oesophagostomum*) may also be a cause of these tumour-like lesions of the human intestine.

Like so many other features of *T. deminutus* infections, the clinical significance of human infection with this species remains to be fully elucidated.

Most reports on human infection with *T. deminutus* have reported the species as generally having a low prevalence and thus giving rise to the opinion that this intestinal nematode is one of minor significance as a human parasite (Cook 1986). However, studies in Zimbabwe have indicated that in that country at least, the species is widespread in both humans and monkeys such as the baboon and the vervet monkey, and also that in some African communities its prevalence may reach 87% (Goldsmid 1982).

Further, the theory of Goldsmid and Jablonski (1982) that the pigment granules around the gut of the worm derive from blood ingested from the host and reinforced by the observation of Goldsmid (1971a) that blood could be seen in the intestinal contents of freshly collected worms (and its nature confirmed using histochemical tests), supported the belief of Blackie (1932) that the worms were blood-feeders, as opposed to the contention of Boch (1956) that they merely fed on mucous material not blood.

Crompton (1986) has recently reviewed the problem of the contribution of intestinal nematode infections to the aetiology of human malnutrition and has pointed out that more research is needed in the area – a fact which also holds true for *Ternidens* infections where nothing is known at all regarding the role that this species may have in the exacerbation of malnutrition and kwashiorkor in children living on an already deficient diet. This aspect of *Ternidens* infection may be relevant in areas of low protein intake (which includes much of Africa) as a predisposing factor to kwashiorkor.

In this respect, it is interesting to compare the age-related prevalence recorded for hookworm and *Ternidens* in Zimbabwe by Goldsmid (1972a) where it can be seen that although hookworm infections start at an earlier age than do *Ternidens* infections, the prevalence of the latter infections increases rapidly to reach a peak at about 10 years of age before dropping and then levelling out until about 35 years of age and then tapering to 50–55 years of age. In hookworm, it was found that prevalence increased steadily until 25–35 years of age, before tailing off. For both infections, children were commonly infected, thus potentiating a clinical correlation with the effects of malnutrition (Fig. 5.1).

Moqbel (1986) has reviewed helminth-induced intestinal inflammation, remarking that 'gastrointestinal inflammation is a prominent feature of the protective reaction in animals immune against helminths'. In *Ternidens* infections, Boch (1956) described in detail the inflammatory responses in the large intestine of infected rhesus monkeys which led to nodule formation of the sort seen in *Oesophagostomum* infections and which have resulted in surgical emergencies in human infections with this species (Anthony & McAdam 1972). The latter authors did, in fact, include one case of *Ternidens*-induced helminthoma in their series and Goldsmid (1971a) found *Ternidens* associated with large bowel ulceration at autopsy in a Zimbabwean.

Studies on immunity to infection with hookworms have indicated that we still do not fully understand the immune response to infection with *N. americanus* and *A. duodenale* and many other intestinal nematodes infecting humans. We do not even know in hookworm infections whether an effective protective immunity is, or can be, developed (Crompton 1987). Thus as Migasena and Gilles (1987) have pointed out for hookworm: 'Longitudinal studies suggest that parasites are regularly lost and subsequently reacquired in the following transmission season and some evidence for herd immunity exists. If hookworms do allocate protective immunity in man, the concept remains to be conclusively proven'. Ball and Bartlett (1969) were able to demonstrate that although antibodies were formed to hookworm infection and were still detectable two years after infection, no protective immunity was developed, although larvae of *N. americanus* incubated in immune sera shown to possess specific antibodies developed precipitates on their surface. Ogilvie et al. (1978), also working with *N. americanus*, concluded that although their studies revealed that the development of antibodies was not related to resistance to infection, it was possible that the antibody responses were important in limiting the severity of the gastrointestinal disturbances. Miller (1979) observed that 'there is no conclusive evidence . . . that man develops a functional immunity to hookworms' despite the fact that 'acquired immunity with a protective function, nevertheless, has been found in other hookworm–host systems when efforts have been made to discover and investigate the subject'. However, despite progress in our understanding of helminth–host relationships in the intervening years, the latest reviews draw the same conclusions today, mainly that people infected with hookworm 'tend to remain susceptible to infection throughout life, even with constant exposure to the infective stages' (Behnke 1987a).

Where does the solution lie regarding the immunological relationship between humans and hookworms? Does the infection persist with no development of protective immunity due to an immunological incompetence on the part of the host to eliminate the parasite and prevent reinfection or are the hookworms successful in somehow evading the immune response as discussed by Behnke (1987a, 1987b). Wakelin

(1986), put forward the concept that humans have an innate capacity to produce effective and protective responses against intestinal nematodes but he also believes that this capacity is 'subject to a number of powerful constraints, arising from parasite evasion, depression of response capacity and deficiency in response capacity'. This view was reinforced by Mitchell (1987) who pointed out rather succinctly that although parasites are immunogenic, 'the variety of immune responses they elicit is matched by their range of mechanisms to evade, subvert or distract these responses'.

Behnke (1987a) also believes that 'humans express a range of response levels', but the fact remains that more work must be done in this area if we are to even consider the concept of developing an effective protective vaccine against the hookworm species and the statement made by Miller (1979) remains a fundamental one, namely that 'substantial evidence of functional protective immunity in the hookworm–man relationship is the most important remaining void in our knowledge of the diseases caused in man by hookworms'.

Details of the immune response, the role of eosinophils (Goetzl & Austin 1972, David et al. 1980, Samster 1980, Butterworth 1984) and mast cells (Lee et al. 1986) as well as the overall concept of cell-mediated damage (Butterworth 1984) all need further study and elucidation in infection, not only with hookworms, but with other parasitic intestinal nematodes as well. The same is true as regards the interaction between these areas and the human complement system (Leid 1988).

The low loads of *Ternidens* recorded to date could suggest a host limitation of load as claimed by Boch (1956) who, working with rhesus monkeys, stated that larval development after infection was inhibited by the presence of adult worms already in the intestine of the host. It is relevant in this regard to note that Herd (1971) recorded a similar histotropic stage in the life cycle of the related species, *Chabertia ovina* in sheep, with larvae being found in the wall of the small intestine. The claim by Boch (1956) for the presence of a similar histotropic stage in the intestinal wall by the larvae of *Ternidens* would be of great significance in terms of clinical load and treatment regimes (Goldsmid 1970) and would suggest a fascinating mechanism of immunological control of the worm burden by the host. Presumably, such control would involve primarily a cell-mediated immune response, although IgA might be involved as suggested for other intestinal helminth infections by Befus (1986); it undoubtedly requires further research.

Rogers and Goldsmid (1977, 1980) and Goldsmid (1982) have demonstrated that infection with *T. deminutus* does result in the mounting of a humoral immune response by the human host and that specific antibodies against *T. deminutus* are formed, resulting in precipitation around infective larvae incubated in sera obtained from other infected humans – a similar result to that obtained for larvae of *N. americanus* in the study of Ball and Bartlett (1969). Rogers and

Goldsmid (1977, 1980) were further able to show that such antibody responses could be used to diagnose *T. deminutus* infections by the indirect fluorescent antibody test using transverse sections of adult worms as antigen.

The protective nature, if any such protection is provided, of such circulating antibody needs further investigation and this type of study remains a priority for other gastrointestinal nematode infections as well (Pritchard 1986).

Certainly it would seem that an immune response against *T. deminutus* infections is mounted by the infected human host. This response comprises a humoral one with specific antibodies being produced and with precipitating antibodies being formed against infecting larvae, findings very similar to those reported by Herd (1971) for the related sheep parasite, *Chabertia ovina*.

Herd (1971) found evidence in *Chabertia* infections of a cell-mediated reaction also and undoubtedly this type of immune response also occurs in *Ternidens* infections and may manifest especially as the histotropic response reported in infected rhesus monkeys by Boch (1956).

This concept of arrested larval development, or hypobiosis, in nematodes is one of increasing interest and one which may be more widespread than has been realized in the past (Gibbs 1986, Armour & Duncan 1987). Thus it has been shown to occur in a range of cattle nematodes and probably occurs also in *A. duodenale* (Pawlowski 1986, Behnke 1987a). Coles (1988) felt that arrested larval development in cattle nematodes should be considered as a possible way in which these helminths can avoid the action of anthelmintics, perhaps this too should be remembered for human nematode species with such characteristics.

Treatment

The last 50 years has seen the development of a range of safe, effective and increasingly broad-spectrum drugs for the treatment of the gastrointestinal nematode parasites of humans. The introduction of pyrantel pamoate and mebendazole heralded a new range of safe, effective and increasingly broad-spectrum anthelmintics – the benzimidazole drugs such as albendazole, cambendazole and flubendazole (Desowitz 1971, Goldsmid 1972c, Cavier & Hawking 1973, Cline 1982, Van den Bossche 1985, Janssens 1985, James & Gillies 1985, Migasena & Gilles 1987, Cavalho 1988, Cook 1989). This trend can be seen from Table 5.1.

The development of safe, effective, broad-spectrum and easily administered anthelmintics such as mebendazole is of major importance where multiple helminth infections are often the rule rather than the exception and where weight-independent dosages are essential for mass treatments with even rudimentary clinic facilities often being

Table 5.1 Anthelmintic efficacy and side-effects.

	Ancylostoma	Necator	Ternidens	Strongyloides	Ascaris	Enterobius	Trichuris	Side effects
Carbon tetrachloride	+	+	−	−	+	−	−	++
Tetrachloroethylene	+	++	−	−	c/i*	−	−	+
Piperazine	−	−	−	−	++	++	−	+
Bephenium hydroxynaphthoate	+++	++	+++	−	++	−	−	+
Phenylene diisothiocyanate	++	++	+	−	+	−	+	+
Thiabendazole	+++	++	+++	++	++	++	+	++
Pyrantel pamoate	+++	++	+++	−	+++	+++	+	+
Mebendazole	+++	++	+++	+	+++	+++	−	−
Albendazole	++++	++++	?	+	+++	+++	++	−

* c/i = contraindicated.

absent in rural areas. The limiting factor to their widespread utilization may, however, be cost, thus necessitating the use of less effective, less safe and less broad-spectrum but cheaper drugs for the treatment of hookworm infections. This cost factor is something which has to be borne in mind in the development of new anthelmintics – it is of little practical value developing drugs, no matter how good, if they are beyond the financial reach of the developing countries which so badly need them.

Mercer (1985), referring to the development of new chemotherapeutic strategies for helminth infections, has called for more basic research on the biochemical metabolic differences between parasites and their hosts to hopefully allow development of 'selective inhibitors or analogues targeted at various aspects of the metabolism and development of the parasite, ideally with little or no deleterious effect on the host'. He went on to suggest that the helminth endocrine system could provide such a target for the new chemotherapeutic agents.

In broad terms, the hookworms are relatively easy to treat although as a generalization, *Necator* is the more difficult species to eradicate. *Ternidens* too, responds well to most, but not all, of the newer, broad-spectrum anthelmintics and the same in all probability applies to *Trichostrongylus* and *Oesophagostomum* – although a significant number of cases of oesophagostomiasis require surgical treatment where helminthomas have formed or where intestinal perforation has occurred. The same applies in some cases of infection with *Ternidens* and even *A. duodenale* (Elmes & McAdam 1954, Gordon *et al.* 1969, Anthony & McAdam 1972).

Strongyloidiasis remains an area where there is a need for the development of new, effective and safe anthelmintics. For *S. stercoralis*, mebendazole has proved to be of only moderate efficacy and has proved particularly disappointing in the treatment of chronic, autoinfecting strongyloidiasis (Rowbottom *et al.* in press, Oliver *et al.* 1989). It would seem, in theory, that mebendazole, with its low rate of absorption from the gut lumen, works to some extent and especially with prolonged use (Mravak *et al.* 1983), in acute strongyloidiasis which is essentially confined to the gut by the immunocompetent host, but that it fails where significant autoinfection is occurring in patients with some degree of immunodeficiency. In these cases, thiabendazole, despite its side-effects, remains the treatment of choice – presumably because it is better absorbed from the gut and can get to the larvae in the tissues (Cavier & Hawking 1973, Bell 1982, Van den Bossche 1985, Grove 1982a, James & Gilles 1985) although Grove (1982b) has recently refuted its larvicidal value in the tissues of the host in experimental laboratory animals infected with *S. stercoralis*.

Albendazole (86% cure rate) and cambendazole (100% cure rate) have recently been reported as promising for the treatment of strongyloidiasis and thus offer renewed hope for disseminated disease (Genta, 1987, Cook 1989).

Goldsmid (1974b) found mebendazole to be effective against *S. fuelleborni* in captive baboons, but its use in humans infected with this helminth (or a very similar species) has not been fully evaluated in either Africa or Papua New Guinea and thiabendazole continues to be the drug of choice for this species as well (Pampiglione & Ricciardi 1972a, Shield 1985, Barnish & Barker 1987). However, with its unacceptably high level of side-effects, thiabendazole is not ideal and it is hoped that the newer benzimidazole drugs such as albendazole will fulfil their initial promise in the treatment of strongyloidiasis and other intestinal nematode infections (Van den Bossche 1985, Janssens 1985; James & Gilles 1985, Barnish & Barker 1987).

Further work is also needed on the effects of histotropic arrest of larval development by adult worms already in the gut as reported by Boch (1956) for *T. deminutus* in order to determine the correct drug regimes for such nematodes, especially where anthelmintics are poorly absorbed from the gut lumen or where they have no larvicidal activity.

In fact, the whole rationale of anthelmintic treatment for intestinal nematodiases needs to be re-evaluated in the light of concepts for worm control rather than simple chemotherapeutic cure for individual clinical cases. As Davis (1983) has pointed out: 'curative treatment . . . is frequently nullified because . . . background conditions lead to reinfection and perpetuation of epidemiologic transmission cycles'.

Control

Mwosu (1983) has considered the role of mass chemotherapy campaigns against parasitic helminth infections and concluded that 'by continuously identifying and treating the diminishing populations of wormy persons over a period of three to four years, it should be possible to reduce the level of infective agents in the environment to a level below the transmission threshold even without latrine hygiene'.

Hookworm control by mass chemotherapy alone will not completely break the parasites' transmission cycle. As Bruyning (1985) has explained, it soon became apparent in early hookworm control programmes that the aim of mass therapy was not the elimination of all the hookworms in the infected people, but rather aimed at a significant reduction of the worm loads, thereby reducing their clinical effects. He went on to outline that as the hookworms have a large egg output, frequent treatment is an essential prerequisite for reduction of transmission and that for overall control, mass treatment must be combined with other control measures such as environmental sanitation and health education.

One must therefore consider the role of sanitation and hygiene in any overall control programme for intestinal nematode infections. It has long been acknowledged that sanitation would seem to be the most appropriate way to control intestinal helminth infections. However, as pointed out by Cairncross (1987), the mere owning of a latrine is not sufficient, it must be used! While seeming simple in theory, the value of sanitation and hygiene in practice is difficult to assess – or to improve. Mwosu (1983) went on to comment on this as well, saying: 'The role of sanitation in reducing infections is well known . . . but the possibility of altering human behaviour to minimize contact with infecting agents and thereby reduce infection has been received with some cynicism'. Nelson (1972) has also drawn attention to the need for detailed studies on human behaviour in parasitic disease control, commenting for schistosomiasis (although it is equally true in principle for hookworm) that 'perhaps too much attention has been given to snail control and to chemotherapy and not sufficient to studying human behaviour' and he went on to observe that, with few exceptions, there had 'been no deliberate scientific observations on defecation habits'. In this regard, Mwosu (1983) has drawn attention to recent studies in Iran as a model for 'the crucial role of ecological-cultural factors' in the control of intestinal helminth infections.

Cairncross (1987) too has emphasized the lack of available behavioural information, although many studies on latrine hardware have been done. He has pointed out that where latrines are to be utilized for parasite control, total community involvement is essential to make the whole plan workable. In fact, when looking at community control of infection with hookworms and related nematode species in Africa, is mere load reduction enough? In terms of immunity to reinfection we have no evidence that the retention of a small number of worms is of protective value, and large numbers of people with low loads can still serve to reinfect the environment in the absence of other backup control measures such as sanitation and the use of protective footwear.

Hotez et al. (1987) feel that chemotherapy aimed at heavily infected individuals who seem prone to hookworm anaemia might work as a realistic proposition for control. Hookworm is essentially a one-host system and in theory sanitation and the wearing of shoes should suffice to break the cycle of transmission, and control should thus be relatively easy if combined with a mass chemotherapy programme (Van den Bossche 1985). In practice we know that it is not as simple as this to achieve success (Nelson 1972, Mwosu 1983).

With the other hookworm-like species found in Africa, control is even more difficult, being hampered by their having animal reservoirs (Trichostrongylus, Oesophagostomum, S. fuelleborni); by a lack of understanding of basic epidemiology (Ternidens); by lack of safe and effective chemotherapeutic agents (S. stercoralis) and by diagnostic difficulties and identification ineptitude in the laboratories.

The hookworms and S. stercoralis could theoretically be controlled within the human communities, at least to some extent, by adequate sanitation, but we need more reliable studies for each community on such factors as behaviour, traditional customs and agricultural practices as discussed by Schad *et al.* (1983) and Mwosu (1983). Further, as regards *Ternidens* and *S. fuelleborni*, we need confirmation as to whether the reservoir is even necessary for the maintenance of infection within the community once infection has entered that community.

If all these species are to be controlled and if their clinical effects on humans in Africa are to be reduced or eliminated, then we must have a detailed and reliable understanding of all aspects of their natural history: the geographical distribution of the individual species; their epidemiology and transmission cycles; host–parasite interactions; diagnosis and treatment as well as the interacting influences of social and behavioural factors within the community involved – and this might stretch to a superstitious distrust by the community members in remote rural areas of the investigating team collecting such outrageous things as faeces!

Mathematical models for control have been considered in detail for a range of helminth species, including the hookworm species by Anderson and May (1985) and Crompton (1989) but do we really have sufficient understanding of the basic parasitology of these species in Africa at this stage to apply such concepts in practice?

Bruyning (1985) has emphasized the fact that all mathematical models have deficiencies, due largely to the unavoidable need for omissions and oversimplification to make them manageable. Despite this, he points out, some models, including ones for hookworms, have proved useful, especially as an aid to the critical evaluation of alternative control measures and in the design of control programmes.

Anderson and May (1985), however, feel that mathematical models do have a significant role to play, emphasizing that a community approach to helminth control requires a thorough and detailed understanding of the population dynamics of these helminths – studies which they feel have been neglected to date and which a mathematical approach can benefit. Anderson (1986) has also applied mathematical criteria to the concept of intestinal nematode control – again emphasizing the need for accurate data in such studies.

A new and fascinating concept for control of parasitic nematodes has been supported by studies on chemotactic sex-attractant factors such as the pheromones (MacKinnon 1987). Extending this theme, Haseeb and Fried (1988) feel that: 'Helminthologists are about 30 years behind entomologists in their studies on chemical communication' and that: 'new approaches to the control of helminthiases are needed,' and suggest that: 'a knowledge of pheromones that regulate attraction. . . could lead to immunological and chemotherapeutic means to interfere with pheromonal communication. Prevention of worms from mating would. . . prevent the completion of the life cycle for most helminths'.

Future prospects

Even in today's technically advanced world, helminthiases must be recognized as a major problem in Third World countries, and as a major threat to travellers visiting these countries. Of these helminth infections, the hookworms and related species remain amongst the most significant (Behnke 1987a). Ukola (1984) has commented that, for example, in Nigeria a child in the villages is 'bombarded by parasitic worms from soon after birth throughout his life', and further, multiple worm species infections are more common than single species infections (Akogan 1989).

Vaccines are being seriously and confidently proposed for hookworm control (Hotez *et al.* 1987) – again do we really have the basic scientific foundation of knowledge to know whether they will work? Do we know enough about the immune interactions between the hookworms and their human hosts? Can we succeed where nature apparently seems, on presently available data, to have failed? Leid (1988) has pointed out that most parasites have evolved powerful mechanisms to overcome the damaging effects of the complement pathway and emphasized that we need to know more about these mechanisms if we are to use them to control such infections – especially control utilizing immunological bases. Recently Klei (1986), drew attention to the fact that despite all the research being done, to date the only commercially produced helminth vaccine is that against parasitic bronchitis due to the bovine lungworm, *Dictyocaulus viviparous*.

Even if effective vaccines can be developed, will the countries of Africa have the financial resources to afford their widespread use?

A host of problems remain in the study of these hookworm-like species in Africa. The hookworms themselves are undoubtedly the most clinically important species in the group, but we really need further information on the other members in order to evaluate scientifically their real significance as parasites of humans and to decide whether specific control is warranted and, if it is, how it is to be implemented. In fact basic studies are still urgently required on the transmission of the hookworms and other similar species in different regions with regard to cultural factors, for the formulation of control guidelines (Crompton, 1989).

As more research results are published, the pieces of the jigsaw are gradually being fitted together to allow us a better and more complete assessment of the whole interrelating pattern for these intestinal nematodes and their importance in relation to all the other social and medical problems facing the population of Africa.

We have a long way to go before complete understanding is available and perhaps, as pointed out by Muller (1987), we need to alter our approach. Is mere scientific excellence enough? Do we need, in addition, a more entrepreneurial approach?

Undoubtedly, we must encourage the teaching institutions in Africa to train capable parasitologists. Befus (1986), discussing immunity in intestinal helminth infections has also made this point, stating that 'most importantly research facilities must be established in developing countries to investigate intestinal, immunological, inflammatory and physiological responses during infection. Through such investigations, risk factors for susceptibility and disease severity may be identified and therapeutic and prophylactic strategies developed'.

As can be seen from an overview of the hookworm problem in Africa, much research remains to be done – we should be expanding our investigations on the parasitic diseases of Africa and elsewhere to fill the gaps in our knowledge with studies based upon a solid basis of scientific fact rather than upon speculation alone. For this research to be done, we need trained parasitologists and so it is perhaps fitting in this contribution to a volume written to honour the lifetime of work contributed by George Nelson to the advancement of the science of medical parasitology to end with a quotation from Professor Nelson, when he made the plea: 'If we are to meet the need for training of specialists in the field of parasitology and medical entomology, it is essential that there be continuity of support for the maintenance of the Schools of Tropical Medical and Public Health in Europe and America' and that the provision of adequate funds will ensure the continuance of 'collaborative studies between European and American scientists in areas where parasitic and vector-borne diseases are endemic and as a by-product this will help to produce the future generation of medical parasitologists in both the developed and developing world' (Nelson 1972).

References

Akogan, O. 1989. Parasitic worms in Nigerian folklore. *Parasitol Today* **5**, 39.

Allen, D. & R. Dean 1965. The anaemia of kwashiorkor in Uganda. *Trans. R. Soc. Trop. Med. Hyg.* **59**, 326–41.

Amberson, J. M. & E. Schwarz 1952. *Ternidens deminutus* Railliet and Henry, a nematode parasite of man and primates. *Ann. Trop. Med. Parasitol.* **46**, 227–37.

Anderson, R. M. 1986. The population dynamics and epidemiology of intestinal nematode infections. *Trans. R. Soc. Trop. Med. Hyg.* **80**, 686–96.

Anderson, R. M. & R. M. May 1985. Helminth infections of humans: mathematical models, population dynamics, and control. *Adv. Parasitol.* **24**, 1–101.

Anthony, P. P. & I. W. J. McAdam 1972. Helminthic pseudotumours of the bowel: thirty-four cases of helminthoma. *Gut* **13**, 8–16.

Armour, J. & M. Duncan 1987. Arrested larval development in cattle nematodes. *Parasitol. Today* **3**, 173–6.

Ashford, R. W. & G. Barnish 1987. Strongyloidiasis in Papua New Guinea. In *Intestinal helmintic infections*, Vol. 2 No.3, *Clinical tropical medicine and communicable diseases*, Z. S. Pawlowski (ed.). 765–73. London: Baillière Tindall.

Ashford, R., J. Vince, M Gratten & J. Bana-Koiri 1979. *Strongyloides* infection in a mid-mountain Papua New Guinea community. *P.N.G. Med. J.* **22**, 128–35.

Baker, N. F. & J. R. Douglas 1966. Blood alterations in helminth infection. In *Biology of parasites*. 155–83. New York: Academic Press.

Ball, P. & A. Bartlett 1969. Serological reactions to infection with *Necator americanus*. *Trans. R. Soc. Trop. Med. Hyg.* **63**, 362–9.

Barnish, G. & R. W. Ashford 1989. *Strongyloides fuelleborni* and hookworm in Papua New Guinea: patterns of infection within the community. *Trans. R. Soc. Trop. Med. Hyg.* **83**, 684–8.

Barnish, G. & J. Barker 1987. An intervention study using thiabendazole suspension against *Strongyloides fuelleborni*-like infections in Papua New Guinea. *Trans. R. Soc. Trop. Med. Hyg.* **81**, 60–3.

Beaver, P. C., R. C. Jung & E. W. Cupp 1984. *Clinical Parasitology*. 9th edn. Philadelphia: Lea & Febiger.

Befus, D. 1986. Immunity in intestinal helminth infections: present concepts, future directions. *Trans. R. Soc. Trop. Med. Hyg.* **80**, 735–41.

Behnke, J. M. 1987a. Do hookworms elicit protective immunity in man? *Parasitol. Today* **3**, 200–6.

Behnke, J. M. 1987b. Evasion in immunity by nematode parasites causing chronic infections. *Adv. Parasitol.* **26**, 1–71.

Bell, D. R. 1982. Anthelmintic drug therapy. In *Recent advances in infection* Vol. 2. Reeves, D. & A. Geddes (eds), 179–84. Edinburgh: Churchill-Livingstone.

Biggar, R. J. 1986. The AIDS problem in Africa. *Lancet* **i**, 79–83.

Blackie, W. K. 1932. A helminthological survey of Southern Rhodesia. *Memoir Series No. 5. London Sch. Hyg. Trop Med.*

Boch, J. von 1956. Knotchenwurmbefall *Ternidens deminutus* bei Rhesusaffen. *Z. Angew. Zool.* **2**, 207–14.

Bruyning, C. F. A. 1985. Epidemiology of gastrointestinal helminths in human populations. In *Chemotherapy of gastrointestinal helminths*. H. Van den Bossche, D. Thienpont & P. G. Janssens (eds), 7–66. Berlin: Springer-Verlag.

Bryant, C. & H. A. Flockhart 1986. Biochemical strain variation in parasitic helminths. *Adv. Parasitol.* **25**, 275–319.

Buckley, J. J. C. 1946. A helminthological survey of Northern Rhodesia. *J. Helminthol.* **21**, 111–74.

Butterworth, A. E. 1984. Cell-mediated damage to helminths. *Adv. Parasitol.* **23**, 143–235.

Cairncross, S. 1987. Low-cost sanitation technology for the control of intestinal helminths. *Parasitol. Today* **3**, 94–8.

Cavalho, E. M. 1988. Helminthic enteropathies. In: *Clinical tropical medicine and communicable diseases*, Vol. 3, No. 2, 535–65. London: Baillière Tindall.

Cavier, R. & F. Hawking 1973. *Chemotherapy of helminthiasis*. Vol. 1. Oxford: Pergamon.

Clark, D. T. 1956. Identification of beta ZnS in the intestinal cells of *Strongylus* spp. *J. Parasitol.* **42**, 77–80.

Cline, B. 1982. Current drug regimes for the treatment of intestinal helminth infections. *Med. Clin. N. Amer.* **66**, 721–42.

Coles, G. C. 1988. Arrested larval development and anthelmintic resistance in cattle nematodes. *Parasitol. Today* **4**, 106.

Cook, G. C. 1986. The clinical significance of gastrointestinal helminths – a review. *Trans. R. Soc. Trop. Med. Hyg.* **80**, 675–85.

Cook, G. C. 1989. Parasitic infections of the gastrointestinal tract: a worldwide problem. *Curr. Opinion in Inf. Dis.* **2**, 106–8.

Corachan, M., H. Oomen & F. Sutorius 1981. Parasitic duodenitis. *Trans. R. Soc. Trop. Med. Hyg.* **75**, 385–8.

Croll, N. A. 1983. Human behavior, parasites and infectious diseases. In *Human ecology and infectious diseases*. N. A. Croll, & J. H. Cross (eds), 1–20. New York: Academic Press.

Crompton, D. W. T. 1986. Nutritional aspects of infection. *Trans. R. Soc. Trop. Med. Hyg.* **80,** 697–705.

Crompton, D. W. T. 1987. Human helminthic populations. In *Clinical tropical medicine and communicable diseases*, Vol. 2 No. 3, *Intestinal helmintic infections*, Z. S. Pawlowski (ed.). 489–510, London: Baillière Tindall.

Crompton, D. W. T. 1989. Hookworm disease: current status and new directions. *Parasitol. Today* **5,** 1–2.

David, J. R., M. A. Vadas, A. Butterworth, P. de Brito *et al.* 1980. Enhanced helminthotoxic capacity of eosinophils from patients with eosinophilia. *N. Eng. J. Med.* **303,** 1147–52.

Davis, A. 1983. The importance of parasitic diseases. In *Parasitology: a global perspective.* K. S. Warren & J. Z. Bowers. (eds), 62–74. New York: Springer-Verlag.

de Azevedo, J. F. 1964. Soil-transmitted helminths in the Portuguese Republic (European and African provinces). *Anais Inst. Med. Trop.* **21,** 273–95.

Desowitz, R. S. 1971. Antiparasite chemotherapy. *Ann. Rev. Pharmacol.* **11,** 351–68.

Dunn, A. 1978. *Veterinary Helminthology.* 2nd edn. London: Heinemann.

Durette-Desset, M.C. 1985. Trichostrongyloid nematodes and their vertebrate hosts: reconstruction of the phylogeny of a parasitic group. *Adv. Parasitol.* **24,** 239–306.

Elmes, B. & I. McAdam 1954. Helminthic abscess, surgical complication of oesophago-stomes and hookworms. *Ann. Trop. Med. Parasit.* **48,** 1–7.

Evans, A. C. & J. J. Joubert 1989. Intestinal helminths of hospital patients in Kovongo territory, Namibia. *Trans. R. Soc. Trop. Med. Hyg.* **83,** 681–3.

Farid, Z., S. Bassily, A Schubert, J. Nichols *et al.* 1966. Blood loss in Egyptian farmers infected with *Ancylostoma duodenale*. *Trans. R. Soc. Trop. Med. Hyg.* **60,** 486–7.

Farid, Z., S. Bassily, S. Young & A. Hassan 1973. Tetramisole in the treatment of *Ancylostoma duodenale* and *Ascaris lumbricoides* infections. *Trans. R. Soc. Trop. Med. Hyg.* **67,** 425–6.

Feacham, R., M. Guy, S. Harrison, K. Iwugo *et al.* 1983. Excreta disposal facilities and intestinal parasitism in urban Africa: preliminary studies in Botswana, Ghana and Zambia. *Trans. R. Soc. Trop. Med. Hyg.* **77,** 515–21.

Fleisher, K. M. 1985. Hookworm infection in Kweneng District, Botswana. A prevalence survey and a controlled treatment trial. *Trans. R. Soc. Trop. Med. Hyg.* **79,** 848–51.

Forsyth, D. 1970. Anaemia in Zanzibar (Tanzania). *Trans. R. Soc. Trop. Med. Hyg.* **64,** 601–6.

Foy, H. & A. Kondi 1960 Hookworms in aetiology of tropical iron deficient anaemia. *Trans. R. Soc. Trop. Med. Hyg.* **54,** 419–33.

Foy, H. & G. Nelson 1963. Helminths in the etiology of anaemia in the tropics, with special reference to hookworms and schistosomes. *Exp. Parasitol.* **14,** 240–62.

Galliard, H. 1967. Pathogenesis of *Strongyloides*. *Helminth Abs.* **36,** 247–60.

Gelfand, M. & B. Warburton 1967. The species of hookworm and degree of egg load in Rhodesia. *Trans. R. Soc. Trop. Med. Hyg.* **61,** 538–40.

Genta, R. M. 1984. Immunobiology of strongyloidiasis. *Trop. Geog. Med.* **36,** 223–9.

Genta, R. M. 1987. Strongyloidiasis. In *Intestinal helminthic infections* Vol. 2, No. 3 *Clinical tropical medicine and communicable diseases*, Z. S. Pawlowski (ed.). 645–65, London: Baillière Tindall.

Genta, R. M., E. A. Ottesen, F. A. Neva, P. D. Walzer *et al.* 1983. Cellular responses in human strongyloidiasis. *Amer. J. Trop. Med. Hyg.* **32,** 990–4.

Gibbs, H. C. 1986. Hypobiosis in parasitic nematodes – an update. *Adv. Parasitol.* **25,** 129–74.

Gigase, P., Baeta, S., Kumar, V. & Brandt, J. 1987. Frequency of symptomatic human oesophagostomiasis (helminthoma) in northern Togo. In *Helminth zoonoses with particular reference to the tropics*. S. Geerts, V. Kumar & J. Brandt, (eds), 228–35, Dordrecht: Martinus Nijhoff.

Gildenhuys, J. 1970. Mediese dienste in Ovamboland en Kaokoland. *S. Afr. Med. J.* **44**, 1008–10.

Gilles, H. M., E. J. W. Williams & P. A. J. Ball 1964. Hookworm infection and anaemia: an epidemiological, clinical and laboratory study. *Quart. J. Med.* **33**, 1–24.

Goetzl, E. & K. Austin 1972. Cellular characteristics of the eosinophil compatible with a dual role in host defence in parasitic infections. *Amer. J. Trop. Med. Hyg.* **26**, 142–8.

Goldsmid, J. M. 1965. Ancylostomiasis: a review of its effects on man and incidence in Southern Rhodesia. *C. Afr. J. Med.* **11**, 160–7.

Goldsmid, J. M. 1967. *Ternidens deminutus* Railliet and Henry (Nematoda): a diagnostic problem in Rhodesia. *C. Afr. J. Med.* **13**, 54–8.

Goldsmid, J. M. 1968a. The differentiation of *Ternidens deminutus* and hookworm ova in human infections. *Trans. R. Soc. Trop. Med. Hyg.* **62**, 109–16.

Goldsmid, J. M. 1968b. Studies on intestinal helminths in African patients at Harari Central Hospital, Rhodesia. *Trans. R. Soc. Trop. Med. Hyg.* **62**, 619–29.

Goldsmid, J. M. 1970. Studies on *Ternidens deminutus* Railliet and Henry, 1909, a nematode parasite of man and other primates in Rhodesia. PhD thesis. University of London.

Goldsmid, J. M. 1971a. *Ternidens deminutus*: a parasitological enigma in Rhodesia. Research Lec. Series No. 4. Fac. Med. University of Rhodesia.

Goldsmid, J. M. 1971b. Studies on the life cycle and biology of *Ternidens deminutus* (Railliet and Henry, 1909) (Nematoda: Strongylidae). *J. Helminthol.* **45**, 341–52.

Goldsmid, J. M. 1972a. *Ternidens deminutus* (Railliet and Henry, 1909) and hookworm in Rhodesia and a review of the treatment of human infections with *Ternidens deminutus*. *C. Afr. J. Med.* **18**, Suppl.

Goldsmid, J. M. 1972b. Further studies on the laboratory diagnosis of human infections with *Ternidens deminutus*. *S. Afr. J. Med. Lab. Tech.* **18**, 4–6.

Goldsmid, J. M. 1972c. Thiabendazole in the treatment of human infections with *Ternidens deminutus* (Nematoda). *S. Afr. Med. J.* **46**, 1046–7.

Goldsmid, J. M. 1974a. The intestinal helminthozoonoses of primates in Rhodesia. *Ann. Soc. Belge Med. Trop.* **54**, 87–101.

Goldsmid, J. M. 1974b. The use of mebendazole as a broad-spectrum anthelmintic in Rhodesia. *S. Afr. Med. J.* **48**, 2265–6.

Goldsmid, J. M. 1982. *Ternidens* infection. In *Handbook series in zoonoses*. H. Steele (ed.) Section C Parasitic Zoonoses, Vol. 2. 269–88. Boca Raton, Florida: CRC Press.

Goldsmid, J. M. 1986. *Inorganic elements in adult Ternidens deminutus* (Nematoda: Strongylidae: Oesophagostominae) from humans and baboons. *J. Helminthol.* **60**, 147–8.

Goldsmid, J. M. & W. Jablonski 1982. Demonstration of ZnS in *Ternidens deminutus* using EDAX analysis. *Int. J. Parasitol.* **12**, 145–9.

Goldsmid, J. M. & N. F. Lyons 1973. Studies on *Ternidens deminutus* Railliet and Henry, 1909 (Nematoda). I. External morphology. *J. Helminthol.* **47**, 119–26.

Goldsmid, J. M. & S. Rogers 1978. A parasitological study on the Chacma baboon, *Papio ursinus* from the Northern Transvaal. *J. S. Afr. Vet. Assoc.* **49**, 109–11.

Goldsmid, J. M., J. F. Morrison & C. R. Saunders 1974. The hookworm problem in the Rhodesian lowveld: its importance and treatment. *C. Afr. J. Med.* **20**, 97–100.

Gordon, J. A., C. M. D. Ross & H. Afflick 1969. Abdominal emergency due to an oesophagostome. *J. Trop. Med. Parasitol.* **63**, 161–4.

Grove, D. I. 1982a. Treatment of strongyloidiasis with thiabendazole: an analysis of toxicity and effectiveness. *Trans. R. Soc. Trop. Med. Hyg.* **76**, 114–8.

Grove, D. I. 1982b. *Strongyloides ratti* and *S. stercoralis*: the effects of thiabendazole, mebendazole, and cambendazole in infected mice. *Am. J. Trop. Med. Hyg.* **31**, 469–76.

Halloran, M. E., D. A. P. Bundy & E. Pollitt 1989. Infectious disease and the UNESCO Basic Education Initiative. *Parasitol. Today* **5**, 359–62.

Haseeb, M. A. & B. Fried 1988. Chemical communication in helminths. *Adv. Parasitol.* **27**, 169–207.

Heinz, H. J. 1961. Factors governing the survival of Bushmen worm parasites in the Kalahari. *S. Afr. J. Sci.* **57**, 207–13.

Herd, R. P. 1971. Studies on the immune response of sheep to *Chabertia ovina* (Fabricius, 1788). *Int. J. Parasitol.* **1**, 265–74.

Hira, P. R. & B. G. Patel 1977. *Strongyloides fuelleborni* infections in man in Zambia. *Am. J. Trop. Med. Hyg.* **26**, 640–3.

Hira, P. R. & B. G. Patel 1980. Human strongylodiasis due to the primate species *Strongyloides fuelleborni*. *Trop. Geogr. Med.* **32**, 23–9.

Hira, P. R. & B. G. Patel 1984. Hookworms and the species infecting man in Zambia. *J. Trop. Med. Hyg.* **87**, 7–10.

Hotez, P. J., N. Le Trang & A. Cerami 1987. Hookworm antigens: the potential for vaccination. *Parasitol. Today* **32**, 247–9.

Hutton, W. & K. Somers 1961. A comparison of bephenium hydroxy-naphthoate with tetrachloroethylene in hookworm infection. *Trans. R. Soc. Trop. Med. Hyg.* **55**, 431–2.

James, D. & H. M. Gilles 1985. *Human antiparasitic drugs: pharmacology and usage.* Chichester: Wiley.

Janssens, P. G. 1985. Chemotherapy of gastrointestinal nematodiasis in man. In *Chemotherapy of gastrointestinal helminths.* H. Van den Bossche, D. Thienpont & P. G. Janssens (eds), 183–368. Berlin: Springer-Verlag.

Kilala, C. P. 1971. *Ternidens deminutus* infecting man in southern Tanzania. *E. Afr. Med. J.* **48**, 636–45.

Klei, T. R. 1986. Development of vaccines against equine helminths. *Parasitol. Today* **2**, 80–1.

Kvalsvig, J. D. 1988. The effects of parasitic infection on cognitive performance. *Parasitol. Today* **4**, 206–8.

Kyronseppa, H. & J. M. Goldsmid 1978. Studies on the intestinal parasites in Africans in Owamboland, South West Africa. *Trans. R. Soc. Trop. Med. Hyg.* **72**, 16–21.

Lapage, G. 1935. The second ecdysis of infective nematode larvae. *Parasitol.* **27**, 186–206.

Lavine, N. 1968. *Nematode parasites of domestic animals and of man.* Minneapolis: Burgess.

Lee, T. D. G., M. Swieter & A. D. Befus 1986. Mast cell responses to helminth infection. *Parasitol. Today* **2**, 186–91.

Leid, R. W. 1988. Parasites and complement. *Adv. Parasit.* **27**, 131–68.

Lockwood, D. & J. Weber 1989. Parasite infections in AIDS. *Parasitol. Today* **5**, 310–16.

Lothe, D. 1958. An immature *Oesophagostomum* sp. from an umbilical swelling in an African child. *Trans. R. Soc. Trop. Med. Hyg.* **52**, 12.

Macdonald, F. & J. M. Goldsmid 1973. Intestinal helminth infections in the Burma Valley area of Rhodesia. *C. Afr. J. Med.* **19**, 113–5.

Mackinnon, B. 1987. Sex attractants in nematodes. *Parasitol. Today* **3**, 156–8.

Marcus, L. C. 1982. Oesophagostomiasis. In *Handbook series in zoonoses.* H. Steele (ed.) Vol. 2. 221–23. Boca Raton, Florida: CRC Press.

McCabe, R. J. & J. M. Goldsmid 1976. Protozoan and helminthic infections of the intestines of humans in the Inyanga area of Rhodesia. *S. Afr. Med. J.* **50**, 779–80.

McCullough, F. & B. Friis-Hansen 1961. A parasitological survey in three selected communities in Luapula Province, Northern Rhodesia. *Bull. Wld Hlth Org.* **24**, 213–9.

Meakins, R., P. Harland & F. Carswell 1981. A preliminary survey of malnutrition and helminthiasis among school children in one mountain and one lowland ujamaa village in

Northern Tanzania. *Trans. R. Soc. Trop. Med. Hyg.* **75,** 731–9.

Mercer, J. G. 1985. Developmental hormones in parasitic helminths. *Parasitol. Today* **1,** 96–100.

Migasena, S. & H. M. Gilles 1987. Hookworm infection. In *Intestinal helmintic infections,* Vol. 2, No. 3, *Clinical tropical medicine and communicable diseases,* Z. S. Pawlowski (ed.). 617–27, London: Baillière Tindall.

Miller, T. A. 1965. Pathogenesis and immunity in hookworm infection. *Trans. R. Soc. Trop. Med. Hyg.* **62,** 473–85.

Miller, T. A. 1979. Hookworm infection in man. *Adv. Parasitol.* **17,** 315–84.

Mitchell, G. F. 1987. Injection versus infection: the cellular immunology of parasitism. *Parasitol. Today* **3,** 106–11.

Moqbel, R. 1986. Helminth-induced intestinal inflammation. *Trans. R. Soc. Trop. Med. Hyg.* **80,** 719–27.

Mravak, S., W. Schopp & V. Bienzle 1983. Treatment of strongyloidiasis with mebendazole. *Acta Trop.* **40,** 93–4.

Muller, R. 1975. *Worms and Disease.* London: Heinemann.

Muller, M. 1987. Wanted: entrepreneurial parasitologists. *Parasitol. Today* **3,** 62–3.

Mwosu, A. B. C. 1983. The human environment and helminth infections: a biomedical study of four Nigerian villages. In *Human ecology and infectious diseases.* N. A. Croll & J. H. Cross (eds), 225–52. New York: Academic Press.

Nelson, G. S. 1965. The parasitic helminths of baboons with particular reference to species transmissible to man. In *The baboon in medical research.* Proceedings of the First International Symposium on the Baboon and its Use as an Experimental Animal. H. Vagtborg (ed.). 441–70. Austin: University of Texas Press.

Nelson, G. S. 1972. Human behaviour in the transmission of parasitic diseases. In *Behavioural aspects of parasite transmission.* E. U. Canning & C. A. Wright (eds), 109–22. London: Zoological Journal of the Linnean Society.

Nelson, G. S. 1983. Teaching medical parasitology. In *Parasitology: a global perspective.* K. S. Warren & J. Z. Bowers (eds), 158–68. New York: Springer-Verlag.

Nesheim, M. C., 1987. Intestinal helminth infection and nutrition. In *Intestinal helminthic infections,* Vol. 2, No. 3, *Clinical tropical medicine and communicable diseases,* Z. S. Pawlowski (ed.). 553–71, London: Baillière Tindall.

Nnochiri, E. 1968. *Parasitic diseases and urbanization in a developing community.* London. Oxford University Press.

Ogilvie, B., A. Bartlett, R. Godfrey, J. Turton *et al.* 1978. Antibody responses in self-infections with *Necator americanus. Trans. R. Soc. Trop. Med. Hyg.* **72,** 66–71.

Oliver, N., D. Rowbottom, P. Sexton, J. M. Goldsmid *et al.* 1989. Chronic strongyloidiasis in Tasmanian veterans – clinical diagnosis by the use of a screening index. *Aust. N.Z. J. Med.* **19,** 458–62.

Pampiglione, S. & M. L. Ricciardi 1971. The presence of *Strongyloides fuelleborni* Van Linstowe, 1905, in man in Central and East Africa. *Parassitologia* **13,** 259–69.

Pampiglione, S. & M. L. Ricciardi 1972a. Experimental infestation with human strain *Strongyloides fuelleborni* in man. *Lancet* **1,** 663–5.

Pampiglione, S. & M. L. Ricciardi 1972b. Geographic distribution of *Strongyloides fuelleborni* in humans in tropical Africa. *Parassitologia* **13,** 329–38.

Pampiglione, S., E. Najera, M. L. Ricciardi & L. Junginger 1979. Parasitological survey of pygmies in Central Africa III. Bambuti Group (Zaire). *Riv. Parassitol.* **49,** 187–234.

Pawlowski, Z. S. 1986. Soil-transmitted helminthiases. In *Clin. Trop. Med. & Comm. Dis.* **1,** 3, 617–42.

Pelletier, L. L. 1984. Strongyloidiasis. *Inf. Dis. Newsletter.* **3,** 11–14.

Petithory, J. & F. Derouin 1987. AIDS and strongyloidiasis in Africa. *Lancet* **1,** 921.

Piot, P. & J. M. Mann 1988 (eds). In: *AIDs and HIV infection in the tropics,* Vol. 3, No. 1,

Clinical tropical medicine and communicable diseases, London: Baillière Tindall.

Pritchard, D. I. 1986. Antigens of gastrointestinal nematodes. *Trans. R. Soc. Trop. Med. Hyg.* **80**, 728–34.

Rees, P. H., E. N. Mngola, P. O'Leary & H. O. Pamba 1974. Intestinal parasites. In *Health and disease in Kenya*. Vogel, L. C., A. S. Muller, R. S. Odingo, Z. Onyango *et al.* (eds), 339–46. Nairobi: E. Afr. Lit. Bureau.

Rogers, S. & J. M. Goldsmid 1977. Preliminary studies using the indirect fluorescent test for the serological diagnosis of *Ternidens deminutus* infections in man. *Ann. Trop. Med. Parasitol.* **71**, 503–4.

Rogers, S. & J. M. Goldsmid 1980. Serological reactions to infection with *Ternidens deminutus* in man. *Afr. J. Clin. Exp. Immunol.* **1**, 347–59.

Rowbottom, D., J. M. Goldsmid, K. Thomas & N. Oliver. In press. Chronic strongyloidiasis in Tasmanian veterans. *Ann. Soc. Belge Med. Trop.*

Samster, M. 1980. Eosinophils nominated but not elected. *N. Engl J. Med.* **303**, 1175–6.

Sandground, J. H. 1929. *Ternidens deminutus* (Railliet and Henry) as a parasite of man in Southern Rhodesia; together with observations and experimental infection studies on an unidentified parasite of man from this region. *Ann. Trop. Med. Parasitol.* **23**, 23–32.

Sandground, J. H. 1931. Studies on the life-history of *Ternidens deminutus*, a nematode parasite of man, with observations on its incidence in certain regions of Southern Africa. *Ann. Trop. Med. Parasitol.* **25**, 147–84.

Schad, G. A., T. A. Nawalinski & V. Kocher 1983. Human ecology and the distribution and abundance of hookworm populations. In *Human ecology and infectious diseases*. N. A. Croll & J. H. Cross (eds), 167–87. New York: Academic Press.

Schad, G. A., K. Murrell, R. Fayer, H. el Naggar *et al.* 1984. Paratenesis in *Ancylostoma duodenale* suggests possible meat-borne human infection. *Trans. R. Soc. Trop. Med. Hyg.* **78**, 203–4.

Schofield, C. 1985. Parasitology today: an ambitious project. *Parasitol. Today.* **1**, 2.

Shield, J. 1985. A study on the effectiveness of mebendazole and pyrantel pamoate as a combination anthelmintic in Papua New Guinean children. *P.N.G. Med. J.* **28**, 41–4.

Skaife, S. H. 1953. *African insect life*. Cape Town: Longmans Green.

Stewart, T. & L. C. Gasbarre 1989. The veterinary importance of nodular worms. (*Oesophagostomum* spp.) *Parasitol. Today* **5**, 209–13.

Stoll, N. 1949. This wormy world. *J. Parasitol.* **33**, 1–18.

Stoll, N. 1962. Helminthic infections. In *Drugs, parasites and hosts*. Biol. Council Symp. Goodman, L. G. & R. H. Nimmo-Smith (eds). London: Churchill-Livingstone.

Sturrock, R. F. 1966. Hookworm studies in Tanganyika (Tanzania): investigations at Hombolo in the Dodoma Region. *E. Afr. Med. J.* **43**, 315–22.

Tongson, M. S. & S. L. Eduardo 1982. Trichostrongyloidiasis. In *Handbook series in zoonoses*. H. Steele (ed.). Section C: Parasitic Zoonoses, Vol. 2. 331–7, Boca Raton, Florida: CRC Press.

Topley, E. 1968. Common anaemia in rural Gambia 1. Hookworm anaemia among men. *Trans. R. Soc. Trop. Med. Hyg.* **62**, 579–94.

Ugochukwu, E. & K. Ejimadu 1985. Comparative study on the infestations of three different breeds of dogs by gastrointestinal helminths. *Int. J. Zoonoses* **12**, 318–22.

Ukoli, F. M. 1984. *Introduction to parasitology in tropical Africa*. Chichester: Wiley.

Van den Berghe, L. 1934. L'existance de *Ternidens deminutus* au Katanga. *Ann. Soc. Belge Med. Trop.* **14**, 189.

Van den Bossche, H. 1985. Pharmacology of anthelmintics. In *Chemotherapy of gastrointestinal helminths*. H. Van den Bossche, D. Thienpont & P. G. Janssens (eds), 125–82. Berlin: Springer-Verlag.

Van Oye, E. 1961. The problem of ancylostomiasis in the Congo and Ruanda-Burundi. *Trans. R. Soc. Trop. Med. Hyg.* **55,** 17–19.

Vince, J. D., R. Ashford, M. Gratten & J. Bana-Koiri 1979. *Strongyloides* species infestation in young infants of Papua New Guinea: association with generalized oedema. *P.N.G. Med. J.* **22,** 120–7.

Von Brand, T. 1979. *Biochemistry and physiology of endoparasites*. Amsterdam: Elsevier.

Wakelin, D. 1986. Genetic and other constraints on resistance to infection with gastrointestinal nematodes. *Trans. R. Soc. Trop. Med. Hyg.* **80,** 742–7.

Weinkel, C. S., B. N. Gaynes & J. K. Roche 1988. Diarrhoeal diseases in the immunocompromised patient. In: *Clinical tropical medicine and communicable diseases*. Vol. 3. No. 3. 401–15, London: Baillière Tindall.

WHO 1980. Parasite-related diarrhoeas. *Bull. Wld Hlth Org.* **58,** 819–30.

WHO 1984. Intestinal parasitic infections and how to prevent them. *Wld Health.* 16–17. March 1984.

Woodruff, A. W. 1965. Pathogenicity of intestinal helminthic infections. *Trans. R. Soc. Trop. Med. Hyg.* **59,** 585–606.

Yamaguti, S. 1954. Parasitic worms mainly from the Celebes. X. Nematodes of birds and mammals. *Acta. Med. Okayama.* **9,** 135–59.

6 Onchocerciasis – river blindness

Introduction

Personal perspective

In his review on onchocerciasis published in 1970 in *Advances in Parasitology* George Nelson wrote: 'Throughout the world . . . no more than a dozen scientists are devoted to the study and control of this disease. The present review is not designed for this small group of experts; its aim is to bring onchocerciasis to the attention of biologists, clinicians and public health workers in order to stimulate interest in a fascinating but formidable problem'. Four years later, I was one of the biologists whom he introduced to this 'formidable problem', and in the 20 years since he wrote these lines 'river blindness' has risen from obscurity to join the ranks of the most widely recognized of tropical diseases.

What was it that brought about this change and what has been the impact on the disease? Two major initiatives stand out, and their most important effect has been to decimate the infection in one of its greatest strongholds in the African continent. The first of these initiatives was the inclusion of onchocerciasis, along with filariasis and five other tropical infections, in the World Health Organization (WHO) Special Programme for Research and Training in Tropical Diseases. This was instigated in 1974, the year I arrived in George Nelson's department at the London School of Hygiene and Tropical Medicine, and its significance for me and other new recruits was one of the first things he brought to my attention. At that time George was chairing a WHO Expert Committee on Filariasis (WHO 1974) and it was no doubt high in his mind how much there was to do in research on this sadly neglected group of infections. The second initiative taken that same year was the setting up of the Onchocerciasis Control Programme (OCP) in West Africa under the joint auspices of WHO, UNDP (the United Nations Development Programme), FAO (the Food and Agriculture Organization) and the World Bank. In one of the greatest undertakings in community health to be seen anywhere in Africa, OCP set out to reduce the transmission of onchocerciasis in the Volta river basin (an area of over 764 000 square kilometres) below a threshold of intensity at which it exists as a blinding disease. Through sheer determination and good management this programme has enjoyed enormous success, but its future and with it the health of many West

138

Africans still depends on tools that research has yet to develop. The requirement for progress in the development of drugs led to the formation of the Onchocerciasis Chemotherapy Project (OCT) in 1982, and others have been attracted into the field, such as the Edna McConnell Clark Foundation, which in 1983 instigated an ambitious programme to develop a vaccine for onchocerciasis.

In this chapter I have attempted to give no more than a brief overview of onchocerciasis, choosing instead to focus on aspects of this broad subject most closely related to my own research. For me the fascination has been in the biology of the parasites causing infection, the adaptations they have evolved for parasitism, and how one might manipulate the host–parasite relationship. These interests are reflected in the emphasis given in the following pages to parasite development, to the relevance of animal parasites in human onchocerciasis, and to experimental studies with laboratory models. In large measure, these are an inheritance of my association with George Nelson.

Introduction to the infection

Onchocerciasis in man is caused by the filarial nematode, *Onchocerca volvulus* (Leuckart 1893), which is a member of the family Onchocercidae within the superfamily Filarioidea to which all the medically important filarial species belong. The parasite occurs throughout tropical Africa in a patchy rather than uniform geographic distribution (Fig. 6.1), and on the most recent estimate from WHO infects in excess of 17 million people (WHO 1987). Outside Africa there are foci of transmission in the Yemen and in Central and South America (Mexico, Guatemala, Venezuela, Colombia, Ecuador, Brazil), but the total number of cases is estimated to be no more than 100000 (WHO 1987). Like most helminthiases, onchocerciasis rarely presents itself as an acute clinical condition, parasites persist for many years, reinfection is the norm, and infection is not synonymous with disease. Nevertheless, onchocerciasis is one of the most emotive parasitic infections of Africa, because of the high rate of morbidity leading as it may to ocular pathology and blindness. As a public health problem its significance varies among the African states, but throughout its range it is an affliction of the poor and often remote rural communities. An inevitable consequence of this is that the populations at risk lack a powerful influence on health politics, a factor of much importance in the decision to create the OCP and in the design of its mode of operation.

The popular name 'river blindness' stems from the association of onchocerciasis with riverine habitats, which arises because of the mode of parasite transmission that depends on haematophagous insects of the genus *Simulium* (blackflies, buffalo flies). In different parts of Africa various species or sub-species of *Simulium* are involved, but all have

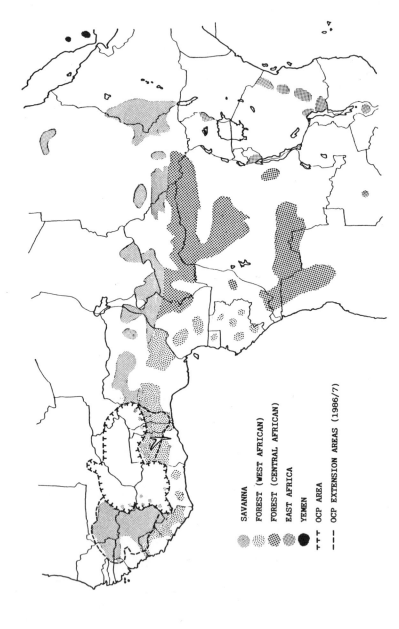

Figure 6.1 Distribution of onchocerciasis in Africa, highlighting some of the geographic variations in the infection. Boundaries of the Onchocerciasis Control Programme (OCP) in West Africa are shown as they were originally established (1974) and as they are today. (Reprinted with kind permission of the authors from Muller and Baker, 1988.)

SAVANNA

FOREST (WEST AFRICAN)

FOREST (CENTRAL AFRICAN)

EAST AFRICA

YEMEN

OCP AREA

OCP EXTENSION AREAS (1986/7)

aquatic larvae and pupae that have a requirement for fast flowing, well aerated water (Fig. 6.2). The flight range of the flies therefore specifies the boundaries of transmission, and transmission intensity is usually a function of distance from breeding sites in the rivers. However, many factors modify the practical importance of particular habitats as places of exposure to infection, and transmission may be seasonal or perennial depending on the conditions prevailing in the local watercourses over the year.

It is customary at this point in articles dealing with parasitological subjects to provide the reader with a short and easily digestible account of the life-cycle. Here I intend to make a break with tradition, by going straight into a discussion of parasite biology from the perspective of a natural historian.

Life-cycle

Blacklock's identification of *Simulium damnosum s.1.* as the vector of *O. volvulus* in Sierra Leone opened the way to our present understanding of the natural history of onchocerciasis (Blacklock 1926). His discovery came 33 years after the original description of *O. volvulus* by Manson (which was subsequently credited to Leuckart who had sent Manson the

Figure 6.2 (a) Patients attending a clinic in a village in Sierra Leone. The man in the foreground is blind, while his companion shows signs of dermal atrophy and depigmentation. (b) Typical breeding site of *S. damnosum s.1.* (mainly *S. soubrense*) in West African forest (near Bo, Sierra Leone). (c) Pupae of *S. soubrense* on vegetation.

material) and followed much speculation among his contemporaries over the mode of transmission. For example, on epidemiological grounds Robles (1919) had suggested simuliids might be vectors in Guatemala, while from East Africa it was reported by Dry (1921) that local people had linked the disease with a biting fly that turned out to be *Simulium neavei*. Today, there are still many aspects of development that are unknown or poorly understood, although in the broad sense most of the chief characteristics of the life cycle have been determined.

Parasite biology in the intermediate host

Throughout Africa there are numerous blackfly species or sub-species that have been incriminated in the transmission of onchocerciasis. This is a crucial aspect of the ecology of *O. volvulus* and of the epidemiology of infection. For the sake of clarity, this will not be entered into in the following description of parasite biology, but is a subject of much importance that will be returned to in the section dealing with the epidemiology of onchocerciasis.

Uptake of microfilariae by blackflies

Because the microfilariae of *O. volvulus* concentrate in the skin, as shown by O'Neill as early as 1875, it follows that their ingestion will be favoured by a haematophagous insect with scarifying mouthparts. This guided Blacklock in his investigations among the Konno villages of Sierra Leone, where in the first instance he had explored the possibility that the Congo floor maggot, *Auchmeromyia luteola*, might be a vector (Blacklock 1926, 1927). However, he quickly recognized the significance of the short, heavily armoured proboscis of female simuliids, designed as it is for creating tissue damage and feeding on the pool of extravasated blood. Blade-like mandibles and maxilla slash through the epidermis and permit the penetration of the hypopharynx just deep enough to pick up microfilariae from the superficial layers of the dermis.

The numbers of microfilariae ingested vary greatly among individual insects, but for a given blackfly species and parasite isolate they are essentially a function of the density in skin. Other factors may influence the weight of infection acquired, such as the length of time spent in engorgement, as a variable independent of the size of the blood meal (Eichler 1971). The phenomenon may be linked to reports that have described the apparent concentration of microfilariae from tissues surrounding the bite site (Strong *et al.* 1934, De Leon & Duke 1966, Duke *et al.* 1967a, Duke 1970, Garms 1973, Omar & Garms 1975), and indeed the feeding of blackflies on patients has in the past been used as a form of xenodiagnosis (Nettel 1945, Wilkinson 1949). No specific mechanism of attraction has been demonstrated or chemotactic substance isolated from blackfly saliva, but the concentration of microfilariae appears to be an authentic phenomenon which exhibits specificity for certain vector–parasite combinations (De Leon & Duke 1966). This suggests that *O. volvulus* exists in more than one form and has developed adaptations to the local vectors, a subject that will be returned to under the heading of the epidemiology of onchocerciasis.

In lymphatic filariasis and loaiasis circadian rhythms of microfilariae in the blood are a well-known adaptation of the parasites to the feeding habits of the mosquito or tabanid vectors (Hawking 1967). In onchocerciasis such periodicity has also been observed (Lartigue 1967, Duke *et al.* 1967b, Duke & Moore 1974, Anderson *et al.* 1975), but the fluctuations in numbers of microfilariae are small and not all investigators are agreed (Picq & Jardel 1974, Tada & Figueroa 1974). Diurnal periodicity has been demonstrated in man (Lartigue 1967, Duke *et al.* 1967b, Anderson *et al.* 1975) and in experimentally infected chimpanzees (Duke & Moore 1974), with peak levels of microfilariae corresponding with the time of day associated with highest biting by the vectors. A seasonal periodicity of *O. volvulus* has also been found (Fuglsang *et al.* 1976, Hashiguchi *et al.* 1981), as it has been for *O. lienalis* in cattle (Ivanov 1964, Eichler 1973) and *Onchocerca cervicalis* in horses (Foil *et al.* 1987). It has been contended that periodicities of *Onchocerca* microfilariae in fact reflect a changing vertical distribution in the skin (Mellor 1973), a view supported by the seasonal shift of the microfilariae into deep layers of the dermis during the winter (Ivanov 1964, Eichler 1973, Foil *et al.* 1987).

Of more biological significance is the spatial distribution of skin microfilariae over the body that results in the highest parasite densities corresponding with those sites most frequently bitten by the vectors. This elegant adaptation is seen at its best among the *Onchocerca* species parasitizing wild and domestic ungulates (Wenk 1976), some of which have very specific localizations of microfilariae that occur distant from the site of predilection of the adults. For example, in horses the adult worms of *O. cervicalis* are restricted to the nuchal ligament that runs through the neck, yet the microfilariae concentrate along the ventral mid-line which is the site that attracts greatest attention from the *Culicoides* vectors (Mellor 1973, 1974). The adaptation may be just as strongly developed in human onchocerciasis, but it is less obvious because the distribution of microfilariae correlates with that of the adults (Kershaw *et al.* 1954). In different geographic regions the spatial pattern varies in a way that is linked to the biting behaviour of the local vectors, such that parasites predominate below the hip in Africa (Kershaw *et al.* 1954, Nelson 1958a) but above the hip in Mexico (Mazzotti 1951). It might be argued that the localization of adult worms, and hence the microfilariae, is specified by the site of deposition of infective larvae and is therefore linked only coincidentally with the biting habits of the vector. This objection is difficult to address based on data from human infections, although from animal experiments there is evidence that *O. volvulus* may have some intrinsic mechanism of localization. In chimpanzees, it was found by Duke (1980) that the adult worms always occurred in the vicinity of the hip, irrespective of the site of infection by syringe inoculation with third stage larvae.

Parasite development in blackflies

Once microfilariae have been ingested by a blackfly they are carried into the midgut and must penetrate through the midgut epithelium within a matter of minutes in order to avoid becoming trapped in the coagulating blood meal. Two principal features of the vector govern the efficiency of infection at this stage, the first of which is the presence and physical characteristics of the cibarial and pharyngeal armatures. These are comb-like structures lining the pharynx that guard the passage into midgut, which are highly elaborate or relatively inconspicuous according to the species of *Simulium*. Physical damage to microfilariae caused by shearing as they pass through the pharynx may affect the viability of a high proportion of the parasites ingested (Omar and Garms 1975). Once in the midgut, a second characteristic of the intermediate host comes into play which is the rate of formation and thickness of the peritrophic membrane. This is secreted by the epithelial cells to protect them from damage shortly after the intake of blood, and in *S. damnosum s.1.* it forms an effective barrier to the escape of many microfilariae (Duke & Lewis 1964). Naturally, these mechanical barriers to infection can greatly reduce the numbers of microfilariae establishing themselves in the vector, a feature of the host–parasite relationship of mutual advantage where this limits blackfly mortality by guarding against excessive worm burdens. As many as 72 infective larvae of *O. volvulus* have been reported from a single female of *S. neavei* by Nelson and Pester (1962), and following experimental infection with *O. lienalis* up to 107 third stages have been recovered from *Simulium ornatum s.1.* (Bianco *et al.* 1989b). However, in natural infections the norm is for far lower numbers of larvae, and below those which immediately threaten fly survival are levels that may lead to reduced flight range, fecundity or both (Cooter 1982, Cheke *et al.* 1982, Ham & Banya 1984).

Microfilariae that have left the midgut travel via the haemocoel to the thorax where they penetrate the indirect flight muscles and orientate along the axis of an individual muscle cell. Within this intracellular location they begin to develop, undergoing morphological changes and two moults in a manner directly comparable with other filariae (Bain 1969). Normal development of larvae will occur in blackflies infected without them feeding on blood (for example, following intrathoracic injection of microfilariae), but a greater proportion of the parasites mature when the insect is provided with blood (Ham & Gale 1984). During differentiation of microfilariae to infective, third-stage larvae there is an approximate doubling of length (from ~300 to ~600 µm) and trebling of width (from ~6 µm to ~19 µm), accompanied by the formation of a rudimentary gut, secretory tissue and genital primordia. The rate of larval development is temperature dependent, commonly being completed within 6 to 9 days at the range of temperatures experienced in the onchocerciasis foci of Africa. In *S. damnosum s.1.* maintained at a constant temperature of 27°C, moults occur on days 4–5 and on day 6 and infective larvae appear in the head after 6–7 days. Below a threshold temperature development will not occur (Nelson & Pester 1962, Wegesa 1966) and this is reported to vary according to the vector between values of 18°C (for *S. woodi*) and 21°C (for *S. damnosum s.1.*).

In simuliids there is little evidence of overt encapsulation or melanization reactions such as occur in ticks and mosquitoes in response to filarial larvae (Nelson 1964). Nevertheless, blackflies do exhibit both humoral and cellular defence mechanisms that can be highly effective in blocking the development of *Onchocerca* larvae. One manifestation of this defence not requiring prior exposure to infection is the apparent toxicity of acellular components of blackfly haemolymph to microfilariae (Ham & Garms 1988). It appears that lectin-like defence proteins, distinct from the well documented antibiotic proteins of the Lepidoptera, may be important molecules specifying a level of resistance in refractory or partially susceptible species of blackflies. A second way in which refractoriness to the parasites may be expressed is through a mechanism of heightened resistance induced by prior infection with larvae. This has been elegantly demonstrated recently using the vector–parasite combination of *S. ornatum s.1.* and *O. lienalis*, in which the influence of a secondary infection with microfilariae was to reduce the survival of larvae of both the primary and secondary infections (P.J. Ham, pers. comm. 1989). Passive transfers of haemolymph confer the protective effect on recipient flies (Ham 1986), but the precise interplay of humoral and cellular components has not been elucidated as yet. Whether this is a significant factor in natural infections with an impact on the vectorial capacity of blackflies is more difficult to assess. However, in the blood-fed fly the maturation of eggs occurs considerably faster than parasite development, so that two and possibly three ovarian cycles may be completed before infective larvae are ready for transmission. This means that a nulliparous fly infected on the occasion of its first blood meal will only be capable of transmitting infective larvae the third or fourth time that it feeds. It is therefore not uncommon for multiple infections to occur, which might lead to a reduced tolerance of developing larvae as demonstrated in the experimental studies.

Almost nothing is known about early development in *Onchocerca* at a molecular level, primarily because of the dearth of parasite material and the failure to obtain differentiation beyond the first stage larva *in vitro*

(Pudney & Varma 1980). Of particular interest are genes expressed at critical points of the life-cycle, such as at the moments immediately preceding and following transition of larvae between the intermediate and definitive hosts. For example, is programmed expression a key factor permitting third stage larvae to make the necessary physiological adjustments to life in the vertebrate host, or do extrinsic factors trigger many of the changes through activation of a cascade of co-ordinated transcription?

In an attempt to address such questions about the *Onchocerca* life-history, a method has recently been described whereby gene expression may be studied in developing larvae (Bianco & Maizels 1989, Bianco *et al.* 1990). To do this, a procedure was devised for labelling proteins synthesized by parasites *within* the vector, involving the micro-injection of radio-isotopic amino acids into the thorax of infected blackflies. Pulse labellings of *O. lienalis* larvae in *S. ornatum s.1.* with [^{35}S]-methionine have revealed the complexity of protein synthesis occurring in development, much of which must be central to growth and structural changes. Also there will be products with a role in general metabolism, but of particular interest are those molecules that specify functions which have evolved as adaptive specializations for parasitism. One intriguing product to be identified during these labelling studies is a major acidic protein of 23 kD which is developmentally expressed almost exclusively by infective, third stage larvae (Bianco *et al.* 1990). Homologous proteins occur in *O. lienalis* and *O. volvulus*, but these exhibit size polymorphisms both among species and individual organisms. The 23 kD molecule continues to be elaborated after terminal differentiation of the parasite in flies, but not by post-infective larvae entering the phase of development in the vertebrate host. A shift in temperature from 26°C to 37°C triggers secretion of the 23 kD molecule as a discrete event 24–72 hours after leaving the vector. The function of this molecule has yet to be determined, but in view of the timing of its expression and secretion it is reasonable to interpret that it may have a role in transmission.

Transmission of infective larvae

There has been little investigation of the mechanisms regulating the actions of infective larvae during transmission, although it was established during the earliest studies with infected blackflies that third stage larvae congregate in the mouthparts (Blacklock 1927). In heavy infections larvae lie in parallel packed in the labrum/epipharynx between the mandibles (Nelson & Pester 1962), but they may also spill over into the palps, other regions of the head, the thorax and abdomen and even halteres and legs (personal observation). However, differences exist in the propensity of larvae to accumulate in the head in various blackflies, so whereas this is a well developed feature of *O. volvulus* in *S. damnosum s.1.* it is not the case in *Simulium ochraceum* (an important Central American vector).

Similarly for *O. lienalis* around 80% of infective larvae accumulate in the head in *S. ornatum s.1.*, yet under identical conditions of infection this drops to 5% in *S. vernum* (Ham & Bianco 1983a). It is unknown what controls the tropism for mouthparts in the migration of third stage larvae, but if this varies for a given parasite population in different insects it implies there may be a host-dependent, biochemical basis.

The trigger for egress of larvae and route of escape that is taken were commented upon by Blacklock (1926), but since that time relatively little investigation of these phenomena has been made. According to Blacklock larvae escape from the mouthparts by puncturing the labellum, but this observation was not confirmed by Nelson and Pester (1962). The extrinsic cues used by larvae to trigger their migration from the fly have been more the subject of conjecture rather than of experimental inquiry. Blacklock (1927) reported that the egress of larvae would not occur in warmed saline, but took place spontaneously in monkey serum or when 'light' pressure was applied to the head. Collins and Jones (1978) noted that the larvae of *O. cervicalis* escaped from midges while feeding through a membrane, but did not establish which of the many factors involved acted as a stimulus for their migration (warmth, biochemical signals, pressure changes and so forth). Based on naturally infected forest cytospecies of *S. damnosum s.1.* from Cameroon, Duke (1973) calculated the efficiency of egress of *O. volvulus* infective larvae during a single blood meal. Nearly 80% of third stage larvae escaped from the flies, but only 40% of the flies lost their entire parasite load. A startling finding to come from studies on *Brugia pahangi* in *Aedes togoi* was that 91% of larvae were transmitted in one feed, of which 57% escaped within the first 5 seconds (Ho & Lavoipierre 1975). Nevertheless, egress from the vector is not on its own sufficient to ensure successful transmission, because larvae may still fail to enter host tissues rapidly enough to avoid desiccation on the skin surface. In lymphatic filariasis this accounts for a high proportion of larvae (Ewert & Ho 1967), but equivalent experiments have not been reported for *O. volvulus* or other species of blackfly-transmitted *Onchocerca* species.

Parasite biology in the definitive host

Parasite development

Until recently nothing was known about the early development of *O. volvulus*. The lack of a convenient animal host and colonies of the vectors (discussed under 'Experimental investigations with laboratory models') ensured progress was slow. However, advances in techniques of *in vitro* cultivation and the use of micropore chambers to implant larvae into animals has provided useful information on processes occurring up to and just beyond the third moult (Lok *et al.* 1984, Strote 1985, 1987,

Pudney *et al.* 1988, Bianco *et al.* 1989a). Neither approach has as yet provided the conditions required for normal development of fourth stage larvae, and it may be significant that the point at which morphogenesis is halted is similar in both the *in vitro* and *in vivo* systems.

Lok *et al.* (1984) reported that moulting of *O. volvulus* occurred after 5–10 days under *in vitro* conditions. However, in micropore chambers implanted into mice, moulting was observed after only 3–6 days (Bianco *et al.* 1989a). Indeed, it is possible that a proportion of larvae enter the ecdysis as early as day 2, as day 3 was the first time point to be examined in the chamber experiments. The rapid moulting of third stage larvae following entry into the vertebrate phase of the life-cycle fits with the limited growth that is apparent in early fourth stage larvae. Parasites recovered from chambers over the period of moulting measured 590–608 μm long by 18–20 μm wide compared with vector-derived larvae of 568 μm long by 17.5 μm wide. Other morphological changes included early development of the spicular primordia, and the first evidence of differentiation of a genital tube.

For further information on the developmental process in *Onchocerca*, it has been necessary to turn to non-human parasites and study development through sequential necropsy examinations of experimentally infected animals. To do this, a project was established on the development of *O. lienalis* in cattle which provided what is currently the most complete picture of the *Onchocerca* life-cycle. A summary of the preliminary findings was provided by Bianco and Muller (1982), and a full account of these investigations is now in preparation (Bianco *et al.* in prep.). In *O. lienalis*, the third moult occurs 2–5 days post infection, at a time when most of the larvae are still located near to the site of inoculation. In these experiments larvae were administered by syringe inoculation, but we infer that the same will be true following natural transmission since the larvae enter a lethargus prior to ecdysis. Recently moulted worms measured a mean of 520 μm long by 20 μm wide, compared with 467 μm long by 16 μm wide in the case of vector-derived third stage larvae. In this and most other respects they resemble the early fourth stage larvae of *O. volvulus*, but they differ in possessing three minute, caudal appendages (one terminal, two sub-terminal) that appear after the third moult. At the ultrastructural level there are major differences in the cuticle before and after the ecdysis, being thick with shallow longitudinal folding in third stage larvae, but much thinner and more deeply corrugated in fourth stage worms. During the moult there is an extensive network of membranous channels communicating between the hypodermis and parasite surface, indicating that even after the initiation of moulting a high level of biosynthetic activity is associated with formation of the new cuticle.

At the molecular level there is a major transformation in the *de novo* synthesis of proteins following the entry of larvae into the vertebrate phase of the life-cycle. This is illustrated by the profile of polypeptides labelled with [^{35}S]-methionine in third stage larvae while still within the blackfly compared with that seen after larvae are placed in culture under conditions that promote further development (Bianco *et al.* 1990). As discussed earlier, certain products synthesized by infective larvae before transmission are secreted at this time into the external environment. One such product of *O. lienalis* has recently been identified as a potent, proteolytic enzyme, with properties of a serine elastase capable of degrading elastin and glycoproteins within a model of cutaneous extracellular matrix (Lackey *et al.* 1989). This accords well with the requirement for larvae to penetrate host tissues following transmission, and parallels findings of proteases in other tissue invasive nematodes, together with helminths such as the cercariae of schistosomes (McKerrow 1989).

Fourth stage larvae of *O. lienalis* migrate widely through the body in the period of intermoult growth, moving through connective tissue planes under the skin and between muscles, ligaments and tendons. There is no apparent selectivity for a particular anatomical location, since worms were recovered from such disparate sites as the head, neck, thorax, abdomen and legs. The fourth moult occurs between 1.5 and 2.5 months post infection, when larvae have attained a length of approximately 5-10 mm. At this time the immature males have developed prominent spicules, and the females are similar in size, in their level of activity, and in having a finely annulated cuticle lacking rugae.

Growth of male worms of *O. lienalis* stabilized at 12–14 mm long from approximately four months post infection when they appeared to be sexually mature. At this time, females measured no more than 17 mm long and had partially differentiated genitalia, but were clearly far from being reproductively active. Since the mature adult females are known to reach in excess of 450 mm long, it is clear that for them there must be a protracted period of growth. However, a curious phenomenon was the recovery of immature females of similar morphology to those at four months from animals infected for as long as two years which was well after they had developed patent infections. Why these females failed to mature, or whether they were capable of completing development, was not ascertained but they do raise the question as to whether they have a specific biological purpose. One exciting possibility is that they represent a discrete population of females, capable of migration unlike their mature counterparts, and therefore able to move in as replacements at sites of loss of fecund females.

Based on skin snip examinations of 23 experimentally infected calves, the mean prepatent period of *O. lienalis* was established to be 11.6 (range 8.9–15.5) months. This compares with a prepatent period of 13–16 months for *O. volvulus* (Cameroon forest form) in experimentally infected chimpanzees (Duke 1980) and around 15–18 months in man based on visitors to endemic areas (Chartres 1955, Diaz 1957). Microfilariae of *O. lienalis* recovered at post mortem from two calves 19 months after infection were distributed with the highest skin densities along the ventral midline, over the belly or at the umbilicus. Longitudinal examinations of animals with patent infections established that densities of microfilariae in the skin increased only slowly and that peak densities varied independently of the original inoculum of infective larvae.

Biology of the adult worms

One of the first features of the host–parasite relationship to be recognized in onchocerciasis was that the adult worms of *O. volvulus* reside within nodules, or onchocercomata. These vary in diameter, from as little as 2 mm to as much as 8 cm (Albiez *et al.* 1988), and may be buried deep within the tissues (Kilian 1988) or more commonly occur in superficial, subcutaneous sites (Nelson 1970). It has been commented upon by numerous investigators that nodules arise with especially high frequency over bony prominences (e.g. hip, knee, ribs, etc.) and several nodules may occur so closely juxtaposed as to create the impression of a single structure. For a recent overview of the available information on the nodule in onchocerciasis, the reader should consult a special edition of *Tropical Medicine and Parasitology* that has been devoted to this subject (*Trop. Med. Parasit.*, **39,** Supplement IV, 329–486, 1988).

Histologically, nodules comprise an outer fibrotic capsule of host tissue, an inner chamber containing diverse inflammatory cells and, usually but not always, one or two adult female worms. Based on examinations of very many nodules digested with collagenase enzyme, it has been calculated for a hyperendemic area with no recent control measures that nodules contain an average of 1.2 live males and 1.6 live females (Schulz-Key 1988). Large nodules may contain more, and sometimes many more females, while males may or may not be present at any particular moment and are assumed by analogy with *Onchocerca flexuosa* to migrate for the purposes of mating (Schulz-Key 1975). Although females store sperm in the spermatheca, reproduction is discontinuous and periodic and a complete reproductive cycle takes in the region of 2–4 months (Schulz-Key & Karam 1986). There are relatively good estimates of the longevity of the adult worms in man, which are put at more than 10 years based on transmigrants or residents of areas that have come under vector control (Nelson & Grounds 1958, Roberts *et al.* 1967, Karam *et al.* 1987). It is believed that there may be a fall off in fecundity of the females with age, so that in the absence of ongoing transmission the levels of microfilariae will fall off more rapidly than the surviving population of adult worms (Karam *et al.* 1987). This phenomenon will be accentuated by parallel changes in the males that reportedly inseminate females with decreased efficiency in an ageing population (Klager 1988).

A highly significant feature of the intensive research on nodules is that there are very few records of juvenile worms being detected. Even in those instances where the presence of 'immature' females has been reported, these have almost always been specimens with the cuticular rugae characteristic of mature adults. An exception to this was described by Schulz-Key (1988) who isolated a juvenile female worm 22 mm long, that was 'without ridges' and may therefore be equivalent to the early migratory stages of *O. lienalis* that have been recovered from cattle (described above). It follows that if the vast majority of juvenile females of *O. volvulus* cannot be detected in nodules, they probably occur free in the body and only as they approach sexual maturity do they induce nodule formation. There are a handful of documented cases where mature females have been found outside nodules at post mortem (Sharp 1927, van den Berghe 1936, Becker 1950), but it is difficult to establish whether these are simply unusual or if they represent a small fraction of a large, but virtually undetectable population.

Biology of microfilariae

Prompted by the requirement to screen drugs for macrofilaricidal or embryotoxic activity, there has been careful evaluation of the process of development of embryos in *O. volvulus* (Fig. 6.3). This has led to the concept of an 'embryogram', a quantitative measure of the various stages of the parasite present in the uterus and oviducts (Schulz-Key *et al.* 1980). From such data it has been deduced that development *in utero* takes 3–4 weeks (Schulz-Key 1988) and previous studies have put the time for migration of microfilariae to the skin at a further two weeks (Duke 1957).

The mechanism whereby microfilariae achieve tissue migration may, like third stage larvae, be dependent on secreted proteases. A 'collagenase'-like enzyme, together with one or possibly two additional proteases, have been identified in *Onchocerca* microfilariae of either human or animal origin (Petralanda *et al.* 1986, Lackey *et al.* 1989). Differences in proteases have been detected between microfilariae from the uterus and the skin (Lackey *et al.* 1989), and these might be biologically significant, reflecting developmental changes that continue in microfilariae even after birth. Indeed, there are several other key changes that take place in newly released microfilariae, which include alterations in body dimensions, biochemistry, antigenicity, and most important, infectivity for the invertebrate host (Mustafa 1983). In onchocerciasis, it is not yet known what governs the capacity of microfilariae to develop in their vectors, but in the case of *Brugia malayi* it has recently been demonstrated to correlate with the appearance of a stage-specific, calcium-binding protein (Fuhrman & Piessens 1989). Once infective for insects *Onchocerca* microfilariae may live for months in the host, and for *O. volvulus* it has been calculated their longevity may be in the order of 2.5 years (Duke 1968a). The detection of microfilariae in skin snips has and remains the principal method of parasitological diagnosis, and

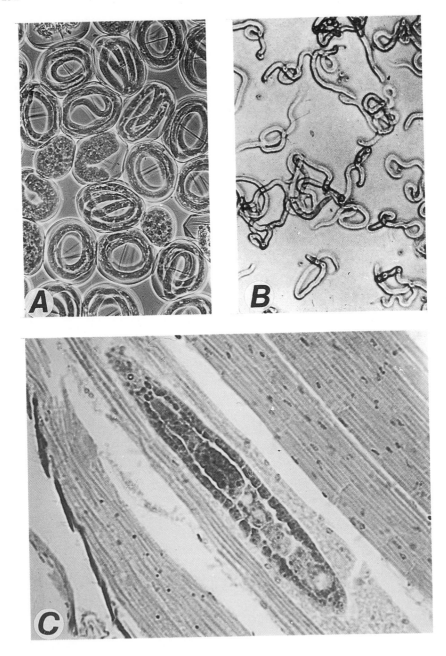

Figure 6.3 (a) Intrauterine stages of *Onchocerca* illustrating a range of developmental forms. (b) Microfilariae recovered from the skin. (c) Developing second stage larva in the thoracic flight muscles of an infected blackfly.

biopsies are now customarily taken with a 'Walser' or 'Holth' corneo-scleral punch (as introduced by Scheffel), superseding the traditional pin and razor blade method (Nelson 1970, Buck 1974).

Pathology

Numerous reviews have been written on the pathology of onchocerciasis. Accordingly, the intention here has been to highlight only the chief manifestations of the disease. Because the immune response can radically influence the outcome of infection, a brief section on aspects of the subject has been included if they have not already been discussed in the context of pathology. For further information the reader is advised to consult the general review of morbidity by Gibson *et al.* (1980), as well as excellent articles by Thylefors (1978) on ocular involvement and by Connor *et al.* (1985) on lesions of the skin. For a rapid introduction to the subject there is a short and beautifully illustrated handbook by Buck (1974).

Clinical manifestations of disease

Unlike lymphatic filariasis, due to the filarial nematodes *Wuchereria bancrofti* and *B. malayi*, it is the microfilariae rather than the adults that

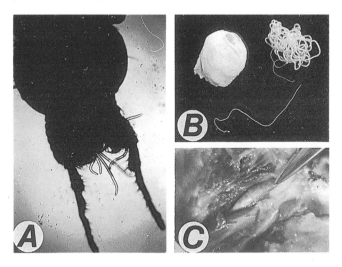

Figure 6.4   (a) *Onchocerca* third stage larvae escaping from the mouthparts of an infected blackfly. (b) Excised *O. volvulus* nodule from a patient (top left) and the adult worms after extraction with collagenase enzyme: shown for comparison are a single female (top left) and male (bottom) worm. (c) Nodule buried in muscle above the hip joint of an experimentally infected chimpanzee (kindly provided by Dr Brian Duke) illustrating the small, slender capsule typical of those formed in this animal.

are the principal cause of pathology in onchocerciasis. Adult worms give rise to localized tissue changes through nodule formation, but although they may be unsightly only rarely do they cause a serious clinical condition (Fig. 6.4).

Involvement of adult worms

Usually nodules are painless unless they are pressing on a joint capsule or occur in an aberrant site, but they may give rise to discomfort during chemotherapy, or if they should perforate as happens particularly with onchocercomata attached to the corium (Albiez et al. 1988). In Africa the nodules are commonest around the pelvis, lying adjacent to the femoral trochanter, coccyx, sacrum and iliac crest. They are also frequently associated with the lateral chest wall. It is not uncommon for them to occur around the knees or lower leg, but only a relatively small fraction are located on the head. Detection of nodules by trained personnel is a rapid procedure estimated to take around three minutes per individual, but two or three times longer if intended as prelude to nodulectomy (Albiez et al. 1988). As a form of diagnosis it is useful but cannot be used as an infallible guide, because many nodules may be too deep for them to be palpated and a proportion may be mistakenly identified. Most likely sources of confusion are superficial lymph nodes and lipomas, which according to a recent analysis accounted for 5% of the total onchocercomata ascribed by clinical diagnosis prior to removal (Albiez et al. 1988). In hyperendemic African foci more than 50% of the population may present with nodules, and the median count of onchocercomata may be in the order of 3–5 per person (Albiez et al. 1988). The clinical benefits of nodulectomy for individuals living in an endemic area can be varied, but for a proportion of those treated there may be a reduction in the absolute level or rate of gain of microfilarial densities in the skin (Albiez 1985). The selective removal of head nodules from a series of patients in Cameroon was claimed to have lowered the local concentrations of microfilariae, and had a beneficial effect on anterior segment ocular lesions (Fuglsang & Anderson 1978).

The histopathology of the nodule is one of the best characterized aspects of onchocercal pathology. Excellent illustrations of histological sections are given in a recent article by Buttner et al. (1988). Briefly, nodules occur in many distinct forms that probably vary as a consequence of age, or as a result of the fate of the worms they contain. Young nodules tend to be small with a thin-walled, collagenous capsule, while older nodules may reach a great size and become inelastic through the infiltration of fibrotic tissue. Large deformable nodules aptly described as 'cystic' also occur, and are filled with necrotic material, caseous fluid and often dead or dying worms (Schulz-Key 1988). In general, the histopathological picture is of a capsule composed of collagen fibres and scar tissue surrounding a cavity containing the worms and a mixed cellular and tissue response. This may include neutrophils, eosinophils, lymphocytes, plasma cells, macrophages and giant cells, together with permeating scar tissue, Russell bodies and immune complexes to varying degrees and in differing proportions. Microfilariae occur throughout the nodule and often do so in great numbers where they may give rise to focal granulomata if they die in this location (Buck 1974).

Involvement of microfilariae

Many of the lesions in the skin and eyes caused by microfilariae are known or suspected to be immunopathological in nature. This has led to the widely held view that, like lymphatic filariasis, the disease presents as a spectrum of clinico-immunological manifestations (Bartlett et al. 1978, Ngu 1978, Ottesen 1984). It follows that such a spectrum may result from at least two independently segregating factors, one being genetic and possibly related to immune response (Ir) genes of the major histocompatibility complex (MHC), and the other stemming from infection exposure (including prenatal exposure) and in consequence having a temporal component.

SKIN DISEASE

The presence of microfilariae in the skin does not in itself predicate dermal pathology, and sometimes remarkably high densities in an individual are associated with no detectable lesions (Nelson 1970, Anderson et al. 1974a) However, dermatitis characterized by itching, altered pigmentation and papules is the most common pathological feature in the early stages of onchocerciasis. The pruritis may be mild, or severe enough to lead to prominent excoriations and can involve any part of the body, although it might first become apparent in one anatomical quarter (Nelson 1966, Buck 1974). The variety of dermatological conditions associated with ongoing onchocercal infections is given in Table 6.1, listed in order of increasing severity of symptoms (Fig. 6.5). In long-standing onchocerciasis the most serious manifestation is advanced dermal atrophy, where the skin assumes a shiny, crushed tissue-paper texture and has lost all of its natural elasticity and resilience (Nelson 1970). As a secondary complication there may be the development of pseudoadenolymphocele or 'hanging groin' (Nelson 1958b), in which enlarged inguinal or femoral lymph nodes hang down in pendulous sacs created by loose redundant folds of skin. The lymphadenitis may ultimately lead to a local obstruction of lymph flow, with the additional risk of elephantiasis commonly affecting the region of the genitals (Connor et al. 1985)

Onchodermatitis of the 'classical' African form is often diffuse in its effects and has come to be known as the 'generalized' form of onchocerciasis. This draws a distinction between a second form of the disease that is termed 'localized' onchocerciasis, in which the skin

Table 6.1 Disease manifestations of onchocerciasis affecting the skin.[1]

Type of condition	Common name(s)	Signs and symptoms
Early onchodermatitis		Pruritis with or without macular or irregular hyperpigmentation and papules.
Papular reaction	Craw craw, Gale filarienne	Papules, large or small, ulcerated and secondarily infected; associated with pruritis, excoriations.
Intradermal oedema	Peau d'orange	Swollen skin resulting from focal lymphoedema.
Lichenification	Lizard skin	Dry, scaly skin with mosaic pattern; associated with pruritis and excoriations.
Depigmentation	Leopard skin	Irregular depigmentation resulting in white areas interrupted by foci of pigmented skin around hair follicles and pore; commonly affects shins.
Dermal atrophy	Tissue-paper skin	Shiny, fragile skin, unevenly pigmented and inelastic; may be wrinkled and hang in folds.
Localized disease	Sowda	Hyperpigmented, thickened, papular and pruritic skin; characteristically affects limbs and unilateral; associated with delayed type hypersensitivity.

1 Based on Nelson 1970, Buck 1974, Engelkirk *et al*. 1982, Connor *et al*. 1985.

pathology is severe and is generally confined to one part of the body, commonly a leg (Fawdry 1957, Anderson *et al*. 1973, Buttner *et al*. 1982, Connor *et al*. 1983). In such cases, few if any microfilariae can be detected in skin-snips, nodules are rarely palpable, and the skin is intensely pruritic, papular, hyperpigmented and thickened. In the Yemen where this condition is particularly prevalent, it is known as 'sowda', the Arabic word for black, in recognition of the hyperpigmentation (Gasparini 1964). In other foci of infection a low percentage of cases present with localized disease, and the reasons for the high prevalence of this manifestation in the Yemen have yet to be determined.

In both forms of the infection, histopathological changes in the skin may be numerous and varied (Buck 1974, Connor *et al*. 1985). One of the most common and serious features is fibrosis, and this may arise comparatively early and develop progressively. Initially there may be nothing more than a mild perivascular infiltration consisting of lympho-cytes, plasma cells, histiocytes, mast cells and eosinophils, together with elevated numbers of fibroblasts among the collagen fibres. There is focal deposition of acid mucopolysaccharide between collagen strands, and melanin appears within phagocytic cells in the upper dermis. Papules

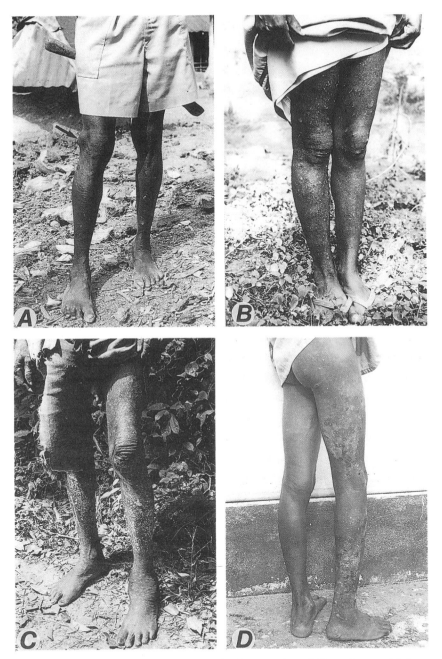

Figure 6.5 Cases of onchocerciasis from the forest region of Sierra Leone. (a) Large *O. volvulus* nodule over the knee (arrowed). (b) Generalized onchodermatitis with prominent papules, excoriations and intradermal oedema. (c) Spotty depigmentation (leopard skins) of the shins and dermal atrophy leading to the wrinkled appearance of skin over the knees. (d) Localized onchodermatitis primarily affecting one limb in a nine-year-old child. There is secondary infection of ulcerated papules resulting from the acute pruritis.

arise as intra-epithelial abscesses which contain interstitial fluid, neutrophils and fibrin surrounding one or more microfilariae. In the late stages of onchodermatitis the epidermis becomes very thin, the rete ridges erode away and much of the dermal architecture is replaced by scar tissue formed from an extensive network of reticulum fibres. Elastic fibres become thickened, disorganized or lost from the dermis, and in cases of depigmentation there is a focal loss of melanin from the epidermis. In localized onchocerciasis there is hyperkeratosis, parakeratosis, acanthosis and papillary elongation of the rete ridges, together with fibrosis of the dermis and a particularly extensive cellular infiltration composed mainly of plasma cells. Oedema of the epidermis and dermis is another notable feature.

Associated with the two basic forms of localized and generalized disease are two distinct histological appearances of the regional lymph nodes draining the areas affected by onchodermatitis (Buck 1974, Connor et al. 1983, 1985). In the generalized disease the nodes may have dilated subcapsular and medulla lymphatics and are hard as a result of fibrosis which may start as minor and then extend throughout the capsule, trabeculae and/or medulla. The sinusoids characteristically show evidence of a histiocytic hyperplasia, while scattered throughout the node are plasma cells, mast cells and eosinophil. Microfilariae frequently arise in the capsule and subcapsular sinusoids, where they may be living, or dead and engulfed in granulomatous reactions. Characteristically, there may be few or no germinal centres, and those that do arise tend to be small. In contrast, the nodes from cases of localized onchocerciasis are greatly enlarged, soft and non-fibrotic, and exhibit marked follicular hyperplasia. Few if any microfilariae are present (Buck 1974), although exceptions have been reported (Abdel-Hameed et al. 1987), and there can be an increase in the numbers of plasma cells bearing surface IgG and IgE (Racz et al. 1983). These features reflect the state of hyper-responsiveness exhibited in localized onchocerciasis, but have not as yet allowed us to establish what conditions predispose individuals to this reaction.

EYE PATHOLOGY

A number of factors have been associated with the development of ocular lesions, such as the duration and intensity of infection, densities of microfilariae near the eyes (i.e. measured at the outer canthus) and the geographic isolate of the parasite (Anderson et al. 1974b, Buck 1974, Garner 1976, Fuglsang & Anderson 1977, Thylefors 1978). Ocular disease does not normally arise rapidly or as an acute condition, although in areas of intense transmission permanent lesions may appear that affect visual acuity in youth (Buck 1974). In the most recent report from WHO, the numbers of Africans blinded by onchocerciasis was estimated to be 326000 (WHO 1987). There are important differences in the aetiology of anterior and posterior segment lesions, so it is more useful to consider these separately rather than to make very many generalizations about ocular pathology.

In the anterior segment, microfilariae may invade the conjunctiva, cornea and anterior chamber and can often been seen with the aid of a slit lamp, wriggling vigorously when they occur in the aqueous humour (Nelson 1970, Anderson & Fuglsang 1973). In the cornea they may also be active when alive and are most frequent immediately beneath Bowman's layer near to the limbus, but do appear elsewhere and at any depth in the stroma (Garner 1980). It is a widely held view that while the parasites are alive they induce no major changes in the anterior segment, but their death triggers host immune mechanisms that lead to a sequence of local tissue reactions (Nelson 1966). Among the cells making up the chronic inflammatory response, T. lymphocytes (CD3$^+$) predominate and include an elevated level of the T suppressor/cytotoxic subset (CD8$^+$) (Chan et al. 1989). Non-lymphoid cells of the conjunctiva and iris show an enhanced Class II MHC antigen expression, indicative of the heightened immunological reactivity within these ocular tissues. Host defence processes are the basis of a range of ocular lesions that have been carefully documented in man and help to explain the chronic and progressive nature of the eye disease (see Table 6.2). Many of the manifestations of onchocercal pathology are not unique to onchocerciasis and expert ophthalmological judgements may be required to distinguish the cause of certain conditions. Depending upon how severe the lesions become they may affect visual acuity and some can lead to blindness, most notably in the case of sclerozing keratitis.

Posterior segment pathology in onchocerciasis has been a controversial issue until comparatively recently. For many years the case incriminating O. volvulus in pathological changes of the fundus rested solely on epidemiological evidence (discussed by Garner 1980). It is now firmly established that microfilariae may induce a variety of profound posterior segment lesions, although there is much yet to be learnt about the way the parasite induces these changes. Microfilariae are thought to gain access to the back of the eye via the vascular and neural channels of the sclera (Neumann & Gunders 1973) and/or by way of the blood (Fuglsang & Anderson 1974). They occur in the vitreous and choroid, but appear to be less common in the retina, which may indicate that the comparatively high frequency of retinal damage results from the indirect actions of parasite toxins or host inflammatory processes. The finding of anti-retinal autoantibodies in serum and ocular fluids has led to the suggestion that these may be instrumental in the pathogenesis of retinal degeneration and optic atrophy (Chan et al. 1987).

Garner (1980) described the pathology as falling into three clinical varieties; pigment disturbance, choroido-retinitis and pigment epithilium/ choroidal atrophy. In contrast, these were regarded as parts of a continuum by Bird et al. (1976), although most experts are agreed that

Table 6.2 Disease manifestations of onchocerciasis affecting the eyes.[1]

Condition	Infection status	Signs and symptoms
Punctuate keratitis	Light/early infection	Opacities approx. 0.5 mm diameter around dead microfilariae in corneal stroma; spontaneously resolve; visual acuity unaffected.
Sclerozing keratitis	Longstanding/heavy infection	Stromal inflammation beginning in nasal or temporal juxtalimbal positions; extends around cornea and then centrally with limbal pigment migration, neovascularization of stroma and opacification; major cause of blindness.
Iridocyclitis	Longstanding/heavy infection	Ranging from mild uveitis to severe plastic inflammation with peripheral anterior and posterior synechiae; may lead to secondary glaucoma and iris pigment atrophy; sight-threatening lesions.
Cataract	Longstanding/heavy infection	May arise secondary to anterior uveitis.
Choroidoretinitis	Longstanding/heavy infection	Focal atrophy of retinal pigment epithelium and choriocapillaris leading to more widespread lesion in posterior pole ('Hisette-Ridley' fundus); macula relatively resistant; cause of tunnel vision.
Postneuritic optic atrophy	Longstanding/heavy infection	Scarring and pigment disturbance around optic disc; optic nerve head inflammation, papillitis and dense sheathing of retinal vessels; major cause of blindness.

1 Based on Buck 1974, Connor & Neafie 1976, Thylefors 1978, Garner 1980.

the damage is primarily either inflammatory or atrophic in nature. A summary of the clinical manifestations of posterior segment lesions is provided in Table 6.2. Most significant of these is the pathology associated with the optic disc and nerve, which it has been estimated accounts for close to 90% of all cases of blindness resulting from ocular onchocerciasis involving the posterior segment (Bird *et al.* 1976).

Immunological responses

There is much yet to be done to characterize the immune response in onchocerciasis, for example in determining which effector mechanisms operate in the pathogenesis of skin and ocular lesions. More ambitious would be to establish which parasite antigens induce the immunopathological response, an area of research that is opening up through the advent of gene cloning techniques. By means of differential screening of the appropriate cDNA libraries it should be possible to identify cloned parasite antigens that are uniquely recognized by antibodies or that stimulate lymphocytes derived from individuals with defined clinical manifestations. Here it will be necessary to consider the isotypic response and T lymphocyte sub-sets, and to look for both positive and negative correlations between antigen recognition and the various disease states. However, there will remain a need to conduct equivalent experiments with native parasite products, because the relevant antigens may not be proteins and will not, therefore, be amenable to recombinant DNA approaches.

For a number of years it has been known that high levels of circulating immune complexes occur in onchocerciasis (Lambert *et al*. 1978) but it is still unclear as to the cause or the full consequences of this phenomenon. Features of the pathogenesis in man, such as the perivascular fibrosis seen in lymph nodes from patients with onchocercal lymphadenitis, are consistent with the idea that deposition of immune complexes contributes to inflammation, fibrosis and lymphatic obstruction (Gibson & Connor 1978). Based on a recent study conducted in the foci of onchocerciasis in northern Sudan, it was reported that levels of *Onchocerca*-specific circulating immune complexes were negatively associated with microfilarial density and positively associated with dermal pathology (Sisley *et al*. 1987). In Cameroon, filarial antigen was found to be associated with immune complexes in the kidneys in 9 of 63 cases of nephropathy among onchocerciasis patients (Ngu *et al*. 1985).

Another important but unresolved issue is the biological significance of antibody dependent cell cytotoxity (ADCC) reactions, which have been demonstrated against microfilariae and infective larvae *in vitro* with serum and cells from people harbouring active *Onchocerca* infections (Mackenzie 1980, Greene *et al*. 1981, Williams *et al*. 1987). According to Mackenzie (1980), more than 70% of cases have antibodies that direct the adherence of granulocytes to infective larvae, yet there is no basis for the belief that such people can resist reinfection if they remain exposed to transmission. Indeed, the only evidence from man of resistance to developing larvae comes from studies that show a low percentage of adults apparently remain free of infection within some endemic foci of onchocerciasis (Ward *et al*. 1988). A small minority of infected individuals develop antibodies that bind the surface of intact, living microfilariae

(Taylor *et al.* 1986) and it would be especially interesting to establish whether this response modifies the host–parasite relationship. This contrasts with the observation that in the vast majority of infections antibodies are induced against the adult worm surface, as measured by immunoprecipitation of the major surface-iodinated glycoprotein of 20 kD (Philipp *et al.* 1984, Bradley *et al.* 1989).

An especially significant feature of the immune response in onchocerciasis is the variation in responsiveness that appears to be the basis of the different manifestations of localized and generalized disease (Bartlett *et al.* 1978, Ngu 1978). Localized onchocerciasis often arises in young people. In the Yemen, patients presenting with generalized onchocerciasis reported experiencing sowda for varying periods in their earlier life (Buttner *et al.* 1982). This fascinating and important observation should be substantiated by a longitudinal study, since it implies there is a changing state of reactivity linked with exposure somewhat reminiscent of hypersensitivity to arthropod bites (Benjamini *et al.* 1961). Certainly, the localized disease appears to be associated with a specific state of heightened reactivity to parasite antigens manifest by delayed hypersensitivity reactions and enlarged lymph nodes containing active germinal centres (Buck 1974, Connor *et al.* 1983, 1985). This contrasts with the picture characteristic of the generalized disease, in which delayed skin reactions are absent, lymph nodes are hypoplastic, and indeed there is evidence of a down-regulation of the cellular arm of the immune response (Bartlett *et al.* 1978, Ngu 1978, Connor *et al.* 1985). Levels of IgG and IgE antibodies are elevated in both forms of onchocerciasis (Greene *et al.* 1985), but these are especially high in the localized disease (Brattig *et al.* 1987) in which it is reported that certain antigens are recognized uniquely by the subclass IgG3 (Parkhouse *et al.* 1987, Cabrera *et al.* 1988). A comparison of haplotypes between groups of patients with either localized or generalized onchocerciasis demonstrated the greatest differences to be among class II (DR) HLA molecules, with some weaker distinctions between those of class I (Brattig *et al.* 1986). Typing of peripheral blood lymphocytes revealed a significantly increased proportion of cells in the localized disease expressing interleukin 2 (IL2) or transferrin receptors, and bearing surface markers for T cells of helper/inducer (CD4$^+$) phenotype. Together with an observed increase in the same patients in numbers of DR antigen-positive cells and of natural killer (NK) lymphocytes, there appears to be a strong case that localized onchocerciasis is associated with significant T cell activation (Brattig *et al.* 1987).

A hypothesis that has been advanced to explain the major difference between localized and generalized onchocerciasis is that the relative quiescence of the cellular response in the latter stems from immunological tolerance established by neonatal exposure to parasite antigens (Connor *et al.* 1985). In defence of this argument it has been established that microfilariae can reach the foetus before birth (Brinkmann *et al.* 1976) and that antigens of *O. volvulus* occur in the breast milk of a high proportion of nursing mothers (Petralanda *et al.* 1988). This is an attractive model but is especially demanding to test, and may well be suited to experimentation using *Onchocerca* parasites in animals (discussed below), as it has been in *Acanthocheilonema viteae* infections in rodents (Haque & Capron 1982).

The state of hyporesponsiveness to parasite antigens that is manifest by T lymphocytes in human onchocerciasis is one of the most intriguing features of the immunology of the infection (Bartlett *et al.* 1978, Ngu 1978, Greene *et al.* 1983, Ward *et al.* 1988). A level of depression in cellular responses to unrelated antigens and mitogens has also been reported (Greene *et al.* 1983). In a comparative study of lymphocytes from age-matched individuals presenting with either active onchocerciasis or without any history of infection, it was established that IL2 production stimulated by *O. volvulus* antigen was specifically depressed in the actively infected group (Ward *et al.* 1988). Longitudinal experiments in cattle experimentally infected with *O. lienalis* have demonstrated that this state of hyporesponsiveness only arises after the onset of patent infections (Bianco & Lloyd, unpublished observations). Lymphocytes from cattle collected during the prepatent infection exhibited a vigorous blastogenic response when stimulated with a PBS-soluble fraction of adult worm extract. The separation of cells into nylon wool adherent and non-adherent populations revealed that specific responsiveness to parasite antigen resided in the adherent population. Following the appearance of microfilariae in the skin, there was a precipitous loss of the antigen-specific proliferative response. However, the treatment of cattle with a combination of diethylcarbamazine and ivermectin to eliminate micro-filariae from the skin resulted in the restoration of T cell responsiveness in more than half the animals tested. Interestingly, microfilariae alone did not appear to be able to ablate the antigen-specific lymphocyte response when administered to uninfected animals immunized with parasite extracts and challenged with inoculations of live microfilariae.

Epidemiology and control within Africa

In the recent report of the WHO Expert Committee, 26 countries in the African region and three in the east of the Mediterranean (Sudan, Yemen and possibly Saudi Arabia) were listed as having foci of onchocerciasis (WHO 1987). Throughout the vast range of the parasite, extending from Senegal in the West to Tanzania in the East, there are many variations in vector–parasite biology, in transmission dynamics, and in the manifestations of infection. In this section some of the variations that have come to light in onchocerciasis will be discussed, and also the impact of animal parasites on the study of the human disease. Because of its importance, a brief description of the control campaign conducted by the OCP in West Africa will also be included.

Geographic variations in transmission and disease

For many years it has been recognized that there are important differences within onchocerciasis with respect to the simuliid vectors in various foci and the types of pathology observed. One major difference is in the vectors between East and West Africa, the former primarily coming from the *S. neavei* species complex and the latter from the *S. damnosum s.1.* complex. Details of the distributions of the various taxa of blackflies involved in transmission are provided by WHO (1987) which lists 42 species, sub-species or forms of the *S. damnosum* complex alone.

Within West Africa it is well established that there are divisions between the forest and savanna habitats, not only with regard to the vector cytospecies, but also with reference to the manifestations of infection in man (see Fig. 6.1). More recently it has become apparent that there may be some basic genetic differences in *O. volvulus* itself, although the extent of infraspecific variation is still impossible to assess. Table 6.3 lists characteristics associated with geographical variation in onchocerciasis as recorded for West Africa alone. Clearly, some of the differences arise from interdependence of particular features, such as the relatively high rates of sclerozing keratitis attributed to savanna parasites which links in with blindness and with the level of pathogenicity in rabbit eyes (discussed in more detail below).

The results of isoenzyme (Cianchi *et al.* 1985) and DNA sequence (Erttmann *et al.* 1987, Unnasch *et al.* 1989) analysis indicate that there has probably been genetic divergence in the parasites that are now associated with different habitats in Africa. This may have arisen through adaptive changes in parasite biology, although as yet we have to identify the genes specifying such phenotypes as infectivity for one cytospecies of *Simulium* over another (Duke *et al.* 1966). One of the less precise manifestations of adaptation to local vectors was reported by Duke *et al.* (1967b) who observed diurnal fluctuations in the levels of microfilariae in the skin that varied between forest and savanna parasites in accordance with the different periods of the day when biting activity was maximal for forest and savanna vector cytospecies. More profound differences were observed in the capacity of forest and savanna microfilariae to develop in blackflies, with far greater efficiency of parasite uptake and development in flies originating from the same habitat (Duke *et al.* 1966). Even following the intrathoracic inoculation of precise doses of microfilariae into these insects, the yields of third stage larvae reflected the difference in their intrinsically greater susceptibility for the local isolate of parasite (Ham & Garms, 1985).

Table 6.3 Geographic variations in onchocerciasis in West Africa.

	Savanna foci	Rain forest foci
Clinical features[1]		
Nodules	++	+++
Densities of microfilariae (mf)	+++	++
Skin atrophy	+++	++
Depigmentation	+	+++
Microfilariae in anterior chamber	+++	+++
Punctuate keratitis	+	++
Sclerozing keratitis	+++	+
Posterior segment lesions	+	+
Blindness	5.1%	2.0%
Pathogenicity in rabbits[2]	+++	+
Susceptibility of chimpanzees[3]	+	+++
Periodicity of microfilariae[4]	Peak density, 1400–1500 hrs	Peak density, 1600–1700 hrs
Isoenzymes[5]	Ldh 100 Hbdh 100	Ldh 110 Hbdh 108
Genomic DNA[6]	pFS1 sequence − ve pSS-1BT sequence + ve	pFS1 sequence + ve pSS-1BT sequence − ve
Principal vectors[7]	*S. damnosum s.s.* *S. sirbanum*	*S. soubrense* *S. soubrense B* *S. sanctipauli* *S. yahense* *S. squamosum*
Typical ATP values[8]	1000–3000	5000 or more (up to 90 000)

1 Based on clinical studies in Cameroon by Anderson *et al.* (1974b).
2 As measured by anterior segment lesions induced by inoculations of microfilariae (Duke & Anderson 1972).
3 Based on densities of microfilariae following experimental infection (Duke 1980).
4 Time of day corresponding with peak densities of microfilariae in the superficial dermis (Duke *et al.* 1967b).
5 Allelic differences at polymorphic loci reported by Cianchi *et al.* (1985).
6 Genomic DNA probes described by Erttmann *et al.* (1987) and Unnasch *et al.* (1989).
7 Simplified breakdown, as in certain circumstances the cytospecies overlap (WHO 1987).
8 ATP = Annual Transmission Potential (Duke 1968b), or theoretical maximum number of larvae received per person/year: figures based on WHO (1987).

One of the perplexing aspects of variation in onchocerciasis is that annual transmission potentials (ATP) (Duke 1968b) can be so much higher in the forest without a proportional increase in worm burdens. In savanna regions of Cameroon where *S. damnosum s.s* or *S. sirbanum* are the vectors, an ATP of 1000–3000 would be considered a typical value. In forest regions of Cameroon, where *S. squamosum* is the vector, ATP figures of 5000 would not be unusual and up to 90 000 have been recorded (WHO 1987). Certainly, there may be an increase in the average number of nodules in the forest, but this is not sufficient to account for the magnitude of difference between the two habitats.

When one considers all forms of onchocerciasis together, the innately greater pathogenicity of savanna microfilariae for the anterior segment of the eye masks the influence that intensity of transmission has on ocular pathology. If one considers, instead, only savanna onchocerciasis, the intensity of transmission is seen to govern the incidence of ocular disease. Permanent eye lesions are primarily restricted to people living in areas of mesoendemicity or hyperendemicity, corresponding with prevalence rates of infection of 40–59% or > 60% respectively (as defined by WHO 1987). In mesoendemic areas vision may be impaired among a proportion of the young adults, but where transmission is hyperendemic even the youths may risk serious disorders and blindness (Buck 1974). According to Prost (1986), cross-sectional data on the prevalence of blindness may underestimate the true magnitude of the problem in the savanna regions of West Africa. High mortality among the blind leads to an erroneously low number of cases being recorded in point prevalence surveys of ocular onchocerciasis. Taking this into account, it was calculated from data collected from hyperendemic villages in Burkina Faso that 46% of males and 35% of females aged 15 are likely to become blind before they die: the corresponding figures for a mesoendemic focus were 14% of males and 10% of females (Prost 1986). Measuring the level of transmission is therefore of much public health importance as a yardstick of the risk to a community from ocular onchocerciasis. One might assume that this should be a comparatively simple procedure, but as we shall see blackflies carry filarial larvae other than *O. volvulus* itself.

Filarial parasites of animals in vectors of human onchocerciasis

Few, if any, of the vectors of *O. volvulus* are strictly anthropophilic, although a detailed knowledge of their host preferences is lacking. In Africa, members of the *S. neavei* and *S. damnosum* complexes exhibit varying degrees of zoophilic behaviour, so not surprisingly some filariae of animals arise in these blackflies (Nelson & Pester 1962, WHO 1976). It is therefore necessary to carefully discriminate between the human and animal parasites if one is to calculate reliable indices of transmission for human onchocerciasis.

However desirable this may be it is a problem that has proved difficult to resolve through a lack of information on the parasite life-cycles and limitations of the taxonomic techniques. The size of the problem is hard to assess, but what evidence exists suggests that it is enough to seriously distort transmission statistics. For example, it was demonstrated in West

Africa in an area within OCP operations that up to 38% of the infective larvae in *S. damnosum s.1.* were not of *O. volvulus* (Omar & Garms 1981). Other instances are described where the proportion has been 40% (Duke 1967) and even up to 70% (Garms & Voelker 1969). This issue is most serious as it affects monitoring of transmission during control, a factor that led WHO to encourage the development of a new taxonomy based on isoenzymes, monoclonal antibodies and DNA probes (discussed below).

Range of filarial species transmitted by simuliids

Onchocerca parasites are predominant among the known genera of filariae to parasitize simuliids, and at present 11 species have been identified that will develop to infective larvae in these insects (Table 6.4). In addition to *O. volvulus* are species from wild and domesticated ungulates, of which *O. cervicalis*, *O. gutturosa*, and *O. ochengi* are common in areas endemic for human onchocerciasis. Not all the parasites in Table 6.4 are known to be carried by blackflies in nature, and indeed *O. cervicalis* and *O. gutturosa* are generally considered to be transmitted by *Culicoides* midges (Mellor 1975, Bain 1979). However, in the laboratory *O. cervicalis* will develop to third stage larvae in some British simuliids, albeit with poor efficiency and after infection by intrathoracic injection or membrane feeding techniques (Bianco *et al.* unpublished observations). Even with this approach, designed as it is to enhance the opportunity for microfilariae to develop, Ham and Garms (1987) failed to obtain infective larvae of *O. gutturosa* in *S. soubrense* and *S. yahense* from Liberia. A report of *O. gutturosa* developing in *Simulium vorax* after feeding on an infected cow in Tanzania (Mwaiko 1981) is an observation that should be confirmed in the light of redescriptions distinguishing *O. gutturosa* and *O. lienalis* (Bain *et al.* 1978).

Of greater importance to human disease is the bovine parasite *O. ochengi*, because this is transmitted in West Africa sympatrically with *O. volvulus* by members of the *S. damnosum* complex (Omar *et al.* 1979). To add to the problem, the infective larvae of *O. ochengi* are indistinguishable from those of *O. volvulus*, creating a major imponderable in the transmission statistics relating to some West African foci. This has placed *O. ochengi* at the top of priorities for development of a sensitive and specific diagnostic probe. Three other genera of filariae, *Splendidofilaria*, *Dirofilaria* and *Tetrapetalonema* also have species that are transmitted by *Simulium* vectors (see Table 6.4). A single member of each genus is currently known to develop in simuliids, and all are New World species so none of them directly contributes to the parasitic fauna of the African blackflies. Nevertheless, they illustrate the range of filariae that occur in these insects, and can therefore be expected to arise when identifications are made of the many types of parasites in African vectors. In addition to named species of filariae there is a long list of unidentified

Table 6.4 Filarial parasites developing to third stage larvae in simuliids.

Parasite	Host	Vectors	Reference
Dirofilaria ursi	Bear	*S. venustum*	Addison 1980
Onchocerca cervipedis	Deer	*Prosimulium imposter*	Weinmann *et al.* 1973
Onchocerca cervicalis	Equines	*S. ornatum s.1.* *S. equinum*	Unpublished results[1]
Onchocerca flexuosa	Deer	*S. ornatum s.1.*	Frank *et al.* 1969
Onchocerca gutturosa	Bovines/ antelope	*S. vorax*	Mwaiko 1981[3]
Onchocerca lienalis	Bovines	*S. ornatum s.1.* *S. jenningsi* several N. American spp. several British spp.	Steward 1937 Lok *et al.* 1983a Lok *et al.* 1983b[2] Ham & Bianco 1983a[2]
Onchocerca ochengi	Bovines	*S. damnosum s.1.* (probably *S. santipauli*)	Omar *et al.* 1979
Onchocerca tarsicola	Deer	*S. ornatum s.1.* *Prosimulium nigripes*	Schulz-Key & Wenk 1981
Onchocerca volvulus	Man	*S. damnosum s.1.* *S. neavei s.1.* *S. albivirgulatum* *S. ochraceum* *S. metallicum s.1.* *S. callidum* *S. exiguum s.1.* *S. guianense* *S. limbatum* *S. oyapockense s.1.*	WHO 1987
Splendidofilaria fallisensis	Duck	*S. croxtoni* *S. latipes* *S. rugglesi* *S. euryadminiculum*	Anderson 1956
Tetrapetalonema ozzardi	Man	*S. amazonicum* *S. sanguineum*	Shelley *et al.* 1980 Tidwell *et al.* 1981

1 Based on laboratory infections by inthrathoracic injection of microfilariae or membrane feeding.
2 Based on intrathoracic injections of microfilariae only.
3 The infective larvae described accord more closely with *O. lienalis* than *O. gutturosa*.

infective larvae that have been recovered from African blackflies and designated by letters or numerals in the absence of any information on other stages in the life-cycles (e.g. Duke 1967, Voelker & Garms 1972). Over and above these are numerous reports of developing larvae from blackflies that are broadly classified as 'distinguishable from *O. volvulus*' and rather beg the question as to how heterogeneous is the 'indistinguishable' category.

Differentiation techniques applied to developing larvae
The simplest and most widely practised method for differentiating infective larvae has been the use of standard morphometrics, coupled with some basic morphological characters (Nelson & Pester 1962, Duke 1967, Garms & Voelker 1969, Voelker & Garms 1972, Bain & Chabaud 1986). The principal morphological features of value include the shape of the caudal extremity and papillae, the position of the anus, the anal ratio (tail length divided by breadth), length and form of the oesophagus (glandular or muscular) and type of buccal cavity (chitinized or not). Nevertheless, the infective larvae of *O. ochengi* and *O. volvulus* cannot be distinguished by morphological characters (Omar *et al.* 1979) and there is a close resemblance with other species such as *O. cervicalis* (Bain & Petit 1978).

These limitations have led to the search for more sensitive taxonomic techniques, and for a period considerable attention was given to histochemical staining procedures. Based on the distribution of acid phosphatase activity in microfilariae it was shown to be possible to differentiate among a range of filarial species (Schillhorn van Veen & Blotkamp 1978) and it was even claimed to detect polymorphisms within populations of *O. volvulus* microfilariae (Braun-Munzinger & Southgate 1977, Omar 1978). With developing larvae from *S. damnosum s.1.* the method also enjoyed a measure of success, distinguishing between larvae of *O. volvulus* and those of 'type D' (Duke 1967), 'type III' (Voelker & Garms 1972) and a hitherto unknown form of infective larva (Omar & Kuhlow 1978, Omar & Garms 1981). However, no differences between forest and savanna forms of *O. volvulus* emerged in the staining patterns of larvae and enthusiasm for the technique waned after it failed to distinguish between *O. ochengi* and *O. volvulus* (Omar & Schulz-Key 1978, Omar *et al.* 1979).

Isoenzyme analysis has also been applied to the microfilariae and adult worms of several *Onchocerca* species (Flockhart 1982, Cianchi *et al.* 1985, Flockhart & Bianco 1985). These techniques have identified differences among species and in some instances between geographic isolates (Cianchi *et al.* 1985), but so far they lack the sensitivity to be applied to the problem of parasite identification in vectors. Naturally infected blackflies harbour small numbers of filariae and may also carry mixed infections, necessitating the development of techniques capable of discriminating amongst the isoenzymes from single third stage larvae.

An attractive approach for the specific identification of individual parasites is to use monoclonal antibodies that are reactive with epitopes unique to *O. volvulus* larvae. A monoclonal antibody of diagnostic value has recently been developed for *B. malayi* (Carlow *et al.* 1987) and such is its specificity that it does not react even with species as closely related as *B. pahangi*. As yet there have been no reports on the production of an *O. volvulus*-specific monoclonal antibody, although one with restricted reactivity for the genus *Onchocerca* has recently been described (Lucius *et al.* 1988a). Another strategy that has been the focus of recent research is to develop DNA probes for taxonomic purposes through *in situ* hybridization. Again, there has been considerable success in the identification of sequences of the requisite specificity for *B. malayi* (McReynolds *et al.* 1986), and more recently there have been encouraging reports from several laboratories on progress with the development of probes for *O. volvulus* (see Proceedings of the O-Now! Symposium on onchocerciasis: recent developments and prospects of control, 1989 Sept. 20–22, Leiden, J.H. van der Kaay (ed.) Institute of Tropical Medicine Rotterdam-Leiden, The Netherlands). A good example of this work is provided by Harnett and colleagues, who have constructed an oligonucleo-tide based on genomic sequence data that is putatively species-specific (Harnett *et al.* 1989). The 60 base pair sequence will hybridize to *O. volvulus* derived from either the forest or the savanna with an intensity 300-fold greater than to any of six other *Onchocerca* spp. tested, including *O. ochengi*. Provided the probe can attain the necessary degree of sensitivity, it may therefore have real potential for the identification of *O. volvulus* larvae in blackflies. Bearing in mind that the estimated total DNA for one third stage larva is only 1.1 ng (Philip McCall pers. comm.), obtaining an adequate signal will depend on high copy number of the target sequence, or amplifying this first to detectable levels using the polymerase chain reaction.

Transmission of Onchocerca *parasites between man and animals*

The presence of infective larvae of animal origin in simuliids collected while attempting to feed on man clearly illustrates there must be an interchange of parasites between animal and human populations. Indeed, there is evidence of zoonotic *Onchocerca* infections in man outside areas of the human disease, based on a small but carefully documented number of case reports (Osborn 1935, Siegenthaler & Gubler 1965, Azarova *et al.* 1965, Beaver *et al.* 1974, Caprioglio 1976, Ali-Khan 1977). There is less evidence, however, of appreciable transmission of *O. volvulus* to animals since only the gorilla has been found to harbour naturally acquired infections (van den Berghe *et al.* 1958). It may be that the wrong animals have been examined or that cryptic infections have been missed, so as the mammalophilic behaviour of the vectors is unravelled it may be that this question should be re-addressed.

If it appears improbable that *O. volvulus* occurs to any significant extent in animals, it does seem more likely that *Onchocerca* parasites of animals are transmitted to humans. Not only simuliids but also ceratopogonids are vectors of animal onchocerciasis of which many are renowned for their biting of man in addition to their animal hosts (Muller 1979). Therefore, the range of species reaching man through the catholic feeding habits of their vectors may be considerable, although difficult to assess until more *Onchocerca* life-cycles have been described. Most reports of zoonotic infections in man are based on worms in histological sections prepared from excised lesions in which interpretation is difficult (e.g. Beaver *et al.* 1974, Ali-Khan 1977). Generic differences have been based on cuticular structure, although identifications to the level of species pose a far greater problem. Detailed comparative studies between the worms from man and *Onchocerca* species from animals have pointed to *O. cervicalis* and *O. gutturosa* as possible sources of these zoonotic infections (Beaver *et al.* 1974, Ali-Khan 1977). A report of *O. volvulus* from an aberrant site in the wall of the aorta in man may be a misidentification of *O. armillata* that inhabits this location in cattle (Meyers *et al.* 1977). The small number of case histories of zoonotic infections is almost certainly a mere fraction of the true frequency in the population. While it may be possible to detect aberrant infections giving rise to gross lesions, any worms lying free in the tissues would be almost impossible to locate. Probably the vast majority of larvae die without undergoing significant development, and only those at least as advanced as immature adults would be recognized as *Onchocerca* parasites even if they were recovered.

Zoonotic infections have not yet been described from Africa, almost certainly because of their cryptic nature and the lack of any immediate or obvious clinical significance. However, some years ago it was proposed by Nelson (1965) that the establishment or course of various infections in man might be influenced by exposure to related parasites of zoonotic origin. The cross-protection exerted in this way was originally termed 'zooprophylaxis', which has recently been reviewed and the term revised to 'the Jennerian principle of cross-protection' (Nelson 1988). Among helminths, the best examples come from schistosomiasis and hookworm infections, but this is mostly on laboratory rather than epidemiological evidence and so does not address the question of the real impact under natural conditions (Nelson 1988). In filarial infections in general, and onchocerciasis in particular, there is meagre data on the possible role of animal parasites in cross-protection. In some of the animal models of filariasis partial immunity has been conferred against infective larvae by heterologous species (Storey & Al-Mukhtar 1982), while cross-protection between microfilariae has been demonstrated in a mouse model of onchocerciasis (Townson *et al.* 1985, Carlow & Bianco 1987a). What is still required is an accurate picture of what man is exposed to by way of zoonotic filariae before we can go on to use epidemiology to test the concept of the Jennerian principle of cross-protection in the context of onchocerciasis.

Current state of immunodiagnosis

One worrying implication stemming from exposure of man to animal parasites is that this has the potential to generate cross-reactive responses which could be picked up in immunodiagnosis. Cross-reactive antigens have been convincingly demonstrated between *O. volvulus* and some of the bovine *Onchocerca* species or other filariae by a range of immunological methods including ELISA, immunoblotting, immuno-precipitation, and the reactivities of monoclonal or monospecific anti-bodies (Weiss 1985, Maizels *et al.* 1985, Cabrera & Parkhouse 1987, Y-M. Kuo pers. comm.). It would therefore be particularly damaging to the usefulness of an immunodiagnostic test if this was incapable of distinguishing between exposure to animal parasites and to infection with *O. volvulus*. Nevertheless, much effort has been put into the development of immunodiagnostic reagents based on antigens from animal *Onchocerca* species, other genera of filariae, and even distantly related nematodes such as *Ascaris*. In the past, clearly this approach has been prompted by the lack of *O. volvulus* material, but with the current availability of parasite gene libraries (Donelson *et al.* 1988) it has largely become an avoidable handicap.

A large number of techniques, ranging from skin tests to 2-site IRMA (immunoradiometric assay) and RAST (radioallergosorption test) have been described and tested in onchocerciasis without any of them gaining widespread acceptance (reviewed by Mackenzie *et al.* 1986). The common denominator is that most of the methods have multiple problems, of sensitivity, specificity, reproducibility, and ease of application to the field. Assays can be broadly divided into those based on antibody or antigen detection, the former being inherently less reliable as indicators of active infection and the latter commonly suffering from sensitivity problems. With the advent of hybridoma technology antigen detection received a good deal of attention, but the majority of monoclonals initially used in this work turned out to be directed against the ubiquitous carbohydrate moiety, phosphorylcholine (reviewed by Weiss 1985). Because filarial nematodes discharge large amounts of material that are rich in phosphorylcholine, these monoclonal antibodies nevertheless revealed remarkable specificity for filarial infections among a range of diseases (Lal & Ottesen 1989). If one could isolate species-specific component(s) associated with the same abundant secretory material, it might be possible to develop an antigen detection assay with the potential to discriminate between onchocerciasis and other filariases.

Among tests based on antibody detection the principle that has emerged from recent experience is that greatest specificity is exhibited by parasite antigens of relatively low molecular weight (Lobos & Weiss 1986, Cabrera & Parkhouse 1987). Based on antibody reactivity with proteins of 33 and 21 kD from *O. volvulus* adult females, Lucius *et al.* (1988a) reported good sensitivity and specificity with sera from onchocerciasis patients. These antigens are the target of an *Onchocerca*-specific monoclonal antibody which was used to identify and clone the corresponding gene from a cDNA library (Lucius *et al.* 1988b). Similar, but non-identical genes were found in *B. malayi* and *Dirofilaria immitis*, which may or may not be of concern when attention is given to the performance of the recombinant antigen in immunodiagnostic assays.

Differential screening of an *O. volvulus* cDNA library with pooled sera from cases of lymphatic filariasis (from Papua New Guinea) or onchocerciasis (from Mali) was used by Bradley and colleagues to isolate a number of clones that encode *Onchocerca*-specific recombinant antigens (Jan Bradley pers. comm. 1989). Rapid re-screening of the clones with panels of individual sera led to the elimination of further cross-reactive antigens (which erroneously appeared specific when evaluated with pooled sera) and to the identification of *Onchocerca*-specific clones encoding immunodominant epitopes. These experiments revealed that few of the antigens were recognized by all of the sera, so in order to achieve 100% sensitivity it will be necessary to create cocktails of the recombinants. A cautionary note is that a high proportion of the clones was shown to be cross-reactive with bovine *Onchocerca* species, a factor that may need to be taken into account during the construction of combined antigen preparations.

The Onchocerciasis Control Programme (OCP) in West Africa

The history of control of onchocerciasis in Africa boasts some remarkable achievements, not least that enjoyed by McMahon and his team who eradicated the infection in Kenya (McMahon *et al.* 1958). But far greater was the challenge posed by onchocerciasis in the massive, virtually contiguous foci of the West, which were those targeted by the OCP when they began vector control operations in 1974.

It was the recognition of a distinction between forest and savanna forms of *O. volvulus* that governed the design of OCP strategy to selectively attack savanna onchocerciasis. First, the savanna parasite was demonstrably more pathogenic in man (see Table 6.3), particularly at the crucial, socio-economical level of being a blinding infection. Secondly, once consideration was given separately to the epidemiology of savanna and forest habitats (as discussed above) it became possible to set a threshold value for the annual transmission potential below which the incidence of blindness would be negligible. Thirdly, and of much importance was the belief that the blackflies responsible for transmission in the forest would be relatively inefficient vectors of the savanna parasite if these invaded from outside the Programme area. As it transpired, blackfly reinvasion proved to be a major concern, but primarily because of the phenomenal migrations on wind currents that it was discovered savanna cytospecies could endure (Cheke & Garms 1983).

At the outset of the Programme there were seven participating countries, but this has expanded to 11 with extension of the boundaries to the south and west of the those originally established (see Figure 6.1). Part of the expansion resulted from a requirement to deal with the sources of breeding of immigrant flies, while more latterly this has come about through redefining the brief of OCP objectives (WHO 1987). The Programme area now incorporates all countries west of Nigeria with foci of the savanna form of onchocerciasis and includes Benin, Burkina Faso, Ivory Coast, Ghana, Guinea, Guinea Bissau, Mali, Niger, Senegal, Sierra Leone and Togo. In the absence of a suitable drug or vaccine for treatment of the human population, the entire plan hinged on the use of larvicide to bring about control of the *Simulium* vectors. In a phased operation, successive sectors of the Programme area were brought under active control by weekly applications of insecticide delivered by a fleet of eight helicopters and one fixed-wing aircraft. Until 1980, exclusive use was made of the organophosphate, temephos, but with the appearance of resistance chlorphoxim was tried (another organophosphate) which in turn became ineffective within a year. Through a massive effort this was replaced with the bacterial agent *Bacillus thuringiensis* H-14 and after a relatively short period blackfly populations reverted to susceptibility to the organophosphates. In order to protect the huge success that has been achieved in controlling the breeding of savanna cytospecies, it has been necessary to establish a network of teams to monitor insecticide resistance as an ongoing activity (WHO 1987).

After 11–15 years of operations, varying according to the year spraying was implemented in a given sector, the annual biting rates and annual transmission potentials have been drastically reduced (Walsh *et al.* 1987, Philippon *et al.* 1989). Transmission has virtually stopped in 80% of the treated areas and in spite of a reduction in non-target aquatic invertebrates the Programme reports there is no evidence of lasting ecological damage. An independent advisory body, the Ecological Group, have from time to time sought to restrict the use of certain insecticides after monitoring ecological impact. Monitoring the effects of control on the incidence of onchocercal infections has shown a reduction to almost nil of new cases within the central controlled area (one case in 5886 children examined) (Ba *et al.* 1987). Elsewhere, where there are problems of reinvasion by migrant blackflies the reduction in incidence is not so dramatic, but is still large at around 80% (calculated from the original data). The effects on ocular pathology have been equally impressive, with a greatly diminished incidence of serious eye lesions and of blindness during the first 7–8 years after control (Dadzie *et al.* 1986).

One of the big concerns now faced by the Programme is to consolidate this success. Inevitably, flies will re-establish breeding within the OCP area soon after the application of insecticide is stopped. Funding is due to end in 1997 when it was anticipated in the original plans that the parasite population within the region would have been eliminated through natural attrition. However, with recent extensions to the Programme not all areas will have been under control for more than 12 years (the estimated longevity of *O. volvulus* in man), and there is always the risk that infection will come back through the movements of people. Reliable, early diagnosis of infection (? immunodiagnosis) will be crucial to monitoring for recrudescences of transmission (Weiss & Karam 1987), and drugs like ivermectin (discussed below) or preferably adulticides are likely to become vital in the very near future. Clearly, there is also a requirement for reliable and predictive mathematical models for onchocerciasis, and we are now beginning to see the emergence of a new generation of computer simulations such as ONCHOSIM which has been developed in association with the OCP (Plaisier *et al.* 1989).

Experimental investigations with laboratory models

For many years there have been no systems in which to study onchocerciasis other than man, a situation that has the beneficial effect of stimulating many scientists to seek insights into the disease directly from human infections. Regrettably, there has also been a penalty for placing so much reliance on clinical research, which is that many questions concerning the host–parasite relationship and approaches to treatment have not been addressed. Until the advent of work with animal models even the most fundamental aspects of parasite development were unknown from the moment that larvae enter the definitive host. The study of ocular pathogenesis, dissection of the host immune response and pre-clinical trials of drugs, are all research areas to have been opened up by the adoption of laboratory models.

The nature of the problems facing development of any form of model for onchocerciasis has meant that most of those now available suffer major drawbacks, reflect only part of the disease process, and may be expensive. A relatively short time has elapsed since investigators took up experimental work with animal models, but it is already clear that through this approach much of value has emerged.

Table 6.5 Attempts to transmit *O. volvulus* to animal hosts.

Host species	Number examined	Source of parasite	Susceptibility	Reference
Primates				
Pan troglodytes	19	Cameroon forest	++	Duke 1980
(chimpanzee)	6	Cameroon savanna	+	Duke 1980
	5	Guatemala	++	Duke 1980
	18	Liberia	++	Greene 1987
Papio doguera	?	Uganda	−	Nelson 1965
Papio anubis	?	Tanzania	−	Wegesa & Lelijveld 1971
Papio anubis	2	Zaire	−	van den Berghe 1941
Papio sp.	?	Cameroon	−	Duke 1982 (pers. comm.)
Mandrillus leocophaeus	7	Cameroon	−	Duke 1962
Mandrillus mandrillus	?	Cameroon	−	Duke 1982 (pers. comm.)
Cercopithecus aethiops	?	Uganda	−	Nelson 1965
Cercopithecus aethiops	?	Tanzania	−	Wegesa & Lelijveld 1971
Cercopithecus preussi	1	Cameroon	−	Duke 1962
Cercopithecus mittis	?	Uganda	−	Nelson 1965
Cercopithecus mittis	?	Tanzania	−	Wegesa & Lelijveld 1971
Cerobus torquatus	1	Cameroon	−	Duke 1962
Cebus albifons	1	Guatemala	−	Kozek & Figueroa 1982
Macaca radiata	4	Guatemala	−	Kozek & Figueroa 1982
Macaca mulatta	4	Guatemala	−	Kozek & Figueroa 1982
Ateles g. villerosus	6	Guatemala	−	Kozek & Figueroa 1982
Ateles paniscus	6	Guatemala	−	Kozek & Figueroa 1982
Aotus trivigatus	?	Tanzania	−	Wegesa & Lelijveld 1971
Galago crasicaudatus	?	Uganda	−	Nelson 1965
Galago crassicaudatus	?	Tanzania	−	Wegesa & Lelijveld 1971
Galago senegalensis	4	Guatemala	−	Kozek & Figueroa 1982
Rodents				
Cavia porcellus	?	Uganda	−	Nelson 1965
Mesocricetus auratus	4	Cameroon	−	Suswillo *et al.* 1977
Mesocricetus auratus	?	Uganda	−	Nelson 1965
Meriones unguiculatus	16	Cameroon	−	Suswillo *et al.* 1977
Meriones unguiculatus	12	Guatemala	−	Kozek & Figueroa 1982
Mus musculus (Swiss)	20	Guatemala	−	Kozek & Figueroa 1982
Mastomy natalensis	13	Guatemala	−	Kozek & Figueroa 1982
Other				
Oryctolagus cuniculus	?	Uganda	−	Nelson 1965
Didelphis marsupialis	2	Guatemala	−	Kozek & Figueroa 1982
Potos flavus	1	Guatemala	−	Kozek & Figueroa 1982
Bos bovis	1	Guatemala	−	Kozek & Figueroa 1982
Capra hircus	1	Cameroon	−	Duke 1962
Sus domestica	?	Cameroon	−	Duke 1982 (pers. comm.)

? = Information not given.

Animal infections with O. volvulus

Naturally, the most desirable model for human onchocerciasis would be the establishment of *O. volvulus* infections in a convenient laboratory animal. Unfortunately, this objective has met with limited success, the chimpanzee proving the only animal so far tested to have developed patent infections (Table 6.5). The use of chimpanzees has provided a unique insight into the course of *O. volvulus* infections (discussed under earlier headings), but on grounds of conservation it is hard to envisage their future role in more than a handful of studies. Competing demands for their allocation to hepatitis or HIV research may well steepen the decline in availability of chimpanzees for onchocerciasis projects.

The search for an experimental host for *O. volvulus* has been a protracted and rather haphazard process that has resulted in the testing for susceptibility of an ad hoc collection of animals (see Table 6.5). In some reports information is scanty on the size of inoculum with infective larvae and on the criteria used to establish whether an infection subsequently developed. Understandably, several investigators concentrated their efforts on primates, but it may be that more ungulates should have been evaluated in view of the predominance of this group of animals as hosts for *Onchocerca* spp. (Muller 1979). While ungulates still leave much to be desired as laboratory hosts, this objection does not apply to rodents that also have been given very little attention. Among the five rodent species that have been infected with *O. volvulus*, there is no published record of the systematic evaluation of the susceptibility of inbred mice. Strains are available that cover diverse genetic backgrounds and immunological defects as profound as the Nu^+Nu^+ (athymic) and scid (severe complete immuno-deficiency) genotypes.

In mice, it may not be appropriate to judge susceptibility on the production of patent infections, but instead to examine survival and growth of larvae at earlier stages of development. The 1–2 year prepatent period approximates to the median life-expectancy of laboratory mice, and the large size of adult female worms would probably restrict the permissible worm burden. It is therefore unlikely that mice could be used for maintenance or passage of infection, yet a strain that was able to support parasite development would still be very useful for a range of studies. Indeed, it has been established that infective larvae within micropore chambers will moult to the fourth stage and survive for several weeks following implantation into CBA HT6T6 mice (Bianco *et al.* 1989a). Unfortunately, fourth stage larvae within chambers subsequently failed to proceed through normal intermoult growth, a problem that has also been encountered with other species, even following implantation into their natural hosts (Abraham *et al.* 1988, Bianco *et al.* 1989a). There is still a place, therefore, for the evaluation of host strain susceptibility by means of systemic administrations of infective larvae.

To circumvent the problem of the long prepatent period in oncho-cerciasis, attempts have been made to transplant *O. volvulus* adult worms into animals either in nodules or after isolation by collagenase digestion (Schulz-Key *et al.* 1977). In chimpanzees, very few microfilariae have appeared in the skin following such transplants and adult worms begin to degenerate after only three weeks (van den Ende *et al.* 1981, Prince *et al.* 1985). In rodents the survival of adult worms was monitored for just 3–5 days, which was inadequate to assess the viability of transplanted parasites or to monitor the accumulation of microfilariae in the skin (Schulz-Key *et al.* 1977). However, when microfilariae are isolated from nodules and these are inoculated into mice they migrate to the skin and subsequently live for up to 12 weeks following injection (Aoki *et al.* 1980). This approach was used initially to infect rodents with the microfilariae of *O. gutturosa* from cattle (Nelson *et al.* 1966), and while very different from an authentic infection with third stage larvae it offers considerable potential as a tool for research on the skin-dwelling microfilariae (discussed below).

It should be recognized that any model of onchocerciasis employing *O. volvulus* itself is currently restricted by availability of the parasites from man. Cyclical maintenance of the life-cycle cannot yet be considered, not only because of the lack of a convenient laboratory host, but also because of the failure to colonize the vectors (Raybould 1981). In the following section we shall consider the merits of *Onchocerca* spp from animals as analogues of the human parasite that may be useful in the construction of a practical model for onchocerciasis.

Onchocerca *infections of animals as models of human onchocerciasis*

There are approximately 30 known species in the genus *Onchocerca* which with very few exceptions occur in ungulate hosts (Muller 1979). Use of these parasites as tools for research in human onchocerciasis has largely been confined to four species, *O. gutturosa*, *O. gibsoni*, *O. cervicalis* and *O. lienalis*. Only *O. lienalis* has been transmitted to laboratory-bred animals (Bianco & Muller 1982), and even in this case patent infections have developed exclusively in the natural, bovine hosts (Townson *et al.* 1981a). As with *O. volvulus*, infective larvae will moult to the fourth stage in diffusion chambers implanted into rodents (Bianco *et al.* 1989a), but growth is retarded. This limits the scope of investigations concerned with developing larvae. Adult worms of various species survive for months following transplantation into mice, but satisfactory results are generally confined to the males that do not require extraction with enzymes to free them from the donor tissues (Townson *et al.* 1981b, El Sinnary *et al.* 1987, Vankan *et al.* 1988). Similarly, microfilariae of *Onchocerca* spp. will survive for days or weeks in mice

(Nelson *et al.* 1966, Beveridge *et al.* 1980a, Townson & Bianco 1982a) and this has been a fertile area for study of immunity, pathology and chemotherapy.

Animals models for studies of parasite biology

At the present time, the parasite of choice for fundamental studies on the *Onchocerca* life-cycle is *O. lienalis* since this species alone may be obtained in all life-cycle forms. The parasite is relatively common and has a widespread distribution (Muller 1979, Engelkirk *et al.* 1982), although it is unclear from existing reports as to whether it occurs in Africa. Microfilariae are readily available from the skin of naturally infected cattle at slaughter (Bianco *et al.* 1980) and, importantly, may be collected without damaging the hide since they accumulate at the umbilicus (Eichler & Nelson 1971). Because blackflies are the natural vectors of *O. lienalis* (Steward 1937), there is also a useful parallel with *O. volvulus* that may be exploited in studies on vector–parasite relationships. Techniques have been developed to streamline the production of third stage larvae (Bianco *et al.* 1989b) and although these are labour intensive, they offer the most practical solution to the requirement for a routine supply of infective stages. One of the biggest drawbacks is that *S. ornatum s.1.* (a major natural vector in Europe) has not been colonized as yet, although some simuliids that serve as less efficient intermediate hosts have been maintained in the laboratory through several generations (Brenner *et al.* 1980, Tarrant *et al.* 1983, Ham & Bianco 1984).

Investigations concerned with the biology of *Onchocerca* in the definitive host necessitate recourse to cattle, and are therefore dependent on availability of the appropriate facilities. This, together with cost, makes the work especially demanding, but it does give far greater freedom in the conduct of experiments than, say, equivalent studies with *O. volvulus* in chimpanzees. For example, the investigation of *O. lienalis* development in calves described in the section dealing with the *Onchocerca* life-cycle would not have been possible in chimpanzees where investigations involving necropsy are considered unacceptable. Moreover, research with primates is always restricted by the numbers of animals available, whereas there is unlimited access to cattle and a longstanding tradition of their use in veterinary research.

Adult worms of *O. lienalis* are to be found in connective tissues of the gastro-splenic omentum (Beveridge *et al.* 1979) and, as with microfilariae, may be recovered from cattle at abattoirs without necessitating mainten-ance of the life-cycle. Thus, even if it is impractical to work with experimental cattle, it is entirely feasible to have supplies of each of the key stages of microfilariae, infective larvae, and adult worms. However, if a primary objective of the research is concerned with adult worm biology, then *O. lienalis* cannot be considered a good choice. The mature females enmeshed in the tissues are virtually impossible to extract alive even with collagenase digestion, and they do not occur in nodules and therefore lack one of the characteristic features of *O. volvulus*. Several *Onchocerca* species of animals do, however, induce the formation of nodules and the most widely used of these is *O. gibsoni* which occurs in cattle from Africa, Asia and tropical Australia (Engelkirk *et al*. 1982). Much of the descriptive biology of adult worms and of the nodule in *O. volvulus* infections has been repeated with *O. gibsoni* and revealed striking similarities between the species (Beveridge *et al*. 1980b, Vankan & Copeman 1988, Franz & Copeman 1988) which has made *O. gibsoni* a particularly valuable model for certain research applications, most notably for drug screening (see below).

One of the relatively common parasites of horses is *O. cervicalis*, and from time to time this has been tipped as a contender to be developed as a model for *O. volvulus* (Jones & Collins 1979). In its favour is that it is transmitted by *Culicoides* midges that have been successfully colonized (*C. nubeculosus* in Europe, *C. variipennis* in North America) and there is evidence that the microfilariae are pathogenic, since ocular lesions similar to those in man have been reported (Lagraulet 1962, Cello 1971, Schmidt *et al*. 1982). Nevertheless, as of the present there have been no reports of large-scale production of infective larvae, or of the transmission to horses of infections in the laboratory.

Irrespective of the species of *Onchocerca* concerned, all pose problems for experimental studies when reliance is placed on natural infections as a source of material. Here, a major advance has been the development of cryopreservation techniques, which permit storage of live parasites in such a way as to free the investigator to plan experimental work with confidence. Procedures have been devised to cryopreserve microfilariae in skin or after extraction (Ham & Bianco 1981, Ham *et al*. 1981), to freeze infective larvae (Lowrie 1983) and even to cryopreserve the adult male and female worms (Townson 1988, Townson *et al*. 1989). One immediate benefit of the work with *Onchocerca* parasites from animals is that the methodologies of cryopreservation developed with these species has found direct application to *O. volvulus* (El Sheikh & Ham 1982, Ham & Bianco 1983b, Townson *et al*. 1989).

Animal models for studies of ocular pathology

One of the critical areas of research in onchocerciasis is the pathogenesis of ocular pathology. However, a major limitation to progress is that human material for histopathological investigations is generally available only from the most advanced cases of ocular pathology. It is not the end point of the disease process but the beginning that will shed light on the underlying mechanisms, so there is an urgent requirement for an animal model that can serve to mimic the salient features of ocular disease. In the pursuit of such a model the first step was to examine *O. volvulus*-infected chimpanzees, but since these showed no evidence of ocular pathology (Duke 1980) attention was shifted to rabbits, guinea pigs, monkeys and, recently, to mice. As mentioned earlier, horses naturally infected with *O. cervicalis* do develop ocular lesions, but without laboratory passage of infections this is not a system that is readily amenable to experimentation.

Studies on the pathogenesis and pathology of rabbit eyes after injection of *O. volvulus* microfilariae into the subconjunctiva and cornea have yielded valuable information about the development of anterior segment lesions. In contrast with man, rabbits rapidly develop corneal lesions in response to microfilariae, but provided there has been no previous exposure to parasite antigens these soon resolve and cause little or no impairment of vision (Duke & Anderson 1972). In rabbits pre-sensitized with living microfilariae, such reactions were more severe and persisted to cause a sclerozing keratitis resembling that occurring in man (Duke & Garner 1975). At a histological level it was established that injection of microfilariae into the bulbar conjunctiva led to severe local inflammation, infiltration of neutrophils and invasion of the corneal stroma by a proportion of the organisms (Garner *et al.* 1973). In possibly the most important experiment to be performed with the rabbit model, it was demonstrated that parasites originating from a savanna focus of onchocerciasis in Cameroon were intrinsically more pathogenic than microfilariae originating from a region in the rain forest (Duke & Anderson 1972, Garner *et al.* 1973). Savanna parasites administered under identical conditions to those from the forest provoked a substantially more damaging corneal reaction, providing strong evidence to support the contention that infraspecific variation exists in *O. volvulus* that contributes to the epidemiology of ocular onchocerciasis (discussed above).

Studies on the pathogenesis of posterior segment lesions in rabbits due to *O. volvulus* microfilariae have demonstrated that direct inoculation of parasites under the retina induces changes that again mimic human disease (Duke & Garner 1976, Garner & Duke 1976). Both atrophic and inflammatory processes occur in the fundus, with mild to moderate uveitis and choroiditis predominating at first, giving way to degenerative changes in the neuroretina and pigment epithelium after some weeks. Optic nerve involvement, which is commonly associated with blindness in cases of posterior segment disease, also arose in the experimental animals manifested by optic disc pallor, atrophic vessels and a mild cellular infiltration.

Successful as they were, these experiments in rabbits were dependent on a good supply of *O. volvulus* microfilariae, which are not available to many investigators outside endemic areas of onchocerciasis. Accordingly, attempts were made to employ *O. lienalis* or *O. cervicalis*, and the initial trials were performed both on rabbits and guinea pigs. Using *O. lienalis*, only 1 of 8 rabbits and 3 of 8 guinea pigs showed appreciable corneal reaction following inoculation of 5000 microfilariae into the perilimbal conjunctiva (Bianco & Garner 1980, Garner 1980). With *O. cervicalis*, each of 6 guinea pigs developed severe anterior segment inflammation, although when the experiment was repeated with cryopreserved rather than fresh parasites this was reduced to 3 of 8 animals (Garner 1980).

Despite the discouraging preliminary results with *O. lienalis* in guinea pigs, this model was adopted by Donnelly and colleagues and used extensively to study the pathogenesis of anterior segment lesions. Using inbred, Strain 2 guinea pigs pre-sensitized by subcutaneous administrations of microfilariae, intense corneal and uveal inflammation was provoked in response to an intracorneal challenge (Donnelly *et al.* 1983, 1984). IgE antibody was detected in the serum and aqueous humour, and levels of the reaginic response and of corneal inflammation increased when diethylcarbamazine was administered. The transfer of spleen cells from immunized donors to naive congenic recipients predisposed the animals to acute inflammation of the cornea following microfilarial challenge. Donnelly *et al.* (1988a) recently reported the detection of autoantibodies to corneal components in immunized guinea pigs, but these were of the IgG class and did not correlate with the appearance of acute corneal reactions. Recently, Gallin *et al.* (1988) have described an attempt to identify parasite antigens that give rise to ocular inflammation, using the approach of inoculating *O. volvulus* material fractionated by gel filtration into the corneas of guinea pigs.

Work on the development of a mouse model for ocular pathology is at an earlier stage but ultimately might offer advantages over the rabbit and guinea pig systems. However, because the orbit is small, and the curvature of the cornea is great, specialized equipment is required to make clinical inspections of the eye lesions in mice. A useful advance made during development of this model has been the avoidance of direct physical trauma to the eye during inoculation of microfilariae. Using *O. cervicalis*, James *et al.* (1986) have obtained rapid invasion of the anterior chamber and corneal stroma by injection of microfilariae over the scalp rather than into the perilimbal conjunctiva. They also report invasion of the retina following injection of microfilariae into the tail vein, a finding of technical value but also one that fits with the hypothesis that the bloodstream is the route into the posterior segment in human onchocerciasis (Fuglsang & Anderson 1974).

Against the trend of working towards a small animal model of ocular pathology are recent reports describing the pathogenesis of experimentally induced onchocercal chorioretinitis and anterior segment lesions in cynomolgus monkeys (Semba *et al.* 1988, Donnelly *et al.* 1986, 1988b). Intravitreal inoculations of *O. lienalis* microfilariae provoked a wide range of pathological changes, most significant of which were retinal pigment epithelial hypertrophy, hyperplasia and loss of pigment (Semba *et al.* 1988). Interestingly, in contrast with the pathogenesis of anterior segment lesions, prior sensitization with live microfilariae did not exacerbate the retinal disturbances, and pathology distinctly preceded antibody or cellular responses to microfilariae (Donnelly et al. 1988b). Clearly, animal models have a major role to play in this field and are revealing critical features of the complex aetiology of ocular pathology in onchocerciasis.

Animal models for studies of chemotherapy

There has been no higher priority in recent research on onchocerciasis than to develop a safe and effective drug to replace diethylcarbamazine and suramin (WHO 1972). This is required not only for the management of individual cases of infection, but also as a tool to limit reintroduction of the parasite into OCP areas in which transmission has been interrupted by vector control. The primary objective was and remains to obtain a new macrofilaricide, but as it has worked out the major advance has been the discovery of the potent microfilaricidal action of ivermectin (Aziz et al. 1982). In an unprecedented gesture Merck, Sharp & Dohme, who have made huge profits from ivermectin in the veterinary market, agreed to provide the drug free for use in the treatment of human onchocerciasis. Distribution of the drug is regulated by the 'Mectizan committee', so-called because of the trade-name under which the formulation designed for humans is registered. A detailed account of the clinical trials and uses of ivermectin in onchocerciasis has recently been provided by Taylor & Greene (1989).

At the outset of the WHO/TDR programme all drug screening was performed with animal models of filariasis, and only recently have systems that utilize Onchocerca parasites come to figure significantly in the early phases of screening compounds. The first of the specific Onchocerca models was O. gibsoni in cattle, and this was adopted with great success as a tertiary screen for drugs (Copeman 1979). Naturally infected animals were used in the trials, and drug activity against adult worms was evaluated from histological sections of nodules. Highly encouraging similarities were found between the responses of O. gibsoni and O. volvulus to drugs, so that much confidence has been placed on the predictive value of the model, which has consequently retained its original role in the screening process. However, one of the obvious disadvantages is the large animal host, not only because of the cost and specialized requirements, but also because of the demands it places on limited supplies of certain test compounds. To meet these objections there have been attempts to create a small animal model by transplant of adult worms of O. gibsoni subcutaneously into mice (Vankan et al. 1988). Only male worms were used in the experiments so far reported, since these could be obtained readily by simple dissection of nodules. The performance of the model against a panel of reference compounds demonstrated several similarities, but also some differences, in drug activity against transplanted adults versus natural infections (Vankan *et al.* 1988). One of the differences, the negligible activity of suramin in mice, may have arisen because of an inadequate interval between drug administration and the recovery of worms (i.e. three weeks). However, even after this comparatively short period there were some parasite losses from controls, either through death (23%) or failure to locate (27%) the organisms during post mortem examinations. Variation among animals in the intensity of encapsulation reactions around worms was observed independently of macrofilaricidal action, and could seriously complicate the evaluation of drugs.

In a modification of this approach taken by Townson and colleagues, adult male worms of *O. gutturosa* have been implanted into mice within micropore chambers (S. Townson, pers. comm., 1989). Using chambers sealed with membranes with a pore size of 5 μm, recoveries of all worms has been achieved for at least seven weeks, making possible the screening of slow-acting compounds such as suramin. This system ensures ease of recovery of parasites and adds to the reliability of locating worms that have died from drug treatments. The condition of worms may then be quantitatively assessed, using one or more of the plethora of assays that have recently been described (reviewed by Comley *et al.* 1988).

Another model for testing drugs, in this instance against *Onchocerca* microfilariae, was described by Bianco *et al.* (1986) using *O. lienalis* in CBA mice. Microfilariae from naturally infected cattle are inoculated subcutaneously into mice and migrate to the skin where they can be quantified in a representative manner by the numbers in the ears (Townson & Bianco 1982a). Treatment of mice with a battery of reference compounds clearly identified all those with known microfilaricidal action, together with some that are considered to be principally macrofilaricides. Similar results were obtained by Townson *et al.* (1988) using an expanded range of the established filaricidal drugs in this model. Interestingly, diethylcarbamazine was relatively poor at reducing the levels of *O. lienalis* microfilariae in mice, whereas ivermectin was by far the most effective compound, killing virtually all microfilariae at a single dose of 50 μg/kg body weight (Bianco *et al.* 1986). It has been proposed that certain drugs may act in synergy with host immune factors, and using the mouse model it was possible to demonstrate an interplay between diethylcarbamazine and the thymus-dependent immune response (Bianco *et al.* 1986). Modes of drug action are difficult to study unless there is a well characterized host, and here the mouse model has a number of obvious advantages.

One of the ways investigators have tackled the development of primary drug screens for other nematode parasites has been to generate *in vitro* systems so as to minimize the costs, increase simplicity and achieve a higher throughput of compounds (Jenkins *et al.* 1980). In onchocerciasis, this approach has achieved a good measure of success, resulting in the founding of screens based on adult males of *O. gutturosa, O. gibsoni*, and *O. volvulus* (Townson *et al*. 1987, 1989, Nowak *et al*. 1987, Comley *et al*. 1989). To a lesser extent attempts have been made to test compounds against adult female worms, but the relative difficulties of recovery of females in a suitable condition for these assays has limited progress (Townson *et al*. 1989). Court *et al*. (1985) described an *in vitro* system for evaluating prospective prophylactic compounds against *O. lienalis* third stage larvae, and although the approach was subsequently used to test ivermectin against developing stages (Lok *et al*. 1987) it does not appear to have been more generally adopted as a method of screening.

Animal models for studies of protective immunity

VACCINATION AGAINST DEVELOPING LARVAE
Studies on immunity engendered against the developing stages of *Onchocerca* have not been described. This is in large part because of the difficulties of transmitting *Onchocerca* infections in the laboratory. However, another constraint lies in evaluating the expression of host resistance, which is conventionally measured by making comparisons between worm recoveries from immunized or control groups of animals. Based on over fifty post mortems of calves infected with known numbers of *O. lienalis* infective larvae (600–1000 larvae per animal), it proved impossible to quantify worm recoveries with any confidence of accuracy (unpublished observations). An alternative strategy for evaluating protection might be to measure the frequency of patent infections, but this criterion is less informative than the quantification of worm recoveries because patent infections may still result if just a few worms were to evade functional immunity. However, provided that groups of sufficent size were used this approach has the merits of being simple and practical, with a bias towards demonstrating sterilizing immunity (the ideal objective of vaccination). Quantifying levels of microfilariae as an index of adult worm burdens would be of questionable value because of the lack of experimental evidence of a correlation between these parameters.

A second approach that is receiving consideration at present is the development of a mouse model for studies on resistance to developing larvae. This is being built on the success of implantations of third stage larvae in micropore chambers (Bianco *et al*. 1989a) and although far from ideal, it does provide a means to study protective immunity within existing limitations. Details of parasite development in chambers has been discussed under previous headings, but the quantitative aspects of this model are equally important for its application to vaccination research. In preliminary experiments in non-immune mice a relatively constant proportion of larvae survived over the first 24 days after implantation (33–58%). Thereafter it was observed that recoveries declined, although some worms were still alive after 96 days (Bianco *et al.* 1989a). It is still too early to say what contribution the model can make, but in the absence of alternatives it is a lead that is important to follow.

Experiments in mice, paralleled with studies in calves, have shown that it is possible to induce high levels of resistance to *O. lienalis* microfilariae. Inbred CBA/H mice sensitized with a primary inoculation of microfilariae exhibit levels of protection exceeding 90% against parasites of a secondary 'infection' (Townson & Bianco 1982a, Townson *et al.* 1984, Carlow *et al.* 1986, 1988). The resistance induced against challenge by a primary infection is not dependent on the route of sensitization and is apparently systemic (Carlow *et al.* 1986). Primary infections of various durations that have been chemically abbreviated with ivermectin confer significant levels of protection (84 %) when of only seven days. As few as 20 live microfilariae administered as the sensitizing inoculum may elicit almost the same level of resistance (80%) as inocula containing 50–10 000 organisms (88–97%). Similar levels of protection have been demonstrated in mice that received single or multiply divided doses of microfilariae as the challenge infection (Carlow *et al.* 1986).

Attempts to immunize mice with non-living microfilariae or parasite extracts have also been successful, but levels of protection have generally been lower than those stimulated by the living parasites (Townson *et al.* 1984, Carlow & Bianco 1987a). There is evidence that the clearance of microfilariae is mediated by a thymus-dependent immune response, although there may be a contribution to protection through T-cell independent mechanisms. The survival of parasites is significantly prolonged in T-cell deprived mice, and these animals no longer retain the ability to express resistance to a secondary infection (Townson & Bianco 1982a, Carlow *et al.* 1988). In comparison, mice of the B-cell deficient strain, CBA/N, exhibit an intermediate degree of protection against parasite challenge (Carlow *et al.* 1988). Resistance can be adoptively transferred to T-cell deprived recipients from immune donors, and requires both T and B cell populations in order to confer significant levels of protection (Carlow & Bianco 1987b). Serum from immune donors may also passively transfer protection, but not efficiently and only when collected within a short period after sensitization (Carlow & Bianco 1987b). Depletion of C3 complement levels in mice by administration of cobra venom factor had no effect on the expression of resistance to microfilariae in immunized hosts (Carlow *et al.* 1988).

Protection against *O. lienalis* microfilariae in mice has been induced by various life-cycle stages (infective larvae, adult males, egg antigens) and heterologous species (*O. gibsoni, O. gutturosa, O. cervicalis* and *O. volvulus*) (Townson *et al.* 1985, Carlow & Bianco 1987a). Extracts of eggs were used in the experiments into cross-protection among species, and the apparent sharing of host protective antigens, or epitopes, is a highly encouraging finding. In cattle, which are the natural hosts of *O. lienalis*, immunization with parasite extracts has also elicited immunity against microfilarial challenge (Townson & Bianco 1982b, Bianco *et al.* in preparation). The protective effect could be conferred by passive transfer of serum, and killing of microfilariae in diffusion chambers was associated with macrophage and granulocyte adherence (Bianco *et al.* in preparation). However, immunized cattle challenged with infective stage larvae developed patent infections, indicating a breakdown of the resistance manifest against challenge with microfilarial injections. Lymphocyte blastogenesis assays conducted with peripheral blood leucocytes from these animals indicated that this may have been related to immuno-suppression expressed at around the onset of patency (Bianco & Lloyd unpublished observations). Because immunosuppression has also been reported in cases of human onchocerciasis (as discussed above), it may be a general problem in any strategy aimed at reducing parasite burdens through vaccination.

The chief significance of these studies on microfilarial immunity is that they have shown for the first time that it is possible to induce protection against any stage of an *Onchocerca* parasite. However, the desirability of provoking immunity against *Onchocerca* microfilariae is controversial, because in onchocerciasis immunopathology is associated with dying or dead microfilariae. On the other hand, it remains to be established whether the same immunological pathways are responsible for immunity and pathogenesis, or whether distinct mechanisms operate to produce these phenomena. Studies are now beginning to highlight the importance of particular antibody isotypes in onchocerciasis (Cabrera *et al.* 1986, Parkhouse *et al.* 1987), but work is only just starting on identifying and characterizing the targets of protective and/or pathological responses (e.g. Cabrera *et al.* 1988). In recent months an approach that has shown promise has been to clone candidate protective antigens from *O. volvulus* cDNA libraries, using as the basis for selection their restricted reactivity with antibodies from cattle immune to microfilariae and not with antibodies from animals harbouring patent infections (Kuo & Bianco, unpublished results). Testing of such recombinant antigens for their ability to induce protection, while monitoring for any associated immunopathology, should be one of the objectives of studies with animal models, since it is now possible to analyse both anti-microfilarial immunity and ocular pathology using inbred mouse models (as discussed above).

Natural infections of rodents with potential as models for onchocerciasis

A small number of parasites belonging to the genera *Monanema* and *Cercopithifilaria* have received attention in recent years as possible models for onchocerciasis. The attraction of these species is that they have skin-dwelling microfilariae and occur naturally in rodents, offering the promise that they may be amenable to maintenance in laboratory rats or mice. However, all of these filariae have ixodid tick vectors which makes the production of infective larvae somewhat laborious, although not so great as for *Onchocerca* spp.

The first of these parasites to have been maintained in the laboratory was *Monanema globulosa*, originally described by Muller and Nelson (1975) from four genera of rodents in Kenya. An investigation into the biology of the parasite revealed the occurrence of microfilariae in the eyes, the development of anterior segment lesions and adverse skin reactions following drug treatment with diethylcarbamazine (Bianco *et al.* 1983, Bianco & Denham 1984). Colonies were established of the natural vector, *Haemaphysalis leachi s1.* and patent infections were produced in striped mice (*Lemniscomys striatus*) and Mongolian jirds (*Meriones unguiculatus*) (Bianco 1984). Unfortunately, no infections matured in laboratory mice and those which were produced in jirds were of low intensity and made serial passage difficult.

More recently there has been similar work on the related species *Monanema nilotica* and *M. martini*, although there is no record of attempts to transmit the parasites to conventional laboratory rodents (Bain *et al.* 1985, 1986). Using *Rhipicephalus sanguineus* or *Hyalomma truncatum* for the production of third stage larvae, patent infections have been produced in laboratory-reared Nile rats (*Arvicanthus niloticus*) and striped mice (*L. striatus*). A third species, *Cercopithifilaria johnstoni*, not only has skin-dwelling microfilariae but also adult worms that occur in subcutaneous tissues, providing a second parallel with *O. volvulus* infections (Spratt & Varughese 1975). This Australian parasite has a wide range of hosts, including *Rattus fuscipes* and *R. lutreolus*, and following experimental transmission has recently been reported to infect *R. norvegicus*, the laboratory rat (Spratt & Haycock 1988). However, much will depend on the ease with which the vector *Ixodes trichosuri* can be maintained, and how readily it will be possible to produce high microfilarial densities in experimental animals.

Acknowledgements

I have benefited in my understanding of onchocerciasis from a great many people, and in particular those of my colleagues who have devoted major periods of their life to work in Africa. I should especially like to extend my thanks to Professor George Nelson, not simply because this book has been produced to mark the occasion of his retirement, but also because of the limitless pleasure I have derived from research with *Onchocerca* parasites and from the discussions this has prompted with him. His contagious enthusiasm for tropical medicine is what originally brought me into this subject and showed me a way for the biologist to make a meaningful contribution to the understanding of human infection.

Several of the photographs used in this article were taken by the author in Sierra Leone while a guest of Drs Jim Whitworth and Adrian Luty to whom I give thanks. I would also like to acknowledge the agencies that have funded my research, the WHO/TDR Special Programme, Edna McConnell Clark Foundation, British Medical Research Council and Wellcome Trust.

References

Abdel-Hameed, A. A., M. S. Noah, J. F. Schacher & S. A. Taher 1987. Lymphadenitis in sowda. *Trop. Geogr. Med.* **39**, 73–6.

Abraham, D., R. B. Grieve, M. Mika-Grieve & B. P. Seibert 1988. Active and passive immunisation of mice against larval *Dirofilaria immitis*. *J. Parasit.* **74**, 275–82.

Addison, E. M. 1980. Transmission of *Dirofilaria ursi* Yamaguti, 1941 (Nematoda: Onchocercidae) of black bears (*Urus americanus*) by blackflies (Simuliidae). *Can. J. Zool.* **58**, 1913–22.

Albiez, E. J. 1985. Effects of a single, complete nodulectomy on nodule burden and microfilarial density two years later. *Trop. Med. Parasit.* **36**, 17–20.

Albiez, E. J., D. W. Buttner & B. O. L. Duke 1988. Diagnosis and extirpation of nodules in human onchocerciasis. *Trop. Med. Parasit.* **39**, 331–46.

Ali-Khan, Z. (1977). Tissue pathology and comparative micro-anatomy of *Onchocerca* from a resident of Ontario and other enzootic *Onchocerca* species from Canada and the USA. *Ann. Trop. Med. Parasit.* **71**, 469–82.

Anderson, J. & H. Fuglsang 1973. Variation in numbers of microfilariae of *Onchocerca volvulus* in the anterior chamber of the human eye. *Trans. R. Soc. Trop. Med. Hyg.* **67**, 544–8.

Anderson, J., H. Fuglsang & A. J. Al Zubaidy 1973. Onchocerciasis in Yemen with special reference to sowda. *Trans. R. Soc. Trop. Med. Hyg.* **67**, 30–1.

Anderson, J., H. Fuglsang, P. J. S. Hamilton & T. F. de C. Marshall 1974a. Studies on onchocerciasis in the United Cameroon Republic. I. Comparison of populations with and without *Onchocerca volvulus*. *Trans. R. Soc. Trop. Med. Hyg.* **68**, 190–208.

Anderson, J., H. Fuglsang, P. J. S. Hamilton & T. F. de C. Marshall 1974b. Studies on onchocerciasis in the United Cameroon Republic. II. Comparison of onchocerciasis in rain-forest and Sudan-savanna. *Trans. R. Soc. Trop. Med. Hyg.* **68**, 209–22.

Anderson, R. C. (1956). The life cycle and seasonal transmission of *Ornithofilaria fallisensis* Anderson, a parasite of domestic and wild ducks. *Can. J. Zool.* **34**, 485.

Anderson, R. I., L. E. Fazen & A. A. Buck 1975. Onchocerciasis in Guatemala. III. Daytime periodicity of microfilariae in the skin. *Am. J. Trop. Med. Hyg.* **24**, 62–5.

Aoki, Y., C. M. M. Recinos, Y. Hashiguchi 1980. Life span and distribution of *Onchocerca volvulus* microfilariae in mice. *J. Parasit.* **66**, 797–801.

Azarova, N. S., O. Y. Miretsky & M. D. Sonin 1965. The first instance of detection of nematode *Onchocerca* Diesing 1841 in a person in the USSR. *Meditzinskaya Parazitologuya i Parazitamii Bolezri* (Moscow) **34**, 156–8.

Aziz, M. A., S. Diallo, I. M. Diop, M. Lariviere *et al.* 1982. Efficacy and tolerance of ivermectin in human onchocerciasis. *Lancet* **2**, 171–3.

Ba, O., M. Karam, J. Remme & G. Zerbo 1987. Role of children in the evaluation of the Onchocerciasis Control Programme in West Africa. *Trop. Med. Parasit.* **38**, 137–42.

Bain, O. 1969. Morphology of larval stages and microfilariae of *Onchocerca volvulus* in *Simulium damnosum* and redescription of the microfilaria. *Ann. Parasit. Hum. Comp.* **44**, 69–81.

Bain, O. 1979. Transmission of *Onchocerca gutturosa* of cattle by *Culicoides*. *Ann. Parasit. Hum. Comp.* **54**, 483–88.

Bain, O. & A. G. Chabaud 1986. Atlas des larves insectantes de filaires. *Tropenmed. Parasit.* **37**, Supplement, 301–40.

Bain, O. & G. Petit 1978. Redescription du stade infestant d'*Onchocerca cervicalis*. *Ann. Parasit. Hum. Comp.* **53**, 315–18.

Bain, O., G. Petit & B. Poulain 1978. Validité des deux espèces *Onchocerca lienalis* et *O. gutturosa* chez les bovins. *Ann. Parasit. Hum. Comp.* **53**, 421–30.

Bain O., G. Petit & A. Gueye 1985. Transmission expérimentale de *Monanema nilotica* El Bihari et coll., 1977. Filaire à microfilaires dermiques parasites de Muridés africains. *Ann. Parasit. Hum. Comp.* **60**, 83–9.

Bain O., C. Bartlett & G. Petit 1986. Une filaire de Muridés africains dans la paroi du colon, *Monanema martini* n. sp. *Ann. Parasit. Hum. Comp.* **61**, 465–72.

Bartlett, A., J. Turk, J. L. Ngu, C. D. Mackenzie *et al.* 1978. Variation in delayed hypersensitivity in onchocerciasis. *Trans. R. Soc. Trop. Med. Hyg.* **72**, 372–7.

Beaver, P. C., G. S. Horner & J. Z. Bilos 1974. Zoonotic onchocerciasis in a resident of Illinois and observations on the identification of *Onchocerca* species. *Am. J. Trop. Med. Hyg.* **23**, 595–607.

Becker, C. K. 1950. Filaires adultes (*Onchocerca volvulus*) libres dans les tissues. *Ann. Soc. Belge Med. Trop.* **30**, 9–10.

Benjamini, E., B. F. Feingold & L. Kartman 1961. Skin reactivity in guinea pigs sensitised to flea bites. The sequence of reactions. *Proc. Soc. Exp. Biol. Med.* **108**, 700–2.

Beveridge, I., E. L. Kummerow & P. Wilkinson 1979. The prevalence of *Onchocerca lienalis* in the gastro-splenic ligament of cattle in North Queensland. *Aust. Vet. J.* **55**, 204–5.

Beveridge, I., E. L. Kummerow & P. Wilkinson 1980a. Experimental infection of laboratory rodents and calves with microfilariae of *Onchocerca gibsoni*. *Tropenmed. Parasit.* **31**, 82–8.

Beveridge, I., E. L. Kummerow & P. Wilkinson 1980b. Observations on *Onchocerca gibsoni* and nodule development in naturally infected cattle in Australia. *Tropenmed. Parasit.* **31**, 75–81.

Bianco, A. E. 1984. Laboratory maintenance of *Monanema globulosa*, a rodent filaria with skin-dwelling microfilariae. *Z. Parasit.* **70**, 255–64.

Bianco, A. E. & D. A. Denham 1984. The action of diethylcarbamazine on the skin-dwelling microfilariae of *Monanema globulosa* (Nematoda: Filarioidea) in rodents. *Tropenmed. Parasit.* **35**, 53–7.

Bianco, A. E. & A. Garner 1980. Ocular complications of onchocerciasis: the search for an animal model. *Trans. R. Soc. Trop. Med. Hyg.* **74**, 109.

Bianco, A. E. & R. M. Maizels 1989. Parasite development and adaptive specialisation. *Parasitology* **99** (Supplement), S113-23.

Bianco, A. E. & R. Muller 1982. Experimental transmission of *Onchocerca lienalis* to calves. In: Parasites – Their World and Ours. Proceedings of the fifth International Congress of Parasitology, August 7–14, Toronto, 1982. *Mol. Biochem. Parasit.* Supplement p. 349.

Bianco, A. E., P. J. Ham, K. El Sinnary & G. S. Nelson 1980. Large-scale recovery of *Onchocerca* microfilariae from naturally infected cattle and horses. *Trans. R. Soc. Trop. Med. Hyg.* **74**, 109–10.

Bianco, A. E., R. Muller & G. S. Nelson 1983. Biology of *Monanema globulosa*, a rodent filaria with skin-dwelling microfilariae. *J. Helm.* **57**, 259–78.

Bianco, A. E., M. Nwachukwu, S. Townson, M. J. Doenhoff *et al.* 1986. Evaluation of drugs against *Onchocerca* microfilariae in an inbred mouse model. *Trop. Med. Parasit.* **37**, 39–45.

Bianco, A. E., M. B. Mustafa & P. J. Ham 1989a. Fate of developing larvae of *Onchocerca lienalis* and *O. volvulus* in micropore chambers implanted into laboratory hosts. *J. Helm.* **63**, 218–26.

Bianco, A. E., P. J. Ham, S. Townson, M. B. Mustafa & G. S. Nelson 1989b. A semi-automated system of intrathoracic injection for the large-scale production of *Onchocerca lienalis* infective larvae. *Trop. Med. Parasit.* **40**, 57–64.

Bianco, A. E., B. D. Robertson, Y-M. Kuo, S. Townson *et al.* 1990. Developmentally-regulated expression and secretion of a polymorphic antigen by *Onchocerca* infective-stage larvae. *Mol. Biochem. Parasit.* **39**, 203–12.

Bird, A. C., J. Anderson & H. Fuglsang 1976. Morphology of the posterior segment lesions of the eye in patients with onchocerciasis. *Brit. J. Ophthalmol.* **60**, 2–20.

Blacklock, D. B. 1926. The development of *Onchocerca volvulus* in *Simulium damnosum*. *Ann. Trop. Med. Parasit.* **20**, 1–48.

Blacklock, D. B. 1927. The insect transmission of *Onchocerca volvulus* (Leuckhart, 1893). The cause of worm nodules in man in Africa. *Brit. Med. J.* **1**, 129–33.

Bradley, J. E., W. Gregory, A. E. Bianco & R. M. Maizels 1989. Biochemical and immunochemical characterisation of a 20-kilodalton complex of surface-associated antigens from adult *Onchocerca gutturosa* filarial nematodes. *Molec. Biochem. Parasit.* **34**, 197–208.

Brattig, N. W., F. W. Tischendorf, F. Reifegerste, E. J. Albiez *et al.* 1986. Differences in the distribution of HLA antigens in localized and generalized forms of onchocerciasis. *Trop. Med. Parasit.* **37**, 271–5.

Brattig, N. W., F. W. Tischendorf, E. J. Albiez, D. W. Buttner *et al.* 1987. Distribution pattern of peripheral lymphocyte subsets in localized and generalized forms of onchocerciasis. *Clin. Immunol. Immunopathol.* **44**, 149–159.

Braun-Munzinger, R. A & B. A. Southgate 1977. Preliminary studies on the histochemical differentiation of strains of *Onchocerca volvulus* microfilariae in Togo. *Bull. Wld Hlth Org.* **55**, 569–75.

Brenner, R. J., E. W. Cupp & M. J. Bernado 1980. Laboratory colonization and life table statistics for geographic strains of *Simulium decorum* (Diptera: Simuliidae). *Tropenmed. Parasit.* **31**, 487–97.

Brinkmann, U. K., P. Kramer, G. T. Presthus & B. Sawadogo 1976. Transmission *in utero* of microfilariae of *Onchocerca volvulus*. *Bull. Wld Hlth Org.* **54**, 708–9.

Buck, A. A. 1974. *Onchocerciasis. Symptomology, pathology, diagnosis.* W.H.O. publication; Geneva. 80pp.

Buttner, D. W., G. von Laer, E. Mannweiler & M. Buttner 1982. Clinical, parasitological and serological studies on onchocerciasis in the Yemen Arab Republic. *Tropenmed. Parasit.* **33**, 201–12.

Buttner, D. W., E. J. Albiez, J. von Essen & J. Erichsen 1988. Histological examination of adult *Onchocerca volvulus* and comparison with the collagenase technique. *Trop. Med. Parasit.* **39**, 390–417.

Cabrera, Z. & R. M. E. Parkhouse 1987. Isolation of an antigenic fraction for diagnosis of onchocerciasis. *Parasit. Immunol.* **9**, 39–48.

Cabrera, Z., M. D. Cooper & R. M. E. Parkhouse 1986. Differential recognition patterns of human immunoglobulin classes to antigens of *Onchocerca gibsoni*. *Trop. Med. Parasit.* **37**, 113.

Cabrera, Z., D. W. Buttner & R. M. E. Parkhouse 1988. Unique recognition of a low molecular weight *Onchocerca volvulus* antigen by IgG3 antibodies in chronic hyperreactive oncho-dermatitis (sowda). *Clin. Exp. Immunol.* **74**, 223–9.

Caprioglio, A. 1976. Parasitic pseudotumors due to filarial nematodes. *Arch. Sci. Med. Torino* **133**, 403–10.

Carlow, C. K. S & A. E. Bianco 1987a. Resistance to *Onchocerca lienalis* microfilariae in mice conferred by egg antigens of homologous and heterologous *Onchocerca* species. *Parasitology* **94**, 485–96.

Carlow, C. K. S & A. E. Bianco 1987b. Transfer of immunity to the microfilariae of *Onchocerca lienalis* in mice. *Trop. Med. Parasit.* **39**, 283–6.

Carlow, C. K. S., R. Muller & A. E. Bianco 1986. Further studies on the development of resistance to *Onchocerca* microfilariae in CBA mice. *Trop. Med. Parasit.* **37**, 276–81.

Carlow, C. K. S., E. Franke, R. C. Lowrie, R.C., F. Partono *et al.* 1987. Monoclonal antibody to a unique surface epitope of the human filaria *Brugia malayi* identifies infective larvae in mosquito vectors. *Proc. Nat. Acad. Sci. USA* **84**, 6914–18.

Carlow, C. K. S., A. R. Dobinson & A. E. Bianco 1988. Parasite specific immune responses to *Onchocerca lienalis* microfilariae in normal or immunodeficient mice. *Parasit. Immunol.* **10**, 309–22.

Cello, R. M. 1971. Ocular onchocerciasis in the horse. *Equine Vet. J.* **3**, 148–54.

Chan, C. C, R. B. Nussenblatt, M. K. Kim, A. G. Palestine *et al.* 1987. Immunopathology of ocular onchocerciasis. 2. Anti-retinal autoantibodies in serum and ocular fluids. *Ophthalmol.* **94**, 439–43.

Chan, C. C, E. A. Ottesen, K. Awadzi, R. Badu *et al.* 1989. Immunopathology of ocular onchocerciasis. 1. Inflammatory cells infiltrating the anterior segment. *Clin. Exp. Immunol.* **77**, 367–72.

Chartres, J. C. 1955. Onchocerciasis-incubation period, clinical course and treatment at first hand. *W. Afr. Med. J.* **4**, 130–5.

Cheke, R. A. & R. Garms 1983. Reinfestations of the southeastern flank of the onchocerciasis control programme area by windborne vectors. *Phil. Trans. R. Soc. London* (B) **302**, 471–83.

Cheke, R. A., R. Garms & M. Kerner 1982. The fecundity of *Simulium damnosum* in northern Togo and infections with *Onchocerca* spp. *Ann. Trop. Med. Parasit.* **76**, 561–68.

Cianchi, R., M. Karam, M. C. Henry, F. Villani *et al.* 1985. Preliminary data on the genetic differentiation of *Onchocerca volvulus* in Africa (Nematoda: Filarioidea). *Acta Tropica* **42**, 341–51.

Collins, R. C. & R. H. Jones 1978. Laboratory transmission of *Onchocerca cervicalis* with *Culicoides variipennis*. *Am. J. Trop. Med. Hyg.* **27**, 46–50.

Comley, J. C. W., M. J. Rees & A. B. O'Dowd 1988. The application of biochemical criteria to the assessment of macrofilarial viability. *Trop. Med. Parasit.* **39**, 456–9.

Comley, J. C. W., T. M. Szpoa, G. Strote, M. Buttner *et al.* 1989. A preliminary assessment of the feasibility of evaluating promising antifilarials *in vitro* against adult *Onchocerca volvulus*. *Parasitology* **99**, 417–25.

Connor, D. H. & R. C. Neafie 1976. Onchocerciasis. In: *Pathology of tropical and extraordinary diseases*. C. H. Binford and H. Connor, eds. Vol II. Armed Forces Institute of Pathology, Washington, DC, pp. 360–72.

Connor, D. H., D. W. Gibson, R. C. Neafie, B. Merighi *et al.* 1983. Sowda-onchocerciasis in North Yemen: a clinicopathologic study of 18 patients. *Am. J. Trop. Med. Hyg.* **32**, 123–137.

Connor, D. H., G. H. George & D. W. Gibson 1985. Pathologic changes of human onchocerciasis: implications for future research. *Rev. Infect. Dis.* **7**, 809–19.

Cooter, R. J. 1982. Studies on the flight of blackflies (Diptera: Simuliidae). I. Flight performance of *Simulium ornatum* Meigen. *Bull. Ent. Res.* **72**, 303–17.

Copeman, D. B. 1979. An evaluation of the bovine *Onchocerca gibsoni*, *Onchocerca gutturosa* model as a tertiary screen against *Onchocerca volvulus* in man. *Tropenmed. Parasit.* **30**, 469–74.

Court, J. P., A. E. Bianco, S. Townson, P. J. Ham *et al.* 1985. Study on the activity of antiparasitic agents against *Onchocerca lienalis* third-stage larvae *in vitro*. *Trop. Med. Parasit.* **36**, 117–19.

Dadzie, K. Y., J. Remme, A. Rolland & B. Thylefors 1986. The effect of 7–8 years of vector control on the evolution of ocular onchocerciasis in West African savanna. *Trop. Med. Parasit.* **37**, 263–70.

De Leon, J. R. & B. O. L. Duke 1966. Experimental studies on the transmission of Guatemalan and West African strains of *Onchocerca volvulus* by *Simulium ochraceum*, *S. metallicum* and *S. callidum*. *Trans. R. Soc. Trop. Med. Hyg.* **60**, 735–52.

Diaz, F. 1957. Notes and observations on onchocerciasis in Guatemala. *Bull. Wld Hlth Org.* **16**, 676–81.

Donelson, J. E., B. O. L. Duke, D. Moser, W. Zeng *et al.* 1988. Construction of *Onchocerca volvulus* cDNA libraries and partial characterisation of the cDNA for a major antigen. *Mol. Biochem. Parasit.* **31**, 241–50.

Donnelly, J. J., J. H. Rockey, A. E. Bianco & E. J. L. Soulsby 1983. Aqueous humor and serum IgE antibody in experimental ocular *Onchocerca* infection of guinea pigs. *Ophthalmic Research* **15**, 61–7.

Donnelly, J. J., J. H. Rockey, A. E. Bianco & E. J. L. Soulsby 1984. Ocular immunopathologic findings of experimental onchocerciasis. *Archives of Ophthalmol.* **102**, 628–34.

Donnelly, J. J., H. R. Taylor, E. Young, M. Khatami *et al.* 1986. Experimental ocular onchocerciasis in cynomolgus monkeys. *Invest. Ophthalmol. Vis. Sci.* **27**, 492–99.

Donnelly, J. J., M. S. Xi, J. P. Haldar, D. E. Hill *et al.* 1988a. Autoantibody induced by experimental *Onchocerca* infection. Effect of different routes of administration of microfilariae and of treatment with diethylcarbamazine citrate and ivermectin. *Invest. Ophthamol. Vis. Sci.* **29**, 827–31.

Donnelly, J. J., R. D. Semba, M. S. Xi, E. Young *et al.* 1988b. Experimental ocular onchocerciasis in cynomolgus monkeys. III. Roles of IgG and IgE antibody and autoantibody and cell mediated immunity in the chorioretinitis elicited by intravitreal *Onchocerca lienalis* microfilariae. *Trop. Med. Parasit.* **39**, 111–16.

Dry, F. W. 1921. Trypanosomiasis in the absence of tsetse and a human disease possibly carried by *Simulium* in Kenya Colony. *Bull. Ent. Res.* **12**, 233–8.

Duke, B. O. L. 1957. The reappearance, rate of increase and distribution of the microfilariae of *Onchocerca volvulus* following treatment with diethylcarbamazine. *Trans. R. Soc. Trop. Med. Hyg.* **51**, 37–44.

Duke, B. O. L. 1962. Experimental transmission of *Onchocerca volvulus* from man to a chimpanzee. *Trans. R. Soc. Trop. Med. Hyg.* **56**, 271.

Duke, B. O. L. 1967. Infective filarial larvae other than *Onchocerca volvulus* in *Simulium damnosum*. *Ann. Trop. Med. Parasit.* **61**, 700–5.

Duke, B. O. L. 1968a. The intake and transmissibility of *Onchocerca volvulus* microfilariae by *Simulium damnosum* fed on patients treated with diethylcarbamazine, suramin and Mel W. *Bull. Wld Hlth Org.* **39**, 169–78.

Duke, B. O. L. 1968b. Studies on factors influencing the transmission of onchocerciasis. IV. The biting cycles, infective biting density and transmission potential of forest *Simulium damnosum*. *Ann. Trop. Med. Parasit.* **62**, 95–106.

Duke, B. O. L. 1970. *Onchocerca–Simulium* complexes. VI. Experimental studies on the transmission of Venezuelan and West African strains of *Onchocerca volvulus* by *Simulium metallicum* and *S. exiguum* in Venezuela. *Ann. Trop. Med. Parasit.* **64**, 421–31.

Duke, B. O. L. 1973. Studies on factors influencing the transmission of onchocerciasis. VIII. The escape of infective *Onchocerca volvulus* larvae from feeding 'forest' *Simulium damnosum. Ann. Trop. Med. Parasit.* **67**, 95–100.

Duke, B. O. L 1980. Observations on *Onchocerca volvulus* in experimentally infected chimpanzees. *Tropenmed Parasit.* **31**, 41–54.

Duke, B. O. L. & J. Anderson 1972. A comparison of the lesions produced in the cornea of the rabbit eye by microfilariae of the forest and Sudan-savanna strains of *Onchocerca volvulus* from Cameroon. I. The clinical picture. *Z. Tropenmed. Parasit.* **23**, 354–68.

Duke, B. O. L. & A. Garner 1975. Reactions to subconjunctival inoculation of *Onchocerca volvulus* microfilariae in pre-immunized rabbits. *Tropenmed. Parasit.* **26**, 435–8.

Duke, B. O. L. & A. Garner 1976. Fundus lesions in the rabbit eye following inoculation of *Onchocerca volvulus* microfilariae into the posterior segment. I. The clinical picture. *Tropenmed. Parasit.* **27**, 3–17.

Duke, B. O. L. & D. J. Lewis 1964. Studies on factors influencing the transmission dynamics of onchocerciasis. III. Observations on the effect of the peritrophic membrane in limiting the development of *Onchocerca volvulus* microfilariae in *Simulium damnosum*. *Ann. Trop. Med. Parasit.* **58**, 83–8.

Duke, B. O. L. & P. J. Moore 1974. The concentration of microfilariae of a Guatemalan strain of *Onchocerca volvulus* in skin snips taken from chimpanzees over 24 hours. *Tropenmed. Parasit.* **25**, 153–9.

Duke, B. O. L., D. J. Lewis & J. P. Moore 1966. *Onchocerca-Simulium* complexes. I. Transmission of forest and Sudan-savanna strains of *Onchocerca volvulus* from Cameroon by *S. damnosum* of various West African bioclimatic zones. *Ann. Trop. Med. Parasit.* **60**, 318–35.

Duke, B. O. L., P. J. Moore, & J. R. De Leon 1967a. *Onchocerca–Simulium* complexes. V. The intake and subsequent fate of a Guatemalan strain of *Onchocerca volvulus* in forest and Sudan-savanna forms of West African *Simulium damnosum*. *Ann. Trop. Med. Parasit.* **61**, 332–7.

Duke, B. O. L., P. O. Scheffel, J. Guyon & P. Moore 1967b. The concentration of *Onchocerca volvulus* microfilariae in skin snips over 24 hours. *Ann. Trop. Med. Parasit.* **61**, 206–19.

Eichler, D. A. 1971. Studies on *Onchocerca gutturosa* (Neumann, 1910) and its development in *Simulium ornatum* (Meigen, 1818). II. Behaviour of *S. ornatum* in relation to the transmission of *O. gutturosa*. *J. Helm.* **45**, 259–70.

Eichler, D. A. 1973. Studies on *Onchocerca gutturosa* (Neumann, 1910) and its development in *Simulium ornatum* (Meigen, 1818). III. Factors affecting the development of the parasite in its vector. *J. Helm.* **47**, 73–88.

Eichler, D. A. & G. S. Nelson 1971. Studies on *Onchocerca gutturosa* (Neumann, 1910) and its development in *Simulium ornatum* (Meigen, 1818). I. Observations on *O. gutturosa* in cattle in south-east England. J. Helm. **45**, 245–58.

El Sheikh, H. & P. J. Ham 1982. Human onchocerciasis: cryopreservation of isolated microfilariae. *Lancet* **1**, 450.

El Sinnary, K. A., A. E. Bianco, & J. F. Williams 1987. Implantation of adult *Onchocerca gutturosa* into laboratory rodents. *Parasitology* **95**, 155–8.

Engelkirk, P. G., J. F. Williams, G. M. Schmidt & R. W. Leid 1982. Zoonotic onchocerciasis. In: CRC Handbook series in Zoonoses, J. H. Steele (ed.). Sect. C: Parasite zoonoses. Vol II (M. G. Schultz ed.). 225–50. Florida: CRC Press.

Erttmann, K. D., T. R. Unnasch, B. M. Greene, E. J. Albiez *et al.* 1987. A DNA sequence specific for forest form *Onchocerca volvulus*. *Nature* **327**, 415–17.

Ewert, A. & B. C. Ho 1967. The fate of *Brugia pahangi* larvae immediately after feeding by infective vector mosquitoes. *Trans. R. Soc. Trop. Med. Hyg.* **61**, 659–62.

Fawdry, A. L. 1957. Onchocerciasis in south Arabia. *Trans. R. Soc. Trop. Med. Hyg.* **51**, 253–6.

Flockhart, H. A. 1982. The identification of some *Onchocerca* spp. of cattle by isoenzyme analysis. *Tropenmed. Parasit.* **33**, 51–6.

Flockhart, H. A. & A. E. Bianco 1985. Differentiation of *Onchocerca gutturosa* and *Onchocerca lienalis* microfilariae by analysis of their GPI isoenzymes. *J. Helm.* **59**, 187–9.

Foil, L. D., T. R. Klei, R. I. Miller, G. E. Church *et al.* 1987. Seasonal changes in density and tissue distribution of *Onchocerca cervicalis* microfilariae in ponies and related changes in *Culicoides variipennis* populations in Louisiana. *J. Parasit.* **73**, 320–6.

Frank, W., P. Wenk & H. Scherb 1969. Studies on *Onchocerca flexuosa* (Nematoda: Filarioidea) a parasite of the red deer (*Cervus elephas*). *Verhandl. Deutsch. Zool. Gesellsch.* (Innsbruck 1968) 540–50.

Franz, M. & D. B. Copeman 1988. The fine structure of male and female *Onchocerca gibsoni*. *Trop. Med. Parasit.* **39**, 466–8.

Fuglsang, H. & J. Anderson 1974. Microfilariae of *Onchocerca volvulus* in blood and urine before, during and after treatment with diethylcarbamazine. *J. Helm.* **48**, 93–97.

Fuglsang, H. & J. Anderson 1977. The concentration of microfilariae in the skin near the eye as a simple measure of the severity of onchocerciasis in a community and as an indicator of the danger to the eye. *Tropenmed. Parasit.* **28**, 63–7.

Fuglsang, H. & J. Anderson 1978. Further observations on the relationship between ocular onchocerciasis and the head nodule, and on the possible benefit of nodulectomy. *Brit. J. Ophthalmol.* **62**, 445–9.

Fuglsang, H., J. Anderson, T. F. de C. Marshall, S. Ayonge *et al.* 1976. Seasonal variation in the concentration of *Onchocerca volvulus* microfilariae in the skin? *Tropenmed. Parasit.* **27**, 365–9.

Fuhrman, J. A. & W. F. Piessens 1989. A stage specific calcium-binding protein from microfiliarae of *Brugia malayi* (Filariidae). *Molec. Biochem. Parasit.* **35**, 249–58.

Gallin, M. Y., D. Murray, J. H. Lass, H. E. Grossniklaus *et al.*1988. Experimental intersitial keratitis induced by *Onchocerca volvulus* antigens. *Arch. Ophthalmol.* **106**, 1447–52.

Garms, R. 1973. Quantatative studies on the transmission of *Onchocerca volvulus* by *Simulium damnosum* in the Bong range, Liberia. *Z. Tropenmed. Parasit.* **24**, 358–72.

Garms, R. & J. Voelker 1969. Unknown filarial larvae and zoophily in *Simulium damnosum* in Liberia. *Trans. R. Soc. Trop. Med. Hyg.* **63**, 676–7.

Garner, A. 1976. Pathology of ocular onchocerciasis: Human and experimental. *Trans. R. Soc. Trop. Med. Hyg.* **70**, 374–7.

Garner, A. 1980. Laboratory research into the pathogenesis of ocular onchocerciasis. *Trans. Opth. Soc. UK* **100**, 179–85.

Garner, A. & B. O. L. Duke 1976. Fundus lesions in the rabbit eye following inoculation of *Onchocerca volvulus* microfilariae into the posterior segment. II. Pathology. *Tropenmed. Parasit.* **27**, 19–29.

Garner, A., B. O. L. Duke & J. Anderson 1973. A comparison of the lesions produced in the cornea of the rabbit eye by microfilariae of the forest and Sudan-savanna strains of *Onchocerca volvulus* from Cameroon. II. The pathology. *Z. Tropenmed. Parasit.* **24**, 385–96.

Gasparini, G. 1964. Problems of onchocerciasis in new suspected areas: sowda and onchocerciasis. *Arch. Ital. Sci. Med. Trop. Parassit.* **45**, 243–60.

Gibson, D. W. & D. H. Connor 1978. Onchocercal lymphadenitis: clinicopathologic study of 34 patients. *Trans. R. Soc. Trop. Med. Hyg.* **72**, 137–54.

Gibson, D. W, C. Heggie & D. H. Connor 1980. Clinical and pathological aspects of onchocerciasis. *Path. Ann.* **15**, 195–240.

Greene, B. M. 1987. Primate model for onchocerciasis research. *Ciba Found. Symp.* **127**, 236–43.

Greene B. M., H. R. Taylor & M. Aikawa 1981. Cellular killing of microfilariae of *Onchocerca volvulus*: Eosinophil and neutrophil-mediated immune serum-dependent destruction. *J. Immunol.* **127**, 1611.

Greene, B. M., M. M. Fanning & J. J. Ellner 1983. Non-specific suppression of antigen-induced lymphocyte blastogenesis in *Onchocerca volvulus* infection in man. *Clin. Exp. Immunol.* **52**, 259–65.

Greene, B. M., A. A. Gbakima, E. J. Albiez & H. R. Taylor 1985. Humoral and cellular immune responses to *Onchocerca volvulus* infection in humans. *Rev. Infect. Dis.* **7**, 789–95.

Ham, P. J. 1986. Acquired resistance to *Onchocerca lienalis* infections in *Simulium ornatum* Meigen and *Simulium lineatum* Meigen following passive transfer of haemolymph from previously infected simuliids (Diptera: Simuliidae). *Parasitology* **92**, 269–77.

Ham, P. J. & A. E. Bianco 1981. Quantification of a cryopreservation technique for *Onchocerca* microfilariae in skin-snips. *J. Helm.* **55**, 59–61.

Ham, P. J. & A. E. Bianco 1983a. Screening of some British simuliids for susceptibility to experimental *Onchocerca lienalis* infection. *Z. Parasit.* **69**, 765–72.

Ham, P. J. & A. E. Bianco 1983b. Development of *Onchocerca volvulus* from cryopreserved microfilariae in three temperate species of laboratory-reared blackflies. *Tropenmed Parasit.* **34**, 137–9.

Ham, P. J. & A. J. Banya 1984. The effect of *Onchocerca* infection on the fecundity and oviposition of laboratory reared *Simulium* sp. (Diptera, Simuliidae). *Trop. Med. Parasit.* **35**, 61–66.

Ham, P. J. & A. E. Bianco 1984. Maintenance of *Simulium wilhelmia lineatum* Meigen and *Simulium erythrocephalum* de Geer through successive generations in the laboratory. *Canad. J. Zool.* **62**, 870–7.

Ham, P. J. & C. L. Gale 1984. Blood meal enhanced *Onchocerca* development and its correlation with fecundity in laboratory reared blackflies (Diptera: Simuliidae). *Tropenmed. Parasit.* **35**, 212–16.

Ham, P. J. & R. Garms 1985. Development of forest *Onchocerca volvulus* in *Simulium yahense* and *Simulium sanctipauli* following intrathoracic injection and ingestion of microfilariae. *Trop. Med. Parasit.* **36** (Suppl. I), 25.

Ham, P. J. & R. Garms 1987. Failure of *Onchocerca gutturosa* to develop in *Simulium soubrense* and *Simulium yahense* from Liberia. *Trop. Med. Parasit.* **38**, 135–6.

Ham, P. J. & R. Garms 1988. The relationship between innate susceptibility to *Onchocerca* and haemolymph attenuation of microfilarial motility *in vitro* using British and West African blackflies. *Trop. Med. Parasit.* **39**, 230–4.

Ham, P. J., S. Townson, E. R. James & A. E. Bianco 1981. An improved technique for the cryopreservation of *Onchocerca* microfilariae. *Parasitology* **83**, 139–146.

Haque, A. & A. Capron 1982. Transplacental transfer of rodent microfilariae induces antigen specific tolerance in rats. *Nature* **299**, 361–3.

Harnett, W., A. E. Chambers, A. Renz, M. J. Worms *et al.* 1989. Development of an oligonucleotide probe specific for *Onchocerca volvulus*. In: O-Now! Symposium on onchocerciasis: recent developments and prospects of control, Sept. 20–22, Leiden, J. H. van der Kaay (ed.) 90, Institute of Tropical Medicine Rotterdam-Leiden, The Netherlands.

Hashiguchi, Y., M. Kawabata, I. Tanaka, T. Okazawa, C. O. Flores *et al.* 1981. Seasonal variation in the microfilarial density of *Onchocerca volvulus* and in the biting activity of *Simulium* spp. in Guatemala. *Trans. R. Soc. Trop. Med. Hyg.* **75**, 839–45.

Hawking, F. 1967. The 24-hour periodicity of microfilariae: biological mechanisms responsible for its production and control. *Proc. Roy. Soc. B.* **169**, 59–76.

Ho, B. C. & M. M. J. Lavoipierre 1975. Studies on filariasis. IV. The rate of escape of the third-stage larvae of *Brugia pahangi* from the mouthparts of *Aedes togoi* during the blood meal. *J. Helm.* **49**, 65–72.

Ivanov, I. V. 1964. Histological changes in skin of cattle infected with *Onchocerca*. *Trudy vses 1st Gel'mint.* **11**, 59–61.

James, E. R., B. Smith & J. Donnelly 1986. Invasion of the mouse eye by *Onchocerca* microfilariae. *Trop. Med. Parasit.* **37**, 359–60.

Jenkins, D. C., R. Armitage & T. S. Carrington 1980. A new primary screening test for anthelmintics utilising the parasitic stages of *Nippostrongylus brasiliensis in vitro*. *Z. Parasitenkd.* **63**, 261–9.

Jones, R. H. & R. C. Collins 1979. *Culicoides variipennis* and *Onchocerca cervicalis* as a model for human onchocerciasis. *Ann. Parasit. Hum. Comp.* **54**, 249–50.

Karam, M., H. Schulz-Key, J. Remme 1987. Population dynamics of *Onchocerca volvulus*. *Acta Tropica* **44**, 445–57.

Kershaw, W. E., B. O. L. Duke & F. H. Budden 1954. Distribution of the microfilariae of *O. volvulus* in the skin. Its relation to the skin changes and the eye lesions and blindness. *Brit. Med. J.* **2**, 724–9.

Kilian H. D. 1988. Deep onchocercomata close to the thigh bones of a Liberian patient. *Trop. Med. Parasit.* **39**, 347–8.

Klager, S. 1988. Investigations on enzymatically isolated *Onchocerca volvulus*: qualitative and quantitative aspects of spermatogenesis. *Trop. Med. Parasit.* **39**, 441–5.

Kozek, W. J. & H. Figueroa Marroquin 1982. Attempts to establish *Onchocerca volvulus* infection in primates and small laboratory animals. *Acta Tropica* **39**, 317–24.

Lackey, A., E. R. James, J. A. Sakanari, S. D. Resnick *et al.* 1989. Extracellular proteases of *Onchocerca*. *Exp. Parasit.* **68**, 176–85.

Lagraulet, J. 1962. L'étude des lésions oculaires dans l'*Onchocerca cervicalis* du cheval peut-elle apporter dans données intéressantes sur la pathogénie de l'onchocercose oculaire humaine. *Bull. Soc. Pth. Exot.* **55**, 417–22.

Lal, R. B. & E. A. Ottesen 1989. Phosphocholine epitopes on helminth and protozoal parasites and their presence in the circulation of infected human patients. *Trans. R. Soc. Trop. Med. Hyg.* **83**, 652–5.

Lambert, P. H., F. J. Dixon, R. H. Zubler *et al.* 1978. A WHO collaborative study for the evaluation of eighteen methods for detecting immune complexes in serum. *Clin. Lab. Immunol.* **1**, 1.

Lartigue, J. J. 1967. Variation in the number of microfilariae of *Onchocerca volvulus* in skin biopsies taken at different hours of the day. *Bull. Wld Hlth Org.* **36**, 491–4.

Lobos, E. & N. Weiss 1986. Identification of non-cross-reacting antigens of *Onchocerca volvulus* with lymphatic filariasis serum pools. *Parasitology* **93**, 389–99.

Lok, J. B., E. W. Cupp & M. J. Bernardo 1983a. *Simulium jenningsi* Malloch (Diptera: Simuliidae): a vector of *Onchocerca lienalis* Stiles (Nematoda: Filarioidea) in New York. *Am. J. Vet. Res.* **44**, 2355–8.

Lok, J. B., E. W. Cupp, M. J. Bernardo & R. J. Pollack, R.J. 1983b. Further studies on the development of *Onchocerca* spp. (Nematoda: Filarioidea) in Nearctic Black flies (Diptera: Simuliidae). *Am. J. Trop. Med. Hyg.* **32**, 1298–305.

Lok, J. B., R. J. Pollack, E. W. Cupp, M. J. Bernardo *et al.* 1984. Development of *Onchocerca lienalis* and *O. volvulus* from the third to fourth larval stage *in vitro*. *Tropenmed. Parasit.* **35**, 209–11.

Lok, J. B., R. J. Pollack & J. J. Donnelly 1987. Studies of the growth-regulating effects of ivermectin on larval *Onchocerca lienalis in vitro*. *J. Parasit.* **73**, 80–4.

Lowrie, R. C. 1983. Cryopreservation of third stage larvae of *Brugia malayi* and *Dipetalonema viteae*. *Am. J. Trop. Med. Hyg.* **32**, 767–71.

Lucius, R., H. Schulz-Key, D. W. Buttner, A. Kern *et al.* 1988a. Characterization of an immunodominant *Onchocerca volvulus* antigen with patient sera and a monoclonal antibody. *J. Exp. Med.* **167**, 1505–10.

Lucius, R., E. Ngozi, A. Kern & J. E. Donelson 1988b. Molecular cloning of an immunodominant antigen of *Onchocerca volvulus*. *J. Exp. Med.* **168**, 1199–204.

Mackenzie, C. D. 1980. Eosinophil leukocytes in filarial infections. *Trans. R. Soc. Trop. Med. Hyg.* **74** (Suppl.), 51.

Mackenzie, C. D., P. J. Burgess & B. M. Sisley 1986. Onchocerciasis. In: *Immunodiagnosis of Parasitic Diseases*, Vol I. *Helminthic Diseases*. K. W. Walls & P. M. Schantz (eds), Florida: Academic Press. 255–301.

McKerrow, J. H. 1989. Parasite proteases. *Exp. Parasit.* **68**, 111–15.

McMahon, J. P., R. B. Highton & H. Goiny 1958. The eradication of *Simulium neavei* from Kenya. *Bull. Wld Hlth Org.* **19**, 75–107.

McReynolds, L. A., S. M. DeSimone & S. A. Williams 1986. Cloning and comparison of repeated DNA sequences from the human filarial parasite *Brugia malayi* and the animal parasite *Brugia pahangi*. *Proc. Nat. Acad. Sci. USA* **83**, 797.

Maizels, R. M., I. Sutanto, A. Gomez-Priego, J. Lillywhite *et al.* 1985. Specificity of surface molecules of adult *Brugia* parasites: cross-reactivity with antibody from *Wuchereria*, *Onchocerca* and other human filarial infections. *Trop. Med. Parasit.* **36**, 233–7.

Mazzotti, L. 1951. Observations based on cutaneous biopsies in onchocerciasis. *Am. J. Trop. Med. Hyg.* **31**, 628–32.

Mellor, P. S. 1973. Studies on *Onchocerca cervicalis* (Railliet and Henry, 1910): I. *Onchocerca cervicalis* in British horses. *J. Helm.* **47**, 97–110.

Mellor, P. S. 1974. Studies on *Onchocerca cervicalis* (Railliet and Henry, 1910): IV. Behaviour of the vector, *Culicoides nubeculosus* in relation to the transmission of *Onchocerca cervicalis*. *J. Helm.* **48**, 283–8.

Mellor, P. S. 1975. Studies on *Onchocerca cervicalis* (Railliet and Henry, 1910): V. The development of *Onchocerca cervicalis* larvae in the vectors. *J. Helm.* **49**, 33–42.

Meyers, W. M., R. C. Neafie & D. H. Connor 1977. Onchocerciasis: invasion of deep organs by *Onchocerca volvulus*. *Am. J. Trop. Med. Hyg.* **26**, 650–7.

Muller, R. 1979. Identification of *Onchocerca*. In: *Problems in the identification of parasites and their vectors*. A. E. R. Taylor, R. Muller (eds), 175–206. Oxford: Blackwell.

Muller, R. & J. R. Baker 1988. The demands of medicine and veterinary science. In: *Prospects in systematics*. D. L. Hawksworth (ed.), 337–95. Oxford: Oxford University Press.

Muller, R. & G. S. Nelson 1975. *Ackertia globulosa sp.n.* (Nematoda: Filarioidea) from rodents in Kenya. *J. Parasit.* **61**, 606–9.

Mustafa, M. B. 1983. Parasitological observations on bovine and equine *Onchocerca* infections. Unpublished PhD. thesis, University of London.

Mwaiko, G. L. 1981. The development of *Onchocerca gutturosa* Neuman to the infective stage in *Simulium vorax* Pomeroy. *Tropenmed. Parasit.* **32**, 277.

Nelson, G. S. 1958a. Onchocerciasis in the West Nile district of Uganda. *Trans. R. Soc. Trop. Med. Hyg.* **52**, 368–76.

Nelson, G. S. 1958b. 'Hanging groin' and hernia complications of onchocerciasis. *Trans. R. Soc. Trop. Med. Hyg.* **52**, 272–5.

Nelson, G. S. 1964. Factors influencing the development and behaviour of filarial nematodes in their arthropodan hosts. In: *Host–parasite relationships in invertebrate hosts*, A. E. R. Taylor (ed.), 2nd. Symp. Brit. Soc. Parasit. 75–119. Oxford: Blackwell.

Nelson, G. S. 1965. Filarial infections as zoonoses. *J. Helm.* **39**, 229–50.

Nelson, G. S. 1966. The pathology of filarial infections. *Helm. Abstracts* **35**, 311–36.

Nelson G. S. 1970. Onchocerciasis. *Adv. Parasit.* **8**, 173–224.

Nelson G. S. 1988. Parasitic zoonoses. In: *The biology of parasitism: a molecular and immunological approach*. P. T. Englund and A. Sher (eds). MBL lectures in Biology, Vol. 9, 13–41. New York: Alan R. Liss.

Nelson, G. S., J. G. Grounds 1958. Onchocerciasis at Kodera. Eleven years after the eradication of the vector. *E. Afr. Med. J.* **35**, 365–8.

Nelson, G. S. & F. R. N. Pester 1962. The identification of infective filarial larvae in Simuliidae. *Bull. Wld Hlth Org.* **27**, 473–81.

Nelson, G. S., M. A. Amin, E. J. Blackie & N. Robson 1966. The maintenance of *Onchocerca gutturosa* microfilariae *in vitro* and *in vivo*. *Trans. R. Soc. Trop. Med. Hyg.* **60**, 17.

Nettel, F. R. 1945. Onchocerciasis: importance of xenodiagnosis and of the study of microfilariae in their transmission phase. *Med. Rev. Mexico* **25**, 194–203.

Neumann, E. & A. E. Gunders 1973. Pathogenesis of the posterior segment of ocular onchocerciasis. *Am. J. Ophthalmol.* **75**, 82–9.

Ngu, J. L. 1978. Immunological studies on onchocerciasis. Varying skin hypersensitivity and leucocyte migration inhibition in a clinical spectrum of the disease. *Acta Trop.* **35**, 269–79.

Ngu, J. L., F. Chatelanat, R. Leke, P. Ndumbe *et al.* 1985. Nephropathy in Cameroon: evidence for filarial derived immune-complex pathogenesis in some cases. *Clin. Nephrol.* **24**, 128–134.

Nowak, M., G. W. Hutchinson & D. B. Copeman 1987. *In vitro* drug screening in isolated male *Onchocerca gibsoni* using motility suppression. *Trop. Med. Parasit.* **38**, 128–30.

Omar, M. S. 1978. Histochemical enzyme staining patterns of *Onchocerca volvulus* microfilariae and their occurrence in different onchocerciasis areas. *Tropenmed. Parasit.* **29**, 462–72.

Omar, M. S. & R. Garms 1975. The fate and migration of microfilariae of a Guatemalan strain of *Onchocerca volvulus* in *Simulium ochraceum* and *S. metallicum* and the role of the buccopharyngeal armature in the destruction of the microfilariae. *Tropenmed. Parasit.* **26**, 183–90.

Omar, M. S. & R. Garms 1981. Histochemical differentiation of filarial larvae developing in *Simulium damnosum* in West Africa. *Tropenmed. Parasit.* **32**, 259–64.

Omar, M. S. & F. Kuhlow 1978. Histochemical differentiation of filarial larvae developing in *Simulium damnosum* in West Africa. WHO/ONCHO/78. 114, 11pp.

Omar, M. S. & H. Schulz-Key 1978. Acid phosphatase activity in the larval stages of *Onchocerca volvulus* developing in the vector *Simulium damnosum*. *Tropenmed. Parasit.* **29**, 359–63.

Omar, M. S., A. M. Denke & J. N. Raybould 1979. The development of *Onchocerca ochengi* (Nematoda: Filarioidea) to the infective stage in *Simulium damnosum s.1.* with a note on the histochemical staining of the parasite. *Tropenmed. Parasit.* **30**, 157–62.

O'Neill, J. 1875. On the presence of a filaria in craw craw. *Lancet* **1**, 265–6.

Osborn, H. A. 1935. Onchocerciasis in England. *Lancet* **229**, 1000.

Ottesen, E. A. 1984. Immunological aspects of lymphatic filariasis and onchocerciasis in man. *Trans. R. Soc Trop. Med. Hyg.* **78** (Suppl.), 9–18.

Parkhouse, R. M., Z. Cabrera & W. Harnett 1987. *Onchocerca* antigens in protection, diagnosis and pathology. *Ciba Found. Symp.* **127**, 125–45.

Petralanda, I., L. Yarzabal & W. F. Piessens 1986. Studies on a filarial antigen with collagenase activity. *Molec. Biochem. Parasit.* **19**, 51–9.

Petralanda, I., L. Yarzabal & W. F. Piessens 1988. Parasite antigens are present in breast milk of women infected with *Onchocerca volvulus*. *Am. J. Trop. Med. Hyg.* **38**, 372–9.

Philipp, M., A. Gomez-Priego, R. M. Parkhouse, M. W. Davies *et al.* 1984. Identification of an antigen of *Onchocerca volvulus* of possible diagnostic value. *Parasitology* **89**, 295–309.

Philippon, B, F. J. Walsh, P. Guillet & D. G. Zerbo 1989. Entomological results of vector control in the onchocerciasis control programme. In: O-Now! Symposium on onchocerciasis: recent developments and prospects of control, Sept. 20–22, Leiden, J. H. van der Kaay (ed.), Institute of Tropical Medicine Rotterdam-Leiden, The Netherlands, p. 25.

Picq J. J. & J. P. Jardel 1974. A method for evaluating the densities of microfilariae of *Onchocerca volvulus* Leuckart, 1893 in patients. Distribution of microfilarial densities, according to the site and level of skin biopsies; variation of microfilarial densities during a 24-hour period. *Bull. Wld Hlth Org.* **51**, 145–53.

Plaisier, A. P., G. L. van Oortmarssen, J. D. F. Habbema & J. Remme 1989.

ONCHOSIM: a simulation model for onchocerciasis transmission and control. In: O-Now! Symposium on onchocerciasis: recent developments and prospects of control, Sept. 20–22, Leiden, J. H. van der Kaay (ed.), Institute of Tropical Medicine Rotterdam-Leiden, The Netherlands, p. 118.

Prince A. M., E. J. Albiez & M. C. van den Ende 1985. *Onchocerca volvulus*: transplantation of adult worms into chimpanzees. *Trop. Med. Parasit.* **36**, 105–8.

Prost, A. 1986. The burden of blindness in adult males in the savanna villages of West Africa exposed to onchocerciasis. *Trans. R. Soc. Trop. Med. Hyg.* **80**, 525–7.

Pudney, M. & M. G. R. Varma 1980. Present state of knowledge of in vitro cultivation of filariae. In: *In vitro cultivation of the pathogens of tropical disease.* 367–89. UNDP/World Bank/WHO Tropical Research Series 3. Basel.

Pudney, M., T. Litchfield, A. E. Bianco & C. D. Mackenzie 1988. Moulting and exsheathment of the infective larvae of *Onchocerca lienalis* (Filarioidea) *in vitro. Acta Tropica* **45**, 67–76.

Racz, P., K. Tenner-Racz, B. Luther, D. W. Buttner *et al.* 1983. Immunopathologic aspects in human onchocercal lymphadenitis. *Bull. Soc. Path. Exot.* **76**, 676–80.

Raybould, J. N. 1981. Colonization of members of the *Simulium damnosum* complex. In: *Blackflies* M. Laird (ed.), 307–18. London: Academic Press.

Roberts, J. M. D., E. Neumann, C. W. Goeckel & R. B. Highton 1967. Onchocerciasis in Kenya 9, 11, and 18 years after elimination of the vector. *Wld Hlth Org. Bull.* **37**, 195–212.

Robles, R. 1919. Human onchocerciasis in Guatemala causing blindness and 'Coastal erysipela'. *Bull. Soc. Path. Exot.* **12**, 442–60.

Schillhorn van Veen, T. W. & J. Blotkamp. 1978. Histochemical differentiation of microfilariae of *Dipetalonema, Dirofilaria, Onchocerca* and *Setaria* spp. of man and domestic animals in the Zaria area. (Nigeria). *Tropemmed. Parasit.* **29**, 33–5.

Schmidt, G. M., J. D. Krehbiel, S. C. Coley & R. W. Leid 1982. Equine ocular onchocerciasis: histopathologic study. *Am. J. Vet. Res.* **43**, 1371–5.

Schulz-Key, H. 1975. Investigations on the filariidae of the Cervidae in southern Germany. I. Development of the nodule, finding of the sexes and production of microfilariae in *Onchocerca flexuosa* (Wedl, 1856), in the red deer. *Tropenmed. Parasit.* **26**, 60–9.

Schulz-Key, H. 1988. The collagenase technique: how to isolate and examine adult *Onchocerca volvulus* for the evaluation of drug effects. *Trop. Med. Parasit.* **39**, 423–40.

Schulz-Key, H. & M. Karam 1986. Periodic reproduction of *Onchocerca volvulus*. *Parasitology Today* **2**, 284–6.

Schulz-Key, H. & P. Wenk 1981. The transmission of *Onchocerca tarsicola* (Filarioidea: Onchocercidae) by *Odagmia ornata* and *Prosimulium nigripes* (Diptera: Simuliidae). *J. Helm.* **55**, 161–6.

Schulz-Key, H., E. J. Albiez & D. W. Buttner 1977. Isolation of living adult *Onchocerca volvulus* from nodules. *Tropenmed. Parasit.* **28**, 428–30.

Schulz-Key, H., B. Jean & E. J. Albiez 1980. Investigations on female *Onchocerca volvulus* for the evaluation of drug trials. *Tropenmed. Parasit.* **31**, 34–40.

Semba, R. D., J. J. Donnelly, J. H. Rockey, J. B. Lok, A. A. Sakla & H. R. Taylor 1988. Experimental ocular onchocerciasis in cynomolgus monkeys. II. Chorioretinitis elicited by intravitreal *Onchocerca lienalis* microfilariae. *Invest. Ophthalmol. Vis. Sci.* **29**, 1642–51.

Sharp, N. A. D. 1927. A new site of *Onchocerca volvulus*. *Lancet* **2**, 1290.

Shelley, A. J., A. P. A. Lunas Dias & M. A. P. Moraes 1980. *Simulium* species of the amazonicum group as vectors of *Mansonella ozzardi* in the Brazilian Amazon. *Trans. R. Soc. Trop. Med. Hyg.* **74**, 784–8.

Siegenthaler, R. von. & R. Gubler 1965. Para-articulaires nematodengranulom (einheimische *Onchocerca*). *Schweiz. Med. Wschr.* **95**, 1102–4.

Sisley, B. M., C. D. Mackenzie, M. W. Steward, J. F. Williams *et al.* 1987. Associations between clinical disease, circulating antibodies and C1q-binding immune complexes in human onchocerciasis. *Parasit. Immunol.* **9**, 447–63.

Spratt D. M. & P. Haycock 1988. Aspects of the life history of *Cercopithifilaria johnstoni* (Nematoda: Filarioidea). *Int. J. Parasit.* **18**, 1087–92.

Spratt D. M. & G. Varughese 1975. A taxonomic revision of filarial nematodes from Australasian marsupials. *Austral. J. Zool.* **35** (Suppl.) 1–99.

Steward, J. S. 1937. The occurence of *Onchocerca gutturosa* Neumann in cattle in England, with an account of its life-history and development in *Simulium ornatum* Mg. *Parasitology* **29**, 212–19.

Storey, D. M. & A. S. Al-Muktar 1982. Vaccination of jirds, *Meriones unguiculatus* against *Litomosoides carinii* and *Brugia pahangi* using irradiated larvae of *L. carinii*. *Tropenmed. Parasit.* **33**, 23–4.

Strong, R. P., J. H. Sandground, J. C. Bequaert & M. M. Ochoa 1934. Onchocerciasis with special reference to the Central American form of the disease. Contributions from the Dept. Trop. Med. & Inst. Trop. Biol. & Med. No. VI. Harvard University Press.

Strote, G. 1985. Development of infective larvae of *Onchocerca volvulus* in diffusion chambers implanted into *Mastomys natalensis*. *Trop. Med. Parasit.* **36**, 120–2.

Strote, G. 1987. Morphology of third and fourth stage larvae of *Onchocerca volvulus*. *Trop. Med. Parasit.* **38**, 73–4.

Suswillo, R.R., G. S. Nelson, R. Muller, P. B. McGreevy et al. 1977. Research note: Attempts to infect jirds (*Meriones unguiculatus*) with *Wuchereria bancrofti, Onchocerca volvulus, Loa loa loa* and *Mansonella ozzardi*. *J. Helm.* **51**, 132–4.

Tada, I. & M. H. Figueroa 1974. The density of *Onchocerca volvulus* microfilariae in the skin at different times of the day in Guatemala. *Jpn. J. Parasit.* **23**, 220–5.

Tarrant, C., S. Moobola, G. Scoles & E. W. Cupp 1983. Mating and oviposition of laboratory reared *Simulium vittatum* (Diptera: Simuliidae). *Can. Entomol.* **115**, 319–23.

Taylor, D. W., J. M. Goddard & J. E. McMahon 1986. Surface components of *Onchocerca volvulus*. *Molec. Biochem. Parasit.* **18**, 283–300.

Taylor, H. R. & B. M. Greene 1989. The status of ivermectin in the treatment of human onchocerciasis. *Am. J. Trop. Med. Hyg.* **41**, 460–6.

Thylefors, B. 1978. Ocular onchocerciasis. *Bull. Wld Hlth Org.* **56**, 63–73.

Tidwell, M. A., B. V. Peterson, J. Ramirez-Perez, M. Tidwell et al. 1981. Notes and preliminary keys relating to Neotropical blackflies belonging to the groups of *Simulium amazonicum* and *S. sanguineum* (Diptera: Simuliidae) including the vectors of *Onchocerca volvulus* and *Mansonella ozzardi*. *Bol. Dir. Mal. San. Amb.* **21**, 77–87.

Townson, S. 1988. The development of a laboratory model for onchocerciasis using *Onchocerca gutturosa*: *in vitro* culture, collagenase effects, drug studies and cryopreservation. *Trop. Med. Parasit.* **39**, 475–9.

Townson S. & A. E. Bianco 1982a. Experimental infection of mice with the microfilariae of *Onchocerca lienalis*. *Parasitology* **85**, 283–93.

Townson, S. & A. E. Bianco 1982b. Immunization of calves against the microfilariae of *Onchocerca lienalis*. *J. Helm.* **56**, 297–303.

Townson, S., A. E. Bianco & D. Owen 1981a. Attempts to infect small laboratory animals with the infective larvae of *Onchocerca lienalis*. *J. Helm.* **55**, 247–49.

Townson, S., K. El Sinnary & A. E. Bianco 1981b. Transplantation of *Onchocerca* into mice. *Trans. R. Soc. Trop. Med. Hyg.* **75**, 889.

Townson, S., A. E. Bianco, M. J. Doenhoff & R. L. Muller 1984. Immunity to *Onchocerca lienalis* microfilariae in mice. I. Resistance induced by the homologous parasite. *Tropenmed. Parasit.* **35**, 202–8.

Townson, S., G. S. Nelson & A. E. Bianco 1985. Immunity to *Onchocerca lienalis* microfilariae in mice II. Resistance induced by heterologous parasites. *J. Helm.* **59**, 337–46.

Townson, S., C. Connelly, A. Dobinson & R. Muller 1987. Drug activity against *Onchocerca gutturosa* males *in vitro*: a model for chemotherapeutic research on onchocerciasis. *J. Helm.* **61**, 271–81.

Townson, S., A. Dobinson, C. Connelly & R. Muller 1988. Chemotherapy of *Onchocerca lienalis* microfilariae in mice: a model for the evaluation of novel compounds for the treatment of onchocerciasis. *J. Helm.* **62**, 181–94.

Townson, S., K. E. Shay, A. R. Dobinson, C. Connelly et al. 1989. *Onchocerca gutturosa* and *O. volvulus*: studies on the viability and drug responses of cryopreserved adult worms *in vitro*. *Trans. R. Soc Trop. Med. Hyg.* **83**, 664–9.

Unnasch, T. R., K. D. Erttmann & B. M. Greene 1989. DNA probes as a new tool to differentiate savannah and forest types of onchocerciasis. In: O-Now! Symposium on onchocerciasis: recent developments and prospects of control, Sept. 20–22, Leiden, J. H. van der Kaay (ed.). Institute of Tropical Medicine Rotterdam, Leiden, The Netherlands, p. 44.

van den Berghe, L. 1936. Preliminary note on extranodular localization of *Onchocerca volvulus* in man. *Ann. Soc. Belge Med Trop.* **16**, 549–51.

van den Berghe, L. 1941. Onchocerciasis research in the Belgian Congo. Paper II. Adult worms and their localisation in man. *Ann. Soc. Belge. Med. Trop.* **21**, 167–87.

van den Berghe, L., M. Chardome & E. Peel 1958. *Gorilla gorilla* in Belgian Congo: probable host of *Onchocerca volvulus* of man. *Fol. Scient. Afr. Contre* **4**, 16.

van den Ende, M. C., E. J. Albiez & E. A. Dennis 1981. Transplantation of adult *Onchocerca volvulus* into chimpanzees. *Trans. R. Soc. Trop. Med. Hyg.* **75**, 310–11.

Vankan, D. M. & D. B. Copeman 1988. Reproduction in female *Onchocerca gibsoni*. *Trop. Med. Parasit.* **39**, 469–71.

Vankan, D. M., D. B. Copeman & M. Novak 1988. An evaluation of implanted male *Onchocerca gibsoni* in mice as a screen for macrofilaricides against *Onchocerca volvulus*. *Trop. Med. Parasit.* **39**, 472–4.

Voelker, J. & R. Garms 1972. On the morphology of unknown filarial larvae in *Simulium damnosum* and *Simulium kenyae* in Liberia and their probable final hosts. *Z. Tropenmed. Parasit.* **23**, 285–301.

Walsh, J.F., B. Philippon, J. E. Hendericks & D. C. Kurtak 1987. Entomological aspects and results of the Onchocerciasis Control Programme. *Trop. Med. Parasit.* **38**, 57–60.

Ward, D. J., T. B. Nutman, G. Zea-Flores, C. Portocarrero, A. Lujan & O. E. Ottesen 1988. Onchocerciasis and immunity in humans: enhanced T cell responsiveness to parasite antigen in putatively immune individuals. *J. Inf. Dis.* **157**, 536–43.

Wegesa, P. 1966. Some factors influencing the transmission of *Onchocerca volvulus* by *Simulium woodi*. *Ann. Rep. East Afr. Inst. Mal. Vect. Dis.* 14–17.

Wegesa, P. & J. L. M. Lelijveld 1971. Experimental transmission of *Onchocerca volvulus* to primates. Proc. 2nd. Int. Cong. Parasit. (Washington). In: *J. Parasit.* **57**, II, 53.

Weinmann, C. J., J. R. Anderson, W. M. Longhurst & G. Connolly 1973. Filarial worms of Columbian black-tailed deer in California. 1. Observations in the vertebrate host. *J. Wildlife Dis.* **9**, 213–20.

Weiss, N. 1985. Monoclonal antibodies as investigative tools in onchocerciasis. *Rev. Infect. Dis.* **7**, 826–30.

Weiss, N. & M. Karam 1987. Humoral immune responses in human onchocerciasis: detection of serum antibodies in early infections. *Ciba Found. Symp.* **127**, 180–8.

Wenk, P. 1976. Coevolution of vector and parasite in simuliids and nematodes. *Z. Angewandte Entomologie* **82**, 38–44.

Wilkinson, P. R. 1949. Some observations on microfilariae in Africans at Jinja. *East Afr. Med. J.* **26**, 344–6.

Williams, J. F., H. W. Ghalib, C. D. Mackenzie, M. Y. Elkhalifa et al. 1987. Cell adherence to microfilariae of *Onchocerca volvulus*: a comparative study. *Ciba Found. Symp.* **127,** 146–63.

WHO (World Health Organization) 1972. Chemotherapy of onchocerciasis. Practice, prospects and needs. *Wld Hlth Org. mimeo document* WHO/ONCHO/72.91.

WHO (World Health Organization) 1974. Technical Report Series, No. 542, (WHO Expert Committee on Filariasis, Third Report).

WHO (World Health Organization) 1976. Epidemiology of onchocerciasis. Technical Report Series, WHO, No.597, 94 pp.

WHO (World Health Organization) 1987. Technical Report Series, No. 752, (WHO Expert Committee on Onchocerciasis, Third Report).

Dracunculus in Africa

'Let not the winding worm touch me and wound my feet'
(Attributed to Vysadeva, Rg Veda 7, Hymn 50, ca. 1350 BC)

Introduction

It was in 1966 when I had just been appointed to a lectureship at the London School of Hygiene and Tropical Medicine that George Nelson suggested that a research project on the nematode *Dracunculus medinensis* might yield fruitful results and would certainly provide a field in which there would be little competition for funds. Over twenty years later I hope that the present world campaign for the elimination of the distressing and anachronistic disease caused by *Dracunculus* is being aided in a small way by the studies I carried out in the intervening period, originally inspired by him.

Dracunculiasis is a long established disease of man which was reported from Africa in antiquity (Adamson 1988) and the actual parasites have been found more than once in Egyptian mummies (see Muller 1985a). From the opening quotation it is clear that human infection must be equally ancient in India, and that the method of treatment has not changed in the intervening three thousand years (Fig. 7.1).

There has been a great growth of interest in dracunculiasis in the last few years which is indicated by an increase in the numbers of published papers on the subject from 14 in 1965 to 44 in 1985. The determination to do something about the disease which has been very much promoted and sustained by the efforts of Dr Donald Hopkins (Hopkins 1988) who had previously been involved in the successful campaign to eradicate smallpox from the world, resulted in studies on the feasibility and economic necessity of control with a resolution of the United Nations World Health Assembly in 1986 endorsing the elimination of dracunculiasis as part of the Clean Drinking Water and Sanitation Decade (1981–90).

Public health importance of dracunculiasis

Before the start of the Decade there had been great reluctance by donor agencies to consider the disease as an important problem and it was not mentioned in the original aims of the Decade. This is perhaps not surprising since the disease has such a low case-fatality rate but many studies over the last few years have shown that it does cause severe

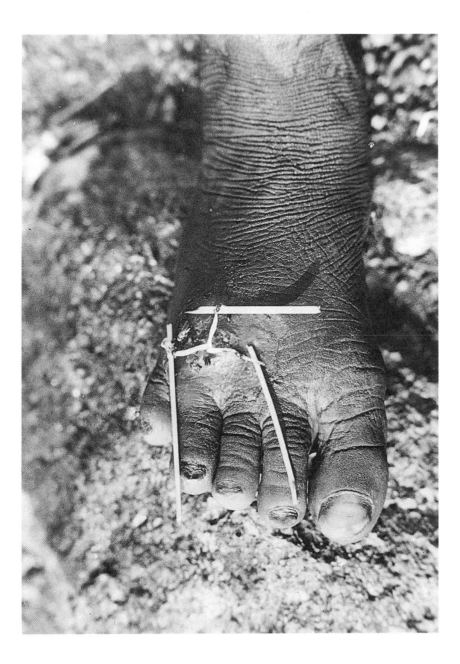

Figure 7.1 Three mature female guineaworms emerging from the foot of a child in Sudan and being wound out on sticks in the traditional manner. Such treatment, if accompanied by bandaging of the lesions, can help to prevent secondary complications and contamination of the water sources. (Courtesy of A. Tayeh.)

disability, often for a long time and at critical periods in the agricultural season. Smith *et al.* (1989), for instance, found that in Imo State, Nigeria, 58% of infected persons suffered severe disability with individuals unable to leave their compounds for a mean of 4.2 weeks and with a mean duration of symptoms of 12.7 weeks. Of the total number of cases 40% were in the 15–49 year working age group, mostly at yam and rice harvest time, while another 40% were in the 5–14 year age group. Ward (1985) has estimated that an attack rate of 35% in a community (a not uncommon figure) would represent a week's work lost for each member of the productive community. The effect on schoolchildren has been evaluated recently by Nwosu *et al.* (1982) and by Ilegbodu *et al.* (1986). Nwosu *et al.* found that the mean absenteeism rate from 13 schools in Anambra State in eastern Nigeria was 13% during most of the school year but rose to 60% at the height of the guineaworm season, while Ilegbodu *et al.* reported that guineaworm-infected children in four primary schools in southwestern Nigeria missed 25% of the school year compared with 2.5% absence for non-infected children and the disease was found to be a major cause of pupils permanently dropping out of school. A similar effect was noted by Edungbola (1983).

Paul *et al.* (1986) have investigated the cost-effectiveness of various approaches to the control of dracunculiasis and by means of models have shown a high benefit-to-cost-ratio, particularly when chemical treatment of ponds is used rather than provision of safe water sources. De Rooy (1987) has estimated that, in the principal rice growing area of Nigeria in Anambra State, rice production is reduced by 11% annually because of disability associated with dracunculiasis which results in an annual economic loss of $20 million, and Guigemde *et al.* (1986) have estimated the annual cost of disability in three villages in Burkina Faso at $29 000.

At present, guineaworm occurs in Africa in 17 countries; former foci in The Gambia, Guinea-Bissau and Somalia appear to have disappeared in the last few years, although a small new focus appeared recently in northern Kenya (Macpherson 1981) and might still be present.

The distribution of the disease in Africa is shown in Figure 7.2, although it is not clear exactly where the disease occurs in the Central African Republic or southern Sudan. The estimated numbers of cases per year and of population at risk for each country are given in Table 7.1. The table is based principally on figures collected by Watts (1987) and also on those provided by delegates to the second African Regional Workshop on Dracunculiasis held in Accra in 1988 (Anon. 1988). The figure for Uganda was obtained from Henderson *et al.* (1988). For many countries, figures were obtained by passive case reporting and these have been multiplied by a factor of 40 (i.e. the assumption has been made that only 2.5% of actual cases are picked up by this procedure). This gives an estimated annual incidence for Nigeria of 2.5 million; for comparison, an active case search carried out from 1 July 1987–30 July 1988 found a

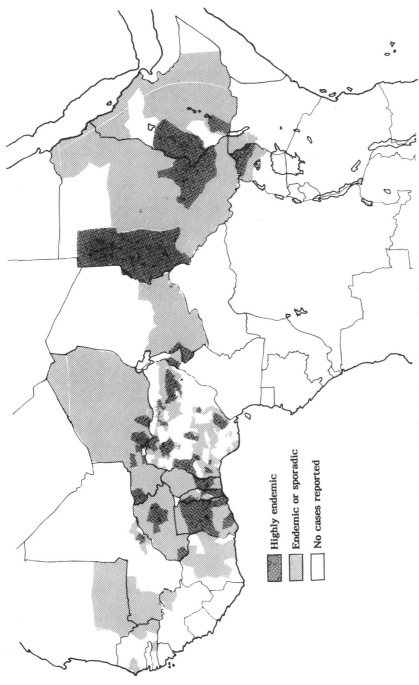

Figure 7.2 Distribution of dracunculiasis in Africa.

Highly endemic

Endemic or sporadic

No cases reported

Table 7.1 Estimated population at risk and annual number of cases of dracunculiasis for endemic countries in Africa.

	1986 population (millions)	1986 population at risk (millions)	Estimated annual incidence
Benin	4.1	2.5	60 000
			13 892 (1988)
Burkina Faso	7.1	5.75	175 000
Cameroon	10.0	1.17	27 640
Central African Republic	2.7	?	?
Chad	5.2	3.17	2200
Ethiopia	43.9	30.4	30 800
Ghana	13.6	9.38	170 000
			71 767 (1988)
Ivory Coast	10.5	4.21	592
Kenya	21.0	0.18	200
Mali	7.9	1.9	20 400
Mauritania	1.9	0.8	50 000
Niger	6.7	5.63	28 000
Nigeria	98.2	30.43	2 500 000
			654 395 (1988)
Senegal	6.9	0.26	500
Sudan	22.9	17.00	60 000
Togo	3.0	1.92	58 240
Uganda	15.2	4.83	120 000

total of 654 395 cases in the country (data presented at the second National Conference on Dracunculiasis held in Lagos in March 1989).

Transmission is usually highly seasonal with maximum transmission, for instance, taking place in the early rainy season in a dry (sahel) savanna zone of Mali with under 80 cm annual rainfall and in the dry season in a humid (guinea) savanna area of southern Nigeria with over 130 cm annual precipitation (Guigemde 1986, Steib & Mayer 1988).

Current situation in Africa

UNICEF allocated $1.55 million to endemic African countries for 1989/90, which together with funds provided by UNDP and WHO should ensure that all countries will have carried out a national active search by the end of 1990 and be in a position to prepare a plan of action and submit proposals for funding to other donor agencies. Useful guidelines for surveillance procedures have been published by the WHO Collaborating Center for Research, Training and Control of Dracunculiasis at the Centers for Disease Control in Atlanta, Georgia (Anon. 1989a).

The impetus for control campaigns in Africa has been reinforced by the success of the eradication campaign in India, which was started in 1981 and is scheduled to eliminate the disease from the country by 1990: by the end of 1988 only 12 123 cases of dracunculiasis were reported from 4278 villages in India, while in 1984 there had been 39 792 cases in 12 840 villages. The same target date was set for elimination of the disease in Pakistan, where only 1110 cases were reported in 1988 (data presented at the EMRO Countries Regional Conference on Guineaworm Eradication held in Islamabad in April 1989).

Country-by-country situation

The situation in various countries is outlined below: up-to-date information is provided by the Guineaworm Wrap-Ups produced regularly by the Centers for Disease Control.

Benin
Background studies on the ecological and socio-economic effects of the disease are being carried out. Dracunculiasis is widespread and the province of Zou has a high percentage of infected villages. A national active case search is in progress.

Burkina Faso
The disease widely distributed and active case reporting was due to begin in all 7600 villages in early 1990 with a national conference planned for then. Many cloth filters have been distributed in two provinces.

Cameroon
Pilot control schemes in the north are going well (Ripert *et al.* 1987) but an active case search has not yet been carried out. One problem is that there is widespread migration of people from the mountains to the plains.

Ethiopia
The disease occurs in areas of the country but no surveys have been undertaken.

Central African Republic
The disease probably exists but there is little information on its exact distribution or on the numbers involved.

Chad
A survey of the endemic regions in the south is due to be carried out in 1990.

Ghana
The disease is widespread and local control efforts are already under way.

A nationwide active case search is being initiated in early 1990. Filters are being widely distributed.

Ivory Coast
Incidence of the disease appears to be decreasing but an active case search is still to be carried out.

Mali
A pilot control project is now being evaluated.

Mauritania and Niger
No active case searches are being carried out, although one is planned for Mauritania soon, but boreholes are being dug to provide safe drinking water (about 2.3% more per year).

Nigeria
There have been two national conferences on the disease (Edungbola 1985) and an eradication programme with an action plan was formulated in 1989. The first active case search was carried out in 1988 and the second was planned for September 1989. There have already been local control schemes in Anambra and Kwara States.

Senegal
A total of 10 cases have been reported over the last three years on the borders with Mali, Mauritania and Guinea but these may have been imported.

Somalia
The disease is not present but there is a possible problem from nomads from south-west Ethiopia introducing it.

Sudan
The disease is widespread and about six million refugees have entered the country in the last few years and the disease appears to be spreading in the Red Sea and Kasala provinces adjacent to Ethiopia. It has not been possible to carry out any recent surveys in the south of the country. Some control efforts as part of primary health care are taking place in Southern Kordofan Province.

Togo
The disease is widespread and it is likely that about 5.5% of the population is infected. There have been successful local control schemes but no national active case search.

Uganda

Seasonal transmission (October–January) occurs in the north but no action has been possible so far.

Control measures

The eradication of dracunculiasis from the world was endorsed by the Steering Committee as a sub-goal of the Clean Drinking Water Supply and Sanitation Decade (1981–90) and was formally adopted as a resolution of the United Nations World Health Assembly in 1986 and by the Regional Committee for Africa in 1988. All endemic African countries have pledged that the disease will be eliminated by 1995 at the latest.

The criterion for certifying elimination in a country is that there has been no confirmed case of viable infection in humans for at least three years in formerly endemic areas despite satisfactory evidence of active surveillance during that period. At the end of this period WHO should be requested for independent assessment of the absence of the disease.

Although it has been known for over a century that *Dracunculus* is transmitted by cyclopid copepods, in the last few years there have been some radical revisions of the group, which could cause confusion in transmission studies. While all intermediate host species were previously usually regarded as belonging to sub-genera of the single genus *Cyclops*, they are now recognized as different genera (Einsle 1970, Steib 1985, Steib *et al.* 1986, Guigemde *et al.* 1987). In particular, species of African *Mesocyclops* have been redescribed (Van der Velde 1984), and it has been determined that what was formerly regarded as a single cosmo-politan, intermediate host species, *Cyclops* (now *Mesocyclops*) *leuckarti* does not even occur in Africa. Instead, specimens recovered from various locations in Africa and originally attributed to this species may belong to one of four different species. Because of this, it is not always possible to determine exactly the intermediate hosts described in all but the most recent literature (Table 7.2).

The life cycle of *Dracunculus is* illustrated in Fig. 7.3.

From the figure it can be seen that there are various possible interventions:

1 Provide a safe drinking water supply;
2 Filter or boil all drinking water;
3 Persuade or prevent infected persons with a patent female worm from entering water;
4 Treat water sources with chemicals to kill cyclops;
5 Treat patients with anthelmintics.

Table 7.2 The species of *Cyclops* which have been found naturally infected with guineaworm in Africa and which can be positively identified.

Valid name	Former name	Country	Authority
Mesocyclops aequatorialis	*Cyclops leuckarti*	Nigeria	Boxshall (unpublished)
Thermocyclops oblongatus	*C. nigerianus*	Nigeria	Muller 1971
T. crassus spp.	*C. hyalinus*	Nigeria	Muller 1971
T. inopinus	—	Burkina Faso	Steib 1985 Oedraogo *et al.* 1986
T. incisus	—	Burkina Faso	Steib & Mayer 1988
Metacyclops margaretae	—	Nigeria	
Mesocyclops kieferi	*C. leuckarti*		

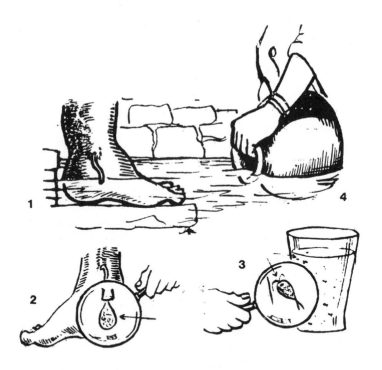

Figure 7.3 The life cycle of *Dracunculus* illustrated by a village education leaflet (from Muller 1985b). 1. The mature female measuring about 70 cm emerges from an ulcer, usually on the foot. 2. The anterior end bursts when immersed in water and releases thousands of microscopic larvae. 3. Larvae are ingested by cyclops, moult twice, and are infective in about two weeks. 4. Infected cyclops are ingested in drinking water and the young worms migrate from the intestine to the subcutaneous tissues; the female worms emerge about one year later.

Provision of safe drinking water

Guineaworm is the only water-associated disease dependent entirely on contamination of drinking water for transmission, unlike other helminthic, protozoal, bacterial or viral diseases where many other factors are likely to be involved to a greater or lesser extent (Table 7.3). Thus, the provision of safe drinking water to all communities throughout the world, which is the ultimate aim of the Decade, would eradicate this disease without specific control measures being required. Also, safe drinking water and adequate sanitation would confer benefits far more wide-ranging and important than just the elimination of guineaworm. UNICEF has estimated that in 1986 there were 1.2 billion people in the world without safe potable water (although the percentage in developing countries with safe water rose from 31% to 36% between 1981 and 1985) and that six million children under the age of five die annually from water-borne diarrhoeal-related diseases.

It is likely that villages in Africa which have guineaworm will have most of the other diseases also and are those requiring clean drinking water the most. Of the total of about 99 000 villages in endemic countries needing safe water, about 7000 also have guineaworm. However, the provision of safe drinking water to all is very much a long-term goal and a strategy of giving guineaworm-infected communities priority in the provision of safe water during the Decade, coupled with specific guineaworm control measures in areas where this is not as yet feasible, can provide an attainable target for eradication of the disease within the next 6–10 years. The cost of eradicating guineaworm from Africa by 1995 has been estimated at 40 million dollars but most of this will be for safe drinking water which will provide much wider benefits, including a lowering of morbidity and mortality due to childhood diarrhoeas, cholera, polio-myelitis, hepatitis and other disease agents as shown in Table 7.3.

Today, a piped water supply is present in virtually all urban areas in Africa so that transmission is confined almost entirely to poor and disadvantaged rural communities. In the past it was also common in towns and cities (hence the Latin name meaning 'little dragon of Medina') and the classical Persian poet Sa'di (1974 translation) tells the tale of a king who was cured when he released all his prisoners and prayed (the disease was eliminated from Iran by 1982).

The provision of tubewells is a major factor in meeting the objectives of providing clean drinking water for all and UNICEF has the aim in suitable geological areas of providing one borehole for every 200 persons. Boreholes are being provided at a fast rate in many African countries but there are precautions that need to be taken, and the author has visited areas where India Mark II pumps which were installed a few months previously were no longer functioning. This was either because the water did not last throughout the dry season or because, in order to serve the

Table 7.3 The possible interventions for water-associated diseases. Guineaworm is the only disease in which transmission depends entirely on the quality of drinking water. Adapted from Balance & Gunn (1984).

Disease	Water quality	Water quantity/ convenience	Personal and domestic hygiene	Waste water disposal/drainage	Excreta disposal	Food sanitation
Viral diarrhoea	00	000	000	—	0	00
Bacterial diarrhoea	000	000	000	—	0	000
Protozoal diarrhoea	0	000	000	—	0	00
Polio and hepatitis A	0	000	000	—	0	00
Guineaworm	000	—	—	—	—	—
Ascaris and *Trichuris*	0	0	0	0	000	00
Hookworm	0	0	0	—	000	—
Schistomiasis	0	0	0	—	000	—
Other helminths with aquatic hosts	—	—	—	—	00	000
Insect transmitted diseases	—	—	—	— to 000	—	—

Value of intervention measure: 0 Low; 00 Medium; 000 High.

maximum number of people, they had been sited between villages and nobody had responsibility for their maintenance. It is very important that a local community should be involved in the siting and operation of a water pump and that members should be able to maintain it with the aid of locally produced materials at minimal cost. Where draw wells are constructed by the local community, they are usually well maintained and provide an adequate source of safe drinking water (Cairncross & Tayeh 1988). The principal expense and labour is involved in digging down by hand and it might be feasible in some areas for the government or other bodies to provide digging equipment to be used over a wide area in order to make the initial holes and the villagers would then be responsible for lining their well, providing a suitable parapet, and keeping it maintained.

The building of large rainwater cisterns as a community project can provide additional sources of safe drinking water, for schools in particular.

In many rural areas labourers drink from many small ponds while working in the fields even if there is a safe water supply in the village and so other specific anti-guineaworm measures may be necessary even when safe water is provided.

Filtration of water

Any material with a mesh of 100 μm or below will filter out all stages of cyclopid copepods likely to be infected with *Dracunculus* larvae. However, local cotton materials have very irregular pore sizes, they tend to smell when wet, and become clogged with debris, leading to very slow filtration rates. Recently, monofilament nylon nets have been recommended (Duke 1984) and these are long lasting, do not clog up so much, are easily washed, and dry quickly. Nets of the correct size for fixing over the neck of a large pottery or calabash vessel kept in the house and secured with a sewn-in elastic band can be distributed in an area. They can certainly be effective if the importance of their use is adequately explained and transmission was completely interrupted in three villages in Burkina Faso after two transmission seasons following their provision to villagers (Gbary *et al.* 1987). Unfortunately, they are expensive and the materials are not usually freely available in local markets. In some parts of Nigeria, material has been given to market women who have prepared and sold the nets but their use as a sole control method has not been as effective as in Burkina Faso. It is intended as part of the Nigerian national plan of action, though, to provide every family living in a village which is still without a safe drinking water supply by July 1991, with a filter. They are also being sold (for the equivalent of 30-45 US cents) in Burkina Faso and Ghana.

Sand filtration is a possibility when water is obtained from the very large artificial ponds which have been dug in many African countries.

This would require fencing of the pond and collection of water at one point after filtration. In Sudan the large ponds (known as hafirs) usually have built-up sides so a gravity system would be possible but in many areas the ponds are below ground level and a pump with a raised water tank would be necessary.

Prevention of water contamination

There have been no detailed studies on the dynamics of transmission so that it is not known for any endemic area how many times during a transmission season larvae are released from emerging female worms into a particular body of water and so infect cyclops. However, many thousands of larvae are released at one time from a single worm (Muller 1971) and it is quite likely that if only one person with an emerging worm should enter the water over the season this will be enough to maintain transmission in a community for the next season, although the incidence may be somewhat less than if many infected people contaminate the water. This supposition is supported by the many reported examples of infection being introduced into a community following a single visit by an infected person.

Since by a comparatively simple change in habits transmission could be completely interrupted, prevention of pond contamination should be a primary health education measure where existing water sources have to continue to be utilized. Kale (1982), for instance, was able to interrupt transmission in villages in western Nigeria following a chemotherapy trial: success was not due to the antiparasitic action of the drugs given but because all lesions were bandaged and patients were then reluctant to enter the water, and consequently contamination was prevented.

However, it is often difficult to obtain water for drinking without entering the pond, apart from the relief of pain obtained by immersing the infected limb. A stone or wooden stepping place, moved as the water level falls in the dry season, is used in some areas.

Treatment of ponds

The use of chemicals to kill cyclops populations in ponds used as drinking water sources is often likely to be necessary in any elimination campaign for various reasons: where the provision of safe drinking water is not feasible because the infected community is too small or remote or because of unfavourable geological conditions; as a short-term measure to reduce or break transmission until permanent sources of safe water can be installed; where a proportion of the population is not willing to comply with health education measures, usually because they do not believe the connection between unsafe drinking water and infection; as an additional safety measure in areas where elimination has almost been achieved.

The disadvantages of chemical treatment are that it does not have any effect on the incidence of other water-borne diseases (see Table 7.3) and that guineaworm infection can always be subsequently reintroduced from another area.

Several pesticides and disinfectants will kill copepods but the most suitable compound is undoubtedly temephos (Abate). This is because it is active at low concentrations ($0.1\,mg\ m^{-1}$ in the laboratory) with good residual action and has been widely used in portable water supplies as a mosquito larvicide for over 20 years, attesting to its lack of toxicity to mammals. Under field conditions the 50% emulsifiable concentrate is most convenient at a concentration of 2 ppm (1 ppm of active compound) when its activity lasts for 4–5 weeks.

The efficacy of chemical treatment of ponds depends on the type of water source and the length of the transmission season. For instance, in the Abakalili zone of Anambra State, Nigeria, which has a population of 1.2 million with an estimated 300 000 cases of dracunculiasis annually, a guineaworm task force was set up in 1986 with the aim of eliminating the disease from the zone. Although some boreholes are being supplied in parts of the zone by UNICEF, JICA (Japanese aid agency) and the State Government, drinking water at present is supplied by 5760 ponds and it will not be possible to provide pumped water for more than a small proportion of the population at risk, so that chemical treatment of ponds is also being undertaken. Most of the ponds are small (less than $500\,m^3$ (Fig. 7.4)) but about 700 are very large (on average $62\,000\,m^3$) and if these were to be treated also it would increase the cost by a factor of nine. It is also not known what importance these large ponds have in the transmission of infection and large water bodies sometimes have religious significance (e.g. they can contain sacred crocodiles) and treatment of them, apart from not being economically feasible, is more likely to meet with local opposition than treatment of small ponds.

In this part of Nigeria transmission occurs primarily in the late dry season from March to May (Nwosu *et al.* 1982), so that chemical treatment will have to be repeated at least three times in a season.

General guidelines for the use of chemicals in copepod control together with procedures and nomograms for calculating the water volume of ponds have been published by the Centers for Disease Control (Anon. 1989b).

Anthelmintic treatment

Various compounds have been recommended for treatment of guineaworm such as thiabendazole, metronidazole, mebendazole and albendazole. If given when a mature female worm begins to emerge, they appear to hasten expulsion and lessen pain; probably through an anti-inflammatory action. However, none has any direct anthelmintic effect on prepatent adults or on contained larvae, so that mass chemotherapy cannot be used as a control strategy. A broad action anthelmintic such as albendazole, which can be given in a single dose and with minimal side-effects, can nonetheless have a useful part to play in ensuring the co-operation of target communities because of its action against intestinal helminths in addition to any action in aiding extraction.

A simple approach to hasten expulsion of worms by covering the ulcer with a permanently moist bandage has been successfully used by members of the Pakistan Guineaworm Eradication Project (M. A. Rab *et al.* unpublished).

In order to discover compounds that are effective against developing worms it will be necessary to test them against infections in experimental animals (see next section) and/or to develop a serological test which can diagnose prepatent infections. Many years ago it was shown that the indirect fluorescent antibody test was positive in experimentally infected rhesus monkeys some months before patency (Muller 1970) and it is likely that the much more convenient ELISA, which shows the presence of specific antibodies in patients with emerging worms (Kliks & Rao 1984) could also be used for diagnosing prepatent infections.

Figure 7.4 Typical infected pond about two metres deep in eastern Nigeria and used as a source of drinking water. There is a water tower in the background which provides piped water but this does not extend outside the town. Each pond has a sacred shrine; here it is situated under the tree on the right.

Dracunculiasis as a zoonosis

The extent to which animals act as reservoirs of human infection is a very puzzling question and one to which it is not possible to give a definitive answer. Emerging female worms belonging to the genus *Dracunculus* have been recovered sporadically from a wide range of mammals and reptiles in many countries around the world including both endemic and non-endemic areas of Africa (reviewed by Muller 1971). Unfortunately, in most of the infections in mammals, only portions of an adult female worm were recovered and specific identification was not possible.

It is clear though, that, since the parasite has such a short patent period which in a fur-covered animal is easily missed by an observer, and what appears to be a rather inefficient mode of transmission, the prevalence of animal infections must be very much higher than records suggest. This is well illustrated by the situation in North America. *Dracunculus* infections in wild carnivores there were probably first recorded by Leidy in 1858 and only about 40 subsequent reports have been published over the subsequent hundred years (see Muller 1971). However, in the early 1970s, Vincent Crichton undertook a PhD project on the subject at the University of Guelph, Canada. He had access to many carcasses of animals killed by fur trappers and found that in southern Ontario over 50% of raccoons and mink were in fact infected, while the parasite was also present in fishers (the marten *Martes pennanti*), weasels, muskrats and opossums (Crichton & Beverley-Burton 1974). In addition a clearly different and previously unrecorded species (which was named *D. lutrae*) was recovered from over 70% of otters in some areas of southern Ontario (Crichton & Beverley-Burton 1973). Recently other workers have shown that guineaworm infection with the same two species is also very common in raccoons and in otters in Arkansas (Tumlison *et al.* 1984).

The species found in raccoons etc. in North America was named *D. insignis* by Chitwood (1950) but the characteristics he used to separate this species from *D. medinensis*, the length of the gubernaculum and the number of pre-anal papillae in the male, were not found to be different when further male specimens of the latter species were examined (Muller 1971). There are thus no morphological differences between the two (apart from the shorter length of females in the former) and it is still an unresolved question whether *D. insignis* is a valid species or is just an animal-adapted strain of *D. medinensis*. There have been few if any autochthonous human infections in North America and attempted cross-infection experiments have not clarified the situation. Beverley-Burton and Crichton (1973) succeeded in experimentally infecting a rhesus monkey with *D. insignis*, showing that it can develop in primates, but they were not able to infect raccoons with *D. medinensis* from man (Berveley-Burton & Crichton 1976). This negative result is not really surprising since the larvae of *D. medinensis* were obtained from a patient from India and the Indian strain of human parasite has not been successfully transmitted to other primates (reviewed in Muller 1971). It was also not possible to infect a rhesus monkey using some of the same batch of larvae sent to Canada and given to the raccoons (Muller, unpublished), although 11 of 14 rhesus monkeys were successfully infected at the first attempt (and the other three after one or two further attempts) using larvae of *D. medinensis* from African patients (Muller 1972). Leiper in 1906 was the first to briefly describe the recovery of male and female worms from a monkey experimentally infected in West Africa and C. W. Daniels (in an unpublished manuscript discovered in the reprint collection of the CAB International Institute of Parasitology) gives some more details of the experiment. It is not clear why a description of the males was never published.

D. insignis (or *D. medinensis*) in ferrets might make a good laboratory model for guineaworm since Eberhard *et al.* (1988) recovered worms from 10 out of 18 ferrets (56%) given infected larvae. An interesting finding from this study which could have relevance to human disease was that male worms did not die after fertilization as reported for *D. medinensis* in rhesus monkeys (Muller 1972) and could conceivably survive for the following transmission season. Fortunately, there is no evidence to suggest that female worms can survive for more than one year as this could have serious implications for control campaigns.

Animal infections with what are assumed to be the human parasite still appear to be widespread even if only occasionally reported and in the last few years worms have been recovered in Argentina from dogs (Riveros *et al.* 1981) and a puma (Rosster *et al.* 1981); in India from dogs (Joseph & Kandazamy 1980, Lalitha & Anandan 1980, Subrahmanyam *et al.* 1976, Tirgari & Radhakishnan 1975) and a horse (Batliwala 1983); in southern USSR from cats, dogs (Ghenis 1972, Velikanov 1984) and wild carnivores (Kairov 1973, Litvinov & Vitvinov 1981, Osmanov *et al.* 1976) and in China from a cat (Hsu & Li 1981). In Africa, infections have been reported in the past from dogs and wild carnivores in Egypt, Ivory Coast, Mali, Sudan, Tanzania and Zambia, and from a cow in West Africa (reviewed by Muller 1971). The lack of knowledge concerning the transmission of *Dracunculus* in animals is illustrated by a recent isolated case of guineaworm infection reported from Japan in a human patient who had not been abroad for many years (Kobayashi *et al.* 1986) and older similar cases from Korea (Hashikura 1927) and Indonesia (Heutsz 1926). It is evident that these occasional human cases from non-endemic areas must represent zoonotic infections; interestingly, in all of them the possibility was raised of freshwater fish which were eaten raw, acting as paratenic hosts.

On the other hand, while in most areas highly endemic for human disease occasional infections are reported from dogs and other domestic animals, there is no evidence that these are capable of providing a reservoir of infection to man. When the disease has been eliminated from any area it has not returned except when infected individuals have come from elsewhere and contaminated the water.

References

Adamson, P. B. 1988. Dracontiasis in antiquity. *Medical History.* **32**, 204–9.

Anon. 1988. Dracunculiasis in Africa. Final report of a workshop. Accra, Ghana 14–18 March, 1988. WHO: Brazzaville.

Anon. 1989a. Guidelines for surveillance in dracunculiasis eradication programs. Atlanta: Centers for Disease Control.

Anon. 1989b. Guidelines for chemical control of copepod populations in dracunculiasis eradication programs. Atlanta: Centers for Disease Control.

Balance, R. C. and R. A. Gunn 1984. Drinking water and sanitation projects: criteria for resource allocation. *Wld Hlth Org. Chron.* **38**, 243–8.

Batliwala, J. C. 1983. Guinea-worm in the horse. *Vet. J.* (Annotations) **36**, 412–3.

Beverley-Burton, M. & V. F. J. Crichton 1973. Identification of guinea-worm species. *Trans. R. Soc. Trop. Med. Hyg.* **67**, 152.

Beverley-Burton, M. & V. F. J. Crichton 1976. Attempted experimental cross infections with mammalian guinea-worms, *Dracunculus spp. (Nematoda: Dracunculoidea). Am. J. Trop. Med. Hyg.* **25**, 704–8.

Cairncross, S & A. Tayeh 1988. Guineaworm and water supply in Kordofan Sudan. *J. Inst. Water Env. Management.* **2**, 268–74.

Chitwood, B.G. 1950. The male of *Dracunculus insignis* (Leidy, 1858) Chandler 1942. *Proc. Hel. Soc. Washington.* **17**, 14–15.

Crichton, V. F. J. & M. Beverley-Burton 1973. *Dracunculus lutrae* n.sp. (Nematoda; Dracunculoidea) from the otter, *Lutra canadensis* in Ontario, Canada. *Canadian J. Zool.* **51**, 521–9.

Crichton, V. F. J. & M. Beverley-Burton 1974. Distribution and prevalence of *Dracunculus* spp. (Nematoda: Dracunculoidea) in mammals in Ontario, Canada. *Canadian J. Zool.* **52**, 163–8.

De Rooy, C. 1987. Guinea worm control as a major contribution to self-sufficiency in rice production in Nigeria. Nigeria: UNICEF.

Duke, B. O. L. 1984. Filtering out the guinea worm. *Wld Hlth* March, 29.

Eberhard, M .L., E. Ruiz-Tiben & S. V. Wallace 1988. *Dracunculus insignis*: experimental infection in the ferret, *Mustela putorius furo. J. Helminthol.* **62**, 265–70.

Edungbola, L. D. 1983. Babana parasitic diseases project. II. Prevalence and impact of dracontiasis in Babana district, Kwara State, Nigeria. *Trans. R. Soc. Trop. Med. Hyg.* **77**, 310–15.

Edungbola L. D. (ed.) 1985. *Dracunculiasis in Nigeria.* Proceedings of the First National Conference on dracunculiasis (guinea worm disease) in Nigeria. Ilorin, Kwara State.

Einsle, U. 1970. Etude morphologiques sur les espèces de Thermocyclops (Crust. Cope.) d'Afrique et d'Europe. *Cahiers ORSTOM Series Hydrobiologia* **4**, 13–38.

Gbary, A. R., T. R. Guigemde & J. B. Ouedraogo 1987. La dracunculose, un fléau éradique dans trois villages du Burkina Faso par l'éducation sanitaire. *Bull. Soc. Path. Exotique,* **80**, 390–5.

Ghenis, D. E. 1972. New cases of *Dracunculuus medinensis* L., 1758 detected in domestic cats and dogs in Kazakhstan. *Med. Parazit. Bol.* **41**, 365.

Guigemde, T. R. 1986. Caractéristiques climatiques des zones d'endémie et modalities épidémiologiques de la dracunculose en Afrique. *Bull. Soc. Path. Exotique,* **79**, 89–95.

Guigemde, T. R., A. R Gbary, J. B. Ouedraogo *et al.* 1986. La dracunculose en Afrique de l'Ouest. *Etudes Médicales OCCGE, Bobo-Dioulasso,* No. 2, 98–140.

Guigemde, T. R., J. B. Ouedraogo, A. R. Gbary & K. Steib 1987. Etude longitudinale des cyclopides, hôte intermédiaries du ver de Guinée en zone Soudano-Sahelienne (Burkina Faso). *Ann. Parasit.* **62**, 484–91.

Hashikura, T. 1927. A case of *Filaria medinensis* in Chosen (Korea). [In Japanese] *Japanese Medical World* **7**, 145–6.

Henderson, P. L., R. E. Fontaine & G. Kyeyvne 1988. Guinea worm disease in northern Uganda: a major public health problem controllable through an effective water programme. *Int. J. Epid.* **17**, 434–40.

Heutsz van, 1926. One indigenous and two imported cases of *Dracunculus* in Java. [In Dutch] *Geneeskundig Tijdschrift Nederlandsch-Indie,* **66**, 25.

Hopkins, D. R. 1988. Dracunculiasis eradication: the tide has turned. *Lancet,* **ii**, 148–50.

Hsu, P. K. & D. N. Li 1981. *Dracunculus medinensis* (Linnaeus, 1788) from a cat in Guangdong. *Ann. Bull. Soc. Parasitol. Guangdong Province,* **3**, 92.

Ilegbodu, V. A., B. C. Christensen, R. A. Wise, O. O. Kale *et al.* 1986. Age and sex differences in new and recurrent cases of guinea worm disease in Nigeria. *Trans. R. Soc. Trop. Med. Hyg.* **81**, 674–6.

Joseph, S. A. & S. Kandazamy 1980. On the occurrence of the guinea-worm, *Dracunculus medinensis* (Linnaeus, 1758) Gallaidant 1773 in an Alsatian dog. *Cheiron,* **9**, 363–5.

Kairov, I. Kh. 1973. Helminthiasis with a natural focal occurrence in the Karakalpak area. [In Russian] *Vopvosy Prirodnoi Ochagovosti Bolezni,* **6**, 131–8.

Kale, O. O. 1982. Fall in incidence of guinea-worm infection in western Nigeria after periodic treatment of infected persons. *Bull. Wld Hlth Org.* **60**, 951–7.

Kliks, M. M. & C. K. Rao 1984. Development of rapid ELISA for early serodiagnosis of dracunculiasis. *J. Comm. Dis.* **16**, 287–94.

Kobayashi, A., K. Katatura, A. Hamada, T. Suzuki *et al.* 1986. Human case of dracunculiasis in Japan. *Am. J. Trop. Med. Hyg.* **35**, 159–61.

Lalitha, C. M. & R. Anandan 1980. Guinea worm infection in dogs. *Cheiron,* **9**, 198–9.

Leiper, R. T. 1906. Some results of the infection of monkeys with guinea worms. *Brit. Assn Adv. Sci.* **76**, 600.

Litvinov, V. F. & V. P. Vitvinov 1981. Helminths of predatory mammals from eastern Azarbaijan SSR, USSR. [In Russian] *Parazitologiya.* **15**, 219–23.

Macpherson, C. N. L. 1981. The existence of *Dracunculus medinensis* (Linnaeus, 1758) in Turkana, Kenya. *Trans. R. Soc. Trop. Med. Hyg.* **75**, 680–1.

Muller, R. 1970. *Dracunculus medinensis*: diagnosis by indirect fluorescent antibody technique. *Exp. Parasitol.* **27**, 357–61.

Muller, R. 1972. *Dracunculus* and dracunculiasis. *Adv. Parasitol.* **9**, 73–151.

Muller, R. 1985a. Bibliography of *Dracunculus*. In *Workshop on opportunities for control of dracunculiasis: contributed papers*. Washington, DC: National Academy Press.

Muller, R. 1985b. Guineaworm eradication -- the end of another old disease? *Parasit. Today.* **1**, 39, 58.

Nwosu, A. B. C., E. O. Ifezulike & A. O. Anya 1982. Endemic dracontiasis in Anambra State of Nigeria: geographic distribution, clinical features, epidemiology and socio-impact of the disease. *Ann. Trop. Med. Parasit.* **76**, 187–200.

Osmanov, S. O., E. Arystanov & M. Ametov 1976. Occurrence of *Dracunculus medinensis* (L., 1758) in a fox near Lake Dautkul (Uzbek SSR) [In Russian] *Vestnik Karakalpakskogo Filiala Akademi Nauk UzSSR* **1**, 90–1.

Ouedraogo, J. B., T. R. Guigemde, A. R. Gbary & K. Steib 1986. Hôtes intermédiaires de *Dracunculus medinensis* (ver de Guinée): determination des espèces vétrices au Burkina Faso. *Document Techniques OCCGE,* No. 8828.

Paul, J. E., R. B. Isley & G. M. Ginsberg 1986. Cost effective approaches to the control of dracunculiasis. *WASH Technical Report* No. 38. Washington, DC: US Agency for International Development.

Ripert, G., B. Roche, B. Couprîce, E. Patuano *et al.* 1987. Epidémiologie de la dracunculose dans les Monts Mandar (Nord-Cameroon): organisation d'une campagne de lutte. *Med. Trop.* **47**, 133–9.

Riveros, C. E., R. A. Moviena, G. M. Bulman & O. J. Lombardero 1981. Dracunculosis en perros de la Provincia de Formosa (Republica Argentina). *Gaceta Vet.* **43**, 255–8.

Rosster, A., C. M. Brunel & G. M. Bulman 1981. Primera cita de dracunculosis en un puma *(Puma concolor concolor).* *Gaceta Vet.* **43**, 164–6.

Sa'di. 1974. The Bustan: Morals pointed and tales adorned. Translated by G.M. Wickas. Tale 12. A holy man cures ruler of sickness, physical and spiritual. 47–51, Leiden: E. J. Brovill.

Smith, G. S., D. Blum, S. R. A. Huttly, N. Okeke *et al.* 1989. Disability from dracunculiasis: effect on mobility. *Ann. Trop. Med. Parasitol.* **83**, 151–8.

Steib, K. 1985. Epidemiologie und vectorokologie der Dracunculose in Obvervolta (Burkina Faso). Doctoral thesis, University of Stuttgart..

Steib, K. & P. Mayer 1988. Epidemiology and vectors of *Dracunculus medinensis* in northwest Burkina Faso, West Africa. *Ann. Trop. Med. Parasitol.* **82**, 189–99.

Steib, K., J. B. Ouedraogo, T. R. Guigemde, A. R. Gbary *et al.* 1986. Les vecteurs du ver de Guinée en Afrique. *Etude Med. OCCGE (Bobo-Dioulasso),* **3**, 87–96.

Subrahmanyam, B., Y. R. Reddy & S. Paul 1976. *Dracunculus medinensis* (guinea worm) infestation in a dog and its treatment with 'Flagyl'. A case report. *Indian Vet. J.* **53**, 637–9.

Tirgari, M. & C. V. Radhakishnan 1975. A case of *Dracunculus medinensis* in a dog. *Vet. Rec.* **96**, 43.

Tumlison, R., R. Smith, J. Hudspeth, P. J. Polechla *et al.* 1984. Dracunculiasis in some Arkansas carnivores. *J. Parasitol.* **70**, 440.

Van der Velde, I. 1984. Revision of the African species of the genus *Mesocyclops* (Copepoda). *Hydrobiologia,* **109**, 3–36.

Velikanov, B.P. 1984. A case of *Dracunculus medinensis* infection in a dog in Turkmenia. [In Russian] *Izvestiya Akademii Nauk Turkamanskoi SSR, Biologicheskikh Nauk* **1**, 64–5.

Ward, W. B. 1985. The impact of dracunculiasis on the household: the challenge of measurement. In *Workshop on opportunities for control of dracunculiasis, contributed papers.* 121–33, Washington, DC: National Academy Press.

Watts, S. 1987. Dracunculiasis in Africa. *Am. J. Trop. Med. Hyg.* **37**, 119–25.

8 Animal reservoirs of schistosomiasis

Introduction

Schistosomiasis, or bilharzia as it is commonly known, is a disease affecting various mammals, including man and domestic livestock, caused by a parasitic trematode of the genus *Schistosoma*. The known history of the disease dates back to the 16th century BC. The Ebers papyrus written at that time contains what is thought to be a reference to its treatment or prevention (Pfister 1912). Ruffer (1910) provided evidence of the earliest records of schistosomiasis in Africa, finding characteristic eggs of *S. haematobium* in mummies dating back to 1250–1000 BC. In 1851 schistosome worms were first recovered from a human body, that being during a post-mortem examination of a patient in Cairo by Theodore Bilharz, who was later responsible for linking infection with these worms to the hematuria in Egyptians passing terminally spined eggs in their urine. Manson (1902) first suggested a possibility of there being more than one species after he had described laterally spined eggs in human faeces in the West Indies. This view was supported by other workers, and in 1907 Sambon designated the laterally spined eggs as *Schistosoma mansoni* after Sir Patrick Manson. Meanwhile another species, *S. japonicum*, which does not occur in Africa, was discovered after Katsurada (1904) recovered adult worms from the portal system of a cat. Its life cycle was worked out by Fujinami (1910), Miyagawa (1912, 1913) and by Miyairi and Suzuki (1913). Their work provided the basis for the research carried out by Leiper (1915–1918) and it was Leiper who finally demonstrated the existence of *S. haematobium* and *S. mansoni* as two distinct species which had morphologically different adult worms and eggs, different distributions in the definitive hosts and a dependence on snails from different genera as intermediate hosts.

Life cycle

All species commonly infecting man have similar life cycles involving sexual generation in the vertebrate host, and an asexual phase in the fresh water molluscan host. Man is the principal host of both *S. haematobium* and *S. mansoni* but *S. japonicum* is a zoonosis. The adult female worm is held in the gynaecophoric canal of the male and the paired worms are found in the mesenteric veins, or veins of the vesicle plexus in the case of

224

S. haematobium. After copulation, the females produce large numbers of eggs daily, some of which pass out in stool or urine. The remainder of the eggs become trapped in the liver or other organs where they die and result in the production of granulomas causing pathology.

The excreted eggs hatch when they reach fresh water and the released miracidia swim freely for several hours before they die. If before then they succeed in locating and penetrating an appropriate snail intermediate host they develop into mother then daughter sporocysts and finally cercariae. When stimulated by light and/or warmth cercariae emerge from snails and swim freely. If a suitable vertebrate host is located within 24 hours cercariae penetrate the skin, transform into schistosomulae and develop into adult worms within 30–40 days. The adult worms migrate to the mesenteric veins or veins of the bladder and start laying eggs to repeat the cycle.

It is believed that both *S. haematobium* and *S. mansoni* originated somewhere in Africa, possibly in central Africa (Wright 1966, Nelson *et al.* 1962) and were spread around the world during the time of the slave trade.

Out of an estimated 200 million people infected in 74 countries (WHO 1985), 150 million infections occur in Africa alone (WHO 1965). The disease is endemic in nearly all African countries. High intensities of infection associated with morbidity have been recorded in Egypt (Adel-Wahab *et al.* 1980, Abdel-Salam & Elsan 1978), Zaire (Gryseels & Polderman 1987), Kenya (Siongok *et al.* 1976), Zambia (Bulsara *et al.* 1985) and Tanzania (Forsyth & Bradley 1966). Although data on morbidity is scanty, the accumulating evidence suggests that schistosomiasis is a public health problem in a number of African countries and particularly those countries where water resources development have been implemented.

Man is also the definitive host for *S. intercalatum* which has a patchy and limited distribution in central and West Africa. A number of species of animals have been infected in the laboratory with *S. intercalatum*, including the rhesus monkey, hamster, opossum gerbil, sheep and goats but only one species of rodent (*Hybomys univitatus*) has been found infected in the wild (Schwetz 1956 – cited in Wright *et al.* 1972). Because zoonotic *S. intercalatum* infections have not been shown to play a role in the transmission and epidemiology of human disease this species is not considered further here.

The purpose of this chapter is to bring together the available information regarding animal infections of *S. haematobium* and *S. mansoni* and the possible role such infections may play in maintaining transmission.

Schistosoma haematobium

Infections in non-human primates

Very few reports of *S. haematobium* infections in non-human primates exist (Table 8.1). Nelson (1960) examined eight baboons (*Papio* sp.), 15 vervet monkeys (*Cercopithecus aethiops*) and 10 sykes monkeys (*C. mitis*) all captured from the Tana River District of Kenya, an area in which *S. haematobium* infections in man were common. Terminal spined eggs or adult worms were found in four of the baboons and in one each of the vervet and sykes monkeys. Most of these infections were classed as *S. mattheei*, but one of the baboons had typical *S. haematobium* eggs in both its bladder and rectum. Only a single male worm was found in the infected vervet, which could have been any one of the terminal spined schistosomes. More recently, Else *et al.* (1982) reported 3 out of 36 sykes monkeys to be infected with typical *S. haematobium* eggs, examined from the same region as Nelson's previous study.

Elsewhere, De Paoli (1965) found *S. haematobium* type eggs in histological preparations from a chimpanzee imported into USA from Sierra Leone and Taylor *et al.* (1972) reported a natural infection in a baboon (*Papio* sp.) captured along the Gambia river in Senegal. A number of worms and eggs were recovered from the baboon, and hatched miracidia successfully infected *Bulinus africanus* snails. The resultant cercariae produced patent infections in experimentally infected baboons 12 weeks after exposure.

In Zimbabwe, Purvis *et al.* (1965) found 2 out of 7 baboons (*Papio rhodesiae*), killed 20 miles from Harare, to harbour adult worms. Although the worms could not be identified, eggs found later in digested brain tissue were terminally spined which suggested that they could have been *S. haematobium*.

Infections in other animals

A single male *S. haematobium* worm was recovered from a buffalo (*Syncerus caffer*) in South Africa (Basson *et al.* 1970) and Mackenzie (1979) reported natural infections in domestic sheep in Zimbabwe. Two reports of *S. haematobium* infections in rodents exist, that of Pitchford (1959), who found infections in wild *Otomys* rats, and Mansour (1973) who found one male worm in one out of 29 Nile rats (*Arvicanthus* sp.). There is an unconfirmed report of a large wild boar (*Sus scrofa*) from Nigeria found to be naturally infected, passing *S. haematobium* eggs (Hill & Onabimiro 1960).

Besides these few examples, the available evidence indicates that *S. haematobium* is not a parasite that normally infects animals other than man. The reports of natural infections are considered to be incidental and are unlikely to play any part in the transmission of the parasite in Africa.

Schistosoma mansoni

Infections in non-human primates

The first record of an animal infection was reported in 1928 by Cameron who found 5 out of 8 vervet monkeys (C. sabaeus) with S. mansoni on the Caribbean island of St Kitts, but nothing is known about the monkey infections except for the report of their occurrence. In Africa, Blackie (1932) examined 5 monkeys and 29 baboons, Van den Berghe (1934) 10 primates in the Congo and Porter (1938) 20 C. aethiops in South Africa, but no schistosome infections were found. Much later, natural infections in preserved tissue from baboons were reported by Miller (1959, 1960) and these reports led Nelson (1960) to investigate the subject further and soon the picture had changed considerably. Nelson returned to the Kibwezi area, from where Miller's infected baboons had originated, and within a week had found infections in 10 out of 12 animals. Overall Nelson examined 144 primates from various locations in East Africa of which 86 were baboons, 39 vervet monkeys, 16 sykes monkeys and 3 were bush babies. The 25 baboons from areas free of human schistosomiasis were all negative, but 35 out of 64 (53%) baboons from S. mansoni endemic areas were found infected with S. mansoni; 16 vervets and 3 sykes from the same endemic areas were also infected.

Nelson considered that his results, combined with Miller's observations showed that baboons were important reservoirs of S. mansoni infection in several areas of East Africa. The high infection rates, together with the location of the worms in the baboons, and the pathology, which were all similar to human infections led Nelson to suggest that a true zoonosis could be present.

This view was not universally accepted and in a detailed review of the literature on reservoir hosts, Barretto (1964) in his English summary concluded that 'while Papio have been found infected in Africa, he was of the opinion . . . that this may merely represent infections acquired from man'. The same author in 1966 stated 'man is the only proved reservoir host of S. mansoni.'

Reviewing the advances in knowledge of schistosomiasis in East Africa at that time, Webbe and Jordan (1966) stated that the overall impression was that baboons were probably capable of maintaining and transmitting S. mansoni. However, areas had not yet been found where it was certain that the infection was being maintained in a baboon community in the absence of infection contracted from man.

Table 8.1 Animal species found to harbour *Schistosoma haematobium* or *S. mansoni* infections in various African countries.

| Animal species | Common name | Country | Schistosome sp. | | Reference |
			S. haematobium	S. mansoni	
Non-human primates					
Pan satyrus	Chimpanzee	Sierra Leone	+		De Paoli 1965
Papio sp.	Senegalese baboon	Tanzania		+	Taylor *et al.* 1972
Papio sp.	Senegalese baboon	Senegal	+		Taylor *et al.* 1972
P. dogera	Dog-faced baboon	Kenya	+		Nelson 1960, Else *et al.* 1982
P. dogera	Dog-faced baboon	Tanzania		+	Fenwick 1969
P. dogera	Dog-faced baboon	Kenya		+	Miller 1959, 1960
P. dogera	Dog-faced baboon	Kenya		+	Nelson 1960, Nelson *et al.* 1962
P. dogera	Dog-faced baboon	Uganda		+	Nelson 1960
P. rhodesiae	Chacma baboon	Zimbabwe	+		Purvis *et al.* 1965
Cercopithecus aethiops	Vervet monkey	Kenya	+		Nelson *et al.* 1962
Cercopithecus aethiops	Vervet monkey	Kenya		+	Nelson *et al.* 1962, Else *et al.* 1982
Cercopithecus aethiops	Vervet monkey	Kenya		+	Cheever *et al.* 1970
Cercopithecus aethiops	Vervet monkey	Tanzania		+	Cheever *et al.* 1970
C. mitis	Sykes monkey	Kenya	+		Nelson 1960, Else *et al.* 1982
C. mitis	Sykes monkey	Kenya		+	Nelson *et al.* 1962, Else *et al.* 1982
C. mitis	Sykes monkey	Ethiopia		+	Fuller *et al.* 1979
Ungulates					
Sus scrofa	Large white boar	Nigeria	+		Hill & Onabamiro 1960
Syncerus caffer	Cape buffalo	South Africa	+		Basson *et al.* 1970
Ovis aries	Sheep	Zimbabwe	+		Mackenzie 1979
Ovis aries	Sheep	Zimbabwe		+	Mackenzie 1979
Kobus ellipsiprymnus	Common waterbuck	South Africa		+	Pitchford *et al.* 1974
	Cattle	Sudan		+	Karoum & Amin 1985

Rodents

Host	Common name	Locality			Reference
Otomys sp.	Wild rat	South Africa	+		Pitchford 1959
Arvicanthus niloticus	Nile rat	Egypt	+		Mansour 1973
Arvicanthus niloticus	Nile rat	Egypt		+	Mansour 1973
Arvicantuus niloticus	Nile rat	Ethiopia		+	Polderman 1974
Gerbillus pyramidum	Egyptian gerbil	Egypt		+	Kuntz 1952
Mastomys sp.	Multimate mouse	Kenya		+	Nelson et al. 1962
Mastomys sp.	Multimate mouse	Kenya		+	Ouma 1987
Mastomys sp.	Multimate mouse	South Africa		+	Pitchford 1959
Mastomys sp.	Multimate mouse	South Africa		+	Pitchford & Visser 1962
Damysmys	African marsh rat	Kenya		+	Nelson et al. 1962
Damysmys	African marsh rat	Zaire		+	Schwetz 1953, 1954, 1956
Lophuromys	Swamp rat	Zaire		+	Schwetz 1956
Oenomys	Rat	Zaire		+	Schwetz 1956
Otomys angoniensis	Vlei rat	South Africa		+	Pitchford 1959
Otomys angoniensis	Vlei rat	South Africa		+	Pitchford & Visser 1962
Otomys sp.	Rat	Kenya		+	Nelson et al. 1962
Pelomys sp.	Creek rat	Kenya		+	Kawashima et al. 1978
Rattus rattus	Black rat	Zaire		+	Schwetz 1956
Rattus rattus	Black rat	Sudan		+	Karoum & Amin 1985
Lemniscomys griselda	Striped field mouse	South Africa		+	Pitchford & Visser 1962
Lemniscomys striatus	Striped field mouse	Kenya		+	Ouma 1987
	Rat	Tanzania		+	McMahon & Baalaway 1967

Insectivores

Host	Common name	Locality			Reference
Crocidura luna	Katanga red musk	Zaire		+	Stijns 1952
C. oliviera	Egyptian shrew musk	Egypt		+	Kuntz 1958

Carnivores

Host	Common name	Locality			Reference
Canis familiaris	Domestic dog	Kenya		+	Nelson et al. 1962
Canis familiaris	Domestic dog	Kenya		+	Ouma 1987
Canis familiaris	Domestic dog	Tanzania			Mango 1971
Canis familiaris	Domestic dog	Sudan		+	Karoum & Amin 1985

In the mid-1960s in their search for animal reservoirs of schistosomiasis in the Mwanza area of Tanzania, McMahon and Baalawy (1967) examined 15 baboons and 14 monkeys but found no evidence to suggest that any of the animals had been infected with either *S. mansoni* or *S. haematobium*. In 1965 an expedition of Cambridge students examined 50 baboons from the Serengeti National Park also in Tanzania, but all were negative.

In the late 1960s a study in the Lake Manyara National Park in northern Tanzania indicated for the first time that baboons might be capable of firstly maintaining infection within their own community, and secondly passing on that infection to human intruders (Fenwick 1969). Over a two-month period a number of tourists became severely infected after swimming in a rock pool within the National Park. Thereafter all human contamination and contact with the pool was stopped. Stool samples collected from the baboons in the areas were positive for *S. mansoni* ova, snails collected from the pool were infected with *S. mansoni*, and when baboons were shot or trapped in the area, adult *S. mansoni* worms were found on dissection. It is speculated that in the first place the infestation could well have been introduced by the labourers who built a house on the banks of the pool, but it was concluded that thereafter the baboons maintained the cycle of infection from baboon to snail to baboon.

The discovery of natural infections in 11 of 47 grivet (vervet) monkeys (*C. aethiops*) from Tanzania and 6 of 18 monkeys from Ethiopia by Cheever *et al.* (1970) suggested that these animals too could make a significant contribution to transmission. An average of 49 adult worm pairs were found in 7 of the Tanzania monkeys and 9 pairs per animal in the Ethiopian animals. The faecal egg output was about 100 eggs per worm pair per day.

From 1974 through 1978 up to 50 tourists visiting the Omo National Park (Mui Game Park, Ethiopia) contracted *S. mansoni* after contact with the Mui River, a tributary of the Omo River (Fuller *et al.* 1979). Of the park rangers, 41% of 86 were found to be infected, and 27% of 184 *B. pfeifferi* collected from water contact sites near the tourist camps were shedding cercariae, of which 37 out of 184 (20%) were identified as *S. mansoni*. Studies among the indigenous people living upstream indicated they were infected. A natural infection was found in 1 of 2 baboons, and a heavy infection was found in 1 of 2 grivet monkeys (*C. aethiops*). Two colobus monkeys dissected were uninfected.

More recently Else *et al.* (1982) collected 40 vervet and 36 sykes monkeys from Lake Naivasha in the Kenya Rift Valley to establish a breeding colony in Nairobi. Faecal examination revealed that 8 of the vervet monkeys (20%) were passing *S. mansoni* type eggs and one animal yielded 152 worms and tissue eggs on perfusion. This finding is doubly important firstly because the vervets have limited territorial range, which suggests infection must be present along the lake at one or more of the tented camps and centres specializing in water sports, and secondly because vervets could contribute to transmission given such infection rates. To date no inter-human schistosomiasis transmission has been reported from around the lake (Pamba & Roberts 1979).

The non-human primates *Papio* sp. and *Cercopithicus* sp. are clearly capable of acquiring natural infections with *S. mansoni*, and under certain circumstances of maintaining that infection among themselves through host snails in water bodies that they frequent. In areas where man and monkeys overlap there is a real possibility that monkeys and baboons will act as reservoir hosts for *S. mansoni*.

Infections in rodents

Kuntz (1952) provided the first record of a rodent (a wild gerbil) naturally infected with *S. mansoni*. This finding stimulated laboratory and field investigations to examine wild rodents as reservoirs of human schistosomiasis infections in Africa. Extensive surveys in South Africa revealed natural infections in *Mastomys* and *Otomys* (Pitchford 1959), but they were thought to be of little importance in transmission. Pitchford and Visser (1962) found a single natural infection in *Lemniscomys griselda* and showed this animal to be a good laboratory host for *S. mansoni*. Earlier studies in Zaire by Schwetz (1953, 1954, 1956) recorded infections of *S. mansoni* var. *rodentorum* in *Dasymys*, *Lophuromys* and *Oenomys*, but according to Teesdale and Nelson (1958) and Pitchford and Visser (1962) these may have been *S. mansoni* infections. Natural infections in Nile rats (*A. niloticus*) have been reported in Egypt (Mansour 1973), Ethiopia (Polderman 1974) and Sudan (Karoum & Amin 1985). In Kenya, Nelson *et al.* (1962) found natural infections in *Otomys* sp., *Mastomys* and *Dasyms*. Nelson (*et al.* 1962, Nelson 1983) argued that unfortunately most of the rodents examined so far were savannah species and stressed the importance of the discovery of natural infections in *D. incomtus* and *L. flavopunctatus*, two species which are closely associated with water. In Taveta in Kenya, 18 out of 41 'creek rats' (*Pelomys* sp.) were found to pass viable eggs in their faeces (Kawashima *et al.* 1978) and the authors concluded that the rodent must be a good candidate as a reservoir host for *S. mansoni* there. Ouma (1987) in Kenya recorded natural infections in 6 out of 119 *Lemniscomys striatus*, 2 out of 271 *Tatera robusta* and 3 out of 222 *M. natalensis*, but suggests that the numbers of infected rodents was too few to be of significance in transmission. Elsewhere in East Africa, McMahon and Baalawy (1967) found three out of 47 rats (unspecified species) from Mwanza, Tanzania with high infections of *S. mansoni*.

While the significance of rodents in transmission of *S. mansoni* in Africa remains unclear and requires further study, recent studies in South

America suggest that rodents may be important hosts in maintaining transmission on that continent (Rollinson *et al.* 1986).

Infections in other wild animals

S. mansoni has been recorded in the insectivore *Crocidura luna* (Stijns 1952) in the Congo and by Kuntz (1958) in Egypt. Pitchford *et al.* (1974) recorded a natural infection of *S. mansoni* in a waterbuck. Besides these cases, no other natural infections with *S. mansoni* in wild animals other than primates and rodents have been reported.

Infection in domestic animals

Domestic animals have long been known as reservoirs of great epidemiological importance for *S. japonicum* (Martins 1958). However, very little is known about the role of domestic animals in the transmission of *S. mansoni* and *S. haematobium*. Barbosa *et al.* (1962) found *S. mansoni* adult worms in four cows in Brazil and ova in one of them. The first report of domestic animal infection in Africa was by Nelson *et al.* (1962) who, working in Kenya, found light infections in 2 out of 9 dogs examined. Mackenzie (1979) reported natural infections of both *S. haematobium* and *S. mansoni* in sheep in Zimbabwe. The following year Mango (1971) working around Mwanza in Tanzania recovered a few eggs from the faeces of 14 (8.8%) out of 160 dogs examined. He also observed that dogs frequently ingested human faeces which contained *S. mansoni* eggs. These dogs subsequently excreted a small number of non-viable eggs. An autopsy of one of the dogs revealed only one dead egg in the liver. He concluded that dogs are refractory to the Mwanza 'strain' of *S. mansoni* and are unlikely to play a significant role in transmission of *S. mansoni* in that area.

Karoum and Amin (1985) working in the Blue Nile Project in the Sudan failed to find any *S. mansoni* eggs in the faeces of 55 stray dogs, but they recovered adult worms and eggs in the tissues of 15 (27.3%) of the dogs examined. It was not possible to hatch the eggs recovered and they too concluded that dogs are unlikely to play a major role in *S. mansoni* transmission. The same authors reported natural *S. mansoni* infections in 2 out of 98 cattle examined, but did not find infections in sheep or goats.

Recent studies in the Machakos District of Kenya (Ouma 1987) revealed that 40.5% of 242 dogs examined were passing ova of *S. mansoni*, none of which were viable. However, experimentally infected puppies were found to pass some viable *S. mansoni* eggs after feeding them on infected human faeces. Dogs in the study area were frequently noticed to eat human faeces and it was concluded that dogs may be important in disseminating *S. mansoni* eggs through coprophagy, but further work is needed to confirm this.

Concluding remarks and recommendation for future studies

Although a few wild and domestic animals have been found to be naturally infected with *S. haematobium*, man remains the only significant maintenance host for this parasite in Africa. However, it is recommended that more extensive searches for other suitable animal hosts should be made.

Man is also the most important definitive host of *S. mansoni*, but primates may also be important in the transmission of this parasite in certain areas of East Africa where significant numbers of these animals still exist. Recent findings suggest that in South America, the West Indies and possibly Africa, rodents may play an important role in maintaining transmission of *S. mansoni*. The true significance of both primates and rodents in the epidemiology of human disease in Africa still remains arcane and merits further study. Specifically, we need to know if primates or rodents can maintain the infection amongst their own communities independent of human infection. Using modern techniques such as biochemical analysis, or DNA probes, there is a need to confirm if the parasites naturally recovered from animals represent the same 'strain' as the parasites found in humans. Such studies should prove to be useful in designing suitable control strategies in the affected areas of Africa and also possibly beyond.

References

Abdel-Salam, E. & A. Elsan 1978. Cystoscopic picture of *Schistosoma haematobium* in Egyptian children correlated to intensity of infection and morbidity. *Am. J. Trop. Med. Hyg.* **35**, 786–90.

Abdel-Wahab, M. F., G. T. Strickland, A., El-Shahly, A. S. Zakaria *et al.* 1980. Schistosomiasis mansoni in an Egyptian village in the Nile Delta. *Am. J. Trop. Med. Hyg.* **29**, 868–74.

Barbosa, F. S., I. Barbosa & F. Arruda 1962. *Schistosoma mansoni* natural infection of cattle in Brazil. *Science* **138**, 831.

Barretto, A. C. 1964. The importance of animals as reservoir hosts of human schistosomiasis. *Arq. Hig. Saúde Públ.* **29**, 95–102.

Barretto, A. C. 1966. The importance of animals as reservoir hosts of human schistosomiasis. *Hosp. Rio J.* **69**, 807–15

Basson, P. A., R. M. McCully, S. P. Kruger, J. W. Van-Niekerk *et al.* 1970. Parasitic and other diseases of the African buffalo in the Kruger National Park. *Onders. J. Vet. Res.* **37**, 11–28.

Blackie, W. K. 1932. A helminthological survey of Southern Rhodesia. *Med. Lon. Sch. Hyg. Trop. Med.* **5**, 1–91.

Bulsara, M. K., T. Y. Sukwa, T. Y. & F. K. Wurapa 1985. Risks of liver enlargement in schistosomiasis mansoni infection in a rural Zambian community. *Trans. R. Soc. Trop. Med. Hyg.* **79**, 535–6.

Cameron, T. W. M. 1928. A new definitive host for *Schistosoma mansoni*. *J. Helminthol.* **6**, 219–22.

Cheever, A. W., R. L. Kirschstein & L. V. Reardon 1970. *Schistosoma mansoni* infection of presumed natural origin in *Cercopithecus* monkeys from Tanzania and Ethiopia. *Bull. Wld Hlth Org.* **42**, 486–90.

De Paoli, A. C. 1965. *Schistosoma haematobium* in the chimpanzee – a natural infection. *Am. J. Trop. Med. Hyg.* **14**, 561–5.

Else, J. G., Statzger, M. & R. F. Sturrock 1982. Natural infections of *Schistosoma mansoni* and *S. haematobium* in *Cercopithecus* monkeys in Kenya. *Ann. Trop. Med. Parasitol.* **76**, 111–2.

Fenwick, A. 1969. Baboons as reservoir hosts of *Schistosoma mansoni*. *Trans. R. Soc. Trop. Med. Hyg.* **63**, 557–67.

Forsyth, D. M. & D. J. Bradley 1966. The consequences of bilharziasis. Medical and public health importance in north-west Tanzania. *Bull. Wld Hlth Org.* **34**, 715–35.

Fujinami, K. 1910. Research on the so-called schistosomiasis japonica in Hiroshima Prefecture. Tokyo Iji Shinski (Tokyo Med. J.) **2**, 10.

Fuller, G. K., A. Lemma & T. Haile 1979. Schistosomiasis in the Omo National Park of south-west Ethiopia. *Am. J. Trop. Med. Hyg.* **28**, 526–30.

Gryseels, B. & A. M. Polderman 1987. The morbidity of schistosomiasis in Maniema (Zaire). *Trans. R. Soc. Trop. Med. Hyg.* **81**, 202–9.

Hill, D. H. & S. D. Onabimiro 1960. Vesical schistosomiasis in the domestic pig. *Brit. Vet. J.* **116**, 145–50.

Karoum, K. O. & M. A. Amin 1985. Domestic and wild animals naturally infected with *Schistosoma mansoni* in the Gezira irrigated scheme, Sudan. *J. Trop. Med. Hyg.* **88**, 83–9.

Katsurada, F. 1904. A new parasite of man by which an endemic disease in various areas of Japan is caused. *Annot. Zool. Japan.* **5**, 146–60.

Kawashima, K., D. Katamine, M. Sakamoto, M. Shimada *et al.* 1978. Investigation on the role of wild rodents as reservoirs of human schistosomiasis in the Taveta area of Kenya, East Africa. *Jap. J. Trop. Med. Hyg.* **6**, 195–203.

Kuntz, R. E. 1952. Natural infection of Egyptian gerbil with *Schistosoma mansoni*. *Proc. Helminthol. Soc. Washington.* **19**, 123–4.

Kuntz, R. E. 1958. *Schistosoma* sp. in shrews in Lower Egypt *Proc. Heminthol. Soc. Washington* **25**, 37–40.

Leiper, R. T. 1915–1918. Report on the results of the bilharziasis mission in Egypt. *J. R. Army. Med. Corps.* **25**, 1–55, 147–92, 253–67.

Mackenzie, R. L. 1979. Investigation into schistosomiasis in sheep in Mashamba land. *Cent. Afr. J. Med. Supp.* **16**, 27–8.

Mango A. M. 1971. The role of dogs as reservoirs in the transmission of *Schistosoma mansoni*. *E. Afr. Med. J.* **48**, 298–306.

Manson, P. 1902. Report on a case of bilharzia from West Indies. *J. Trop. Med.* **5**, 384–5.

Mansour, N. S. 1973. *Schistosoma mansoni* and *S. haematobium* found as natural double infection in the rat, *Arvicanthus niloticus* from human endemic area in Egypt. *J. Parasitol.* **59**, 424.

Martins, A. V. 1958. Non-human vertebrate hosts of *Schistosoma haematobium* and *S. mansoni. Bull. Wld Hlth Org.* **8**, 931–44.

McMahon, J. E. & S. S. Baalawy 1967. A search for animal reservoirs of *Schistosoma mansoni* in Mwanza area of Tanzania. *E. Afr. Med. J.* **44**, 325–6.

Miller, J. H. 1959. Correspondence. *E. Afr. Med. J.* **36**, 56.

Miller, J. H. 1960. *Papio doguera* (dog face baboon) a primate reservoir of *Schistosoma mansoni* in East Africa. *Trans. R. Soc. Trop. Med. Hyg.* **4**, 44–6.

Miyagawa, Y. 1912. Experimental study on the infection of *Schistosoma japonicum* by way of mouth. *Tokyo Iji Shinski* (*Tokyo Med. J.*) No. 1766, 1005.

Miyagawa, Y. 1913. Concerning the cercariae of *Schistosoma japonicum* and its penetration into the body of the host. (Contribution to the knowledge of the prophylaxis of the disease). *Iji Shimbun* (Med. News). No. 891, 1597–608.

Miyairi, K. & M. Suzuki 1913. A contribution to the development of *Schistosoma japonicum. Tokyo Iji Shinski* (*Tokyo Med. J.*) 1936, No. 1836, 1–5.

Nelson, G. S. 1960. Schistosome infections as zoonoses in Africa. *Trans. R. Soc. Trop. Med. Hyg.* **54**, 301–24.

Nelson, G. S. 1983. Wild animals as reservoir hosts of parasitic diseases of man in Kenya. In *Tropical parasitoses and parasitic zoonoses*, J. D. Dunsmore (ed.), 59–72. Tenth Meeting of the World Association for the Advancement of Veterinary Parasitology, Perth, Australia.

Nelson, G. S., C. Teesdale & R. B. Highton 1962. The role of animals as reservoirs of bilharziasis in Africa. In *CIBA Foundation Symposium on Bilharziasis*. G. E. W. Wolstenholme & M. O'Connor (eds), 149–227. London: Churchill.

Nojima, H. M. 1978. Investigation on the role of wild rodents as reservoirs of human schistosomiasis in the Taveta area of Kenya, East Africa. *Jap. J. Trop. Med. Hyg.* **6**, 195–203.

Ouma, J. H. 1987. Transmission of *Schistosoma mansoni* in an endemic area of Kenya with special reference to the role of human defecation behaviour and sanitary practices. PhD thesis, University of Liverpool.

Pamba, H. & J. M. Roberts 1979. Schistosomiasis in and around Lake Naivasha, Kenya, seven years surveillance. *E. Afr. Med. J.* **56**, 255–62.

Pfister, E. 1912. Ueber Die A. A. A. Krankheit Der Papyri Ebers and Brugsch. *Arch. Gesch. Med.* **6**, 12–20.

Pitchford, R. J. 1959. Natural schistosome infection in South Africa rodents. *Trans. R. Soc. Trop. Med. Hyg.* **53**, 213.

Pitchford, R. J. & P. S. Visser 1962. The role of naturally infected wild rodents in the epidemiology of schistosomiasis in the eastern Transvaal. *Trans. R. Soc. Trop. Med. Hyg.* **56**, 126–35.

Pitchford, R. J., P. S. Visser, V. Pienaar, V. de V. & E. R. J. Young 1974. Further observations on *Schistosoma mattheei* Veglia and Le Roux 1929 in the Kruger National Park. *J. Afr. Vet. Ass.* **45**, 211–8.

Polderman, A. M. 1974. The transmission of intestinal schistosomiasis in Begemder province, Ethiopia. *Acta Leid. Res.* **42**, 1–193.

Porter, A. 1938. The larval trematode found in certain South African Mollusca. *S. Afr. Inst. Med. Res.* **8**, 2–20.

Purvis, A. J., I. R. Ellison & E. L. Husting 1965. A short note on the findings of schistosomiasis in baboons (*Papio rhodesiae*) *Cent. Afr. J. Med.* **11**, 368.

Rollinson, D., D. Imbert-Establet & G. C. Ross 1986. *Schistosoma mansoni* from natural infected *Rattus rattus* in Guadeloupe, identification, prevalence and enzyme polymorphism. *Parasitology* **93**, 39–53.

Ruffer, M. A. 1910. Remarks on the histology, pathology and anatomy of Egyptian mummies. *Cairo Sci. J.* **1**, 3–7.

Sambon, L. 1907. Remarks on *Schistosomum mansoni*. *J. Trop. Med. Hyg.* **10**, 303.

Schwetz, J. 1953. On a new schistosome of wild rodents found in the Belgian Congo. *Schistosoma mansoni var. rodentorum* var. nov. *Ann. Trop. Med. Parasit.* **47**, 183–6.

Schwetz, J. 1954. On two schistosomes of wild rodents of the Belgian Congo. *Schistosoma rodheini* Brumpt, 1931 and *Schistosoma mansoni* var. *rodentorum* Schwetz, 1953 and their relationship to *S. mansoni* of man. *Trans. R. Soc. Trop. Med. Hyg.* **48**, 89–100.

Schwetz, J. 1956. Role of wild rats and domestic rats (*Rattus rattus*) in schistosomiasis of man. *Trans R. Trop Med. Hyg.* **50**, 275–82.

Siongok K. T. A., A. A. F. Mahmoud, J. H. Ouma *et al.* 1976. Morbidity in schistosomiasis mansoni in relation to intensity of infection study of a community in Machakos, Kenya. *Am. J. Trop. Med. Hyg.* **25**, 273–84.

Stijns, J. 1952. Sur les rongeus hôtes naturals de schistosoma rodhaini Brumpt. *Ann. Parasitol. Human Comp.* **27**, 385–6.

Taylor, M. G., G. S. Nelson & B. J. Andrews 1972. A case of natural infection of *Schistosoma haematobium* in a Senegalese baboon (*Papio* sp.). *Trans. R. Soc. Trop. Med. Hyg.* **66**, 16–17.

Teesdale, C. & G. S. Nelson 1958. Recent work on schistosomes and snails in Kenya. *E. Afr. Med. J.*, **35**, 427–38.

Van den Berghe 1934. Les schistosomes humaine et animals au Katanga (Congo Belge). *Ann. Soc. Belg. Med. Trop.* **14**, 313–71.

Webbe, G. & P. Jordan 1966. Recent advances in knowledge of schistosomiasis in East Africa. *Trans. R. Trop. Med.* **60**, 279–312.

WHO 1965. Monograph series No 50. World Health Organization, Geneva.

WHO 1985. The control of schistosomiasis. WHO Technical report series No. 728, World Health Organization, Geneva.

Wright, C. A. 1966. Relationship between schistosomes and their molluscan hosts in Africa. *J. Helminthol.* **40**, 403–12.

Wright, C. A., V. R. Southgate & R. J. Knowles 1972. What is *Schistosoma intercalatum* Fisher, 1934? *Trans. R. Soc. Trop. Med. Hyg.* **66**, 28–64.

9 Baboons, bovines and bilharzia vaccines

Introduction

The starting point for all our research on bilharzia vaccines was Nelson's work on 'heterologous immunity' in schistosomiasis. This concept originated with Le Roux, who was Nelson's immediate predecessor as Reader in the Department of Medical Helminthology at the London School, and a prominent schistosomiasis researcher. Le Roux had noted in 1961 that in areas like Sardinia, Corsica and Sicily, where *Schistosomiasis bovis* occurs in cattle, there was no *S. haematobium* infection in man, even though the local population of *Bulinus truncatus*, the intermediate host of *S. bovis*, was highly susceptible to experimental infection with *S. haematobium* from Egypt. To explain this, Le Roux suggested that cercariae from animals may immunize man against *S. mansoni* or *S. haematobium*. Conversely, it seemed possible that animals exposed to the human infections may develop some immunity against their own species of schistosomes. This theory was applied to Africa by Nelson *et al.* (1962), who noted that there were many species of schistosomes in domestic and wild animals in Africa, and suggested that, although most of these failed to reach maturity in man, constant exposure to their cercariae might interfere with the development of *S. mansoni* and *S. haematobium*, resulting in a marked modification in the pathogenicity of these infections. Nelson called this phenomenon of modification of disease resulting from natural exposure to relatively non-pathogenic infections 'zooprophylaxis', and proceeded to look for evidence of it, using experimental animal models.

Heterologous immunity – the first experimental evidence

At the time when Nelson started this work, some experiments had already been done on heterologous immunity between different species of human schistosomes in animals. Vogel and Minning (1953) had shown that *S. japonicum*-infected rhesus monkeys developed a strong resistance to heterologous challenge with *S. mansoni*, and Meleney and Moore (1954) had reported that previous infections of a rhesus monkey with *S. haematobium* gave some protection against *S. mansoni*. Later, Hsu and Hsu (1961, 1963) and Hsu *et al.* (1965) looked at the interaction between different geographical strains of one species, rather than between different species, and found that previous infections of rhesus monkeys with cercariae of a zoophilic strain of *S. japonicum* gave a high degree of resistance to challenge with the human strain. Subsequently, the first solid evidence for 'zooprophylaxis' was provided by Hsu *et al.* (1966), when they demonstrated that rhesus monkeys repeatedly exposed to *S. bovis* were highly resistant to *S. haematobium* challenge.

In their initial experiments, Nelson, together with his PhD students M. A. Amin (from Sudan) and M. F. Saoud (from Egypt), and his former co-worker from Kenya, C. Teesdale, carried out a wide range of heterologous resistance experiments in mice, as a prelude to large animal experiments (Nelson *et al.* 1968). In line with the 'zooprophylaxis' theory, they found that a single initial exposure to either *S. bovis*, *S. mattheei*, or *S. rodhaini* conferred a high degree of resistance to *S. mansoni* challenge, as manifested by reductions in both worm and tissue egg counts, and reductions in worm fecundity. Indeed, they detected a strong reciprocal resistant response, with significant reductions produced in adult worm and egg counts of both 'immunizing' and 'challenge' infections. This seemed at variance with the results of Smithers and Terry (1965), working on homologous resistance to *S. mansoni* in rhesus monkeys, who had come to the far-reaching conclusion that the immune mechanisms involved in preventing reinfection did not affect established worms, or their fecundity, although immunity was mainly induced by the adult worms (an example of 'concomitant immunity'). These experiments on mice were then extended to rhesus monkeys, and it was shown that these hosts, too, can develop resistance to *S. mansoni* after exposure to either *S. bovis* or *S. mattheei* cercariae (Amin *et al.* 1968).

Although it was the prevailing belief that the adult worms were the main stimulus to immunity, these experiments suggested that neither the presence of adult worms (nor of eggs in the tissues) were absolute requirements for the development of schistosome immunity, because no *S. bovis* or *S. mattheei* adult worms were recovered at perfusion and because very few eggs were found in the faeces or tissue of highly resistant monkeys. It was conceded that, as suggested by Smithers and Terry (1965), it might be that a marked immunity will only develop if at least some of the worms reached the adult stage. This indeed did seem to hold for heterologous immunity in the mouse model, because it was found in subsequent experiments (Amin & Nelson 1969) using the *S. mattheei/S. mansoni* system that an interval of at least six weeks between initial infection and challenge was necessary for significant heterologous resistance to develop, and that the degree of resistance induced was increased when the immunizing period was extended. It was also shown that the presence of eggs did not seem to be essential for the induction of heterologous resistance in mice, since immunization with male cercariae conferred a marked degree of heterologous resistance (although less than that stimulated by a mixture of male and female cercariae). In these experiments, the main manifestation of immunity was a reduction in tissue egg counts; whereas there was only a marginal difference in the number of adult worms recovered, the number of eggs deposited in the tissues of vaccinated mice was only 10% of the number in the controls.

Since it is now known from studies on resistance to reinfection with *S. mansoni* in mice (Harrison *et al.* 1982) that the presence of supernumery unpaired worms can depress worm fecundity by some non-immunological means, these 'single-sex' experiments seem difficult to interpret in the light of current knowledge. At the time, it was concluded by Amin and Nelson (1969) that eggs were not necessary for the induction of resistance, and this was in line with the conclusions of Smithers (1968), working on homologous immunity to *S. mansoni*, who had shown that single-sex adult worms transferred directly into the hepatic portal system of rhesus monkeys stimulated a marked degree of resistance, in contrast to the absence of resistance in monkeys inoculated with several thousand living eggs. Compatible results were obtained during more recent studies on acquired resistance in bovine bilharziasis (Bushara *et al.* 1983) in which it was shown that the transfer of adult male *S. bovis* from chronically infected cattle into naïve recipients engendered partial protection against challenge infections in the latter. Thus cattle are similar to some other host species in their ability to mount resistance to bilharziasis in the absence of eggs, and migratory stages, of the parasites.

Reverse zooprophylaxis

With two other Sudanese PhD students, M. F. Hussein and A. A. Saeed, Nelson next extended this work into the veterinary field, carrying out an experiment to see whether calves could be protected against *S. mattheei* by prior exposure to the human parasite *S. mansoni*. Interestingly, it was found that calves could develop patent infections with *S. mansoni* in the laboratory, suggesting that they might be involved in the transmission of this parasite to man under natural conditions. Furthermore, although *S. mansoni* had little pathogenicity in calves, it did confer on them some degree of heterologous resistance against subsequent infection with *S. mattheei* (Hussein *et al.* 1970). Heterologous resistance was thus 'reciprocal', since in the experiments with monkeys, *S. bovis* had protected against *S. haematobium* (Hsu *et al.* 1966) and *S. mattheei* had protected against *S. mansoni* (Amin *et al.* 1968). In addition to the reductions in the worm and egg loads, immunization of the calves with *S. mansoni* had apparently led to stunting of the 'challenge' worms. Another finding in these heterologous immunity experiments was the occurrence of cross-pairing between heterologous species, which might have influenced the egg production of the females (see also Taylor *et al.* 1969, Taylor 1970).

In a similar study in sheep, carried out with Nelson's PhD students J. M. Preston and A. A. Saeed, it was later demonstrated that sheep infected with *S. mansoni* were likewise partially resistant to challenge with *S. mattheei* (Preston *et al.* 1972). The reduction in adult worm loads was not as marked as in the calf, but the degree of resistance demonstrated by the reductions in tissue egg counts and adult worm lengths was considerable, and there was also evidence of a reduction in fecundity of the challenge-derived worms. As in the previous experiment, in which calves were immunized with *S. mansoni* cercariae and challenged with *S. mattheei*, the sheep in this experiment all passed some viable *S. mansoni* eggs in their faeces, again suggesting a role for ruminant reservoirs of *S. mansoni* infection (see also Ch. 8).

In subsequent experiments (Massoud & Nelson 1972) these observations were extended by investigating heterologous immunity between animal and human schistosomes in Iran, with a view to assessing their possible importance in nature. In preliminary experiments in mice, the cattle parasite *Ornithobilharzia turkestanicum* was used as an immunizing agent. In subsequent experiments, observations were made on cross-resistance between *O. turkestanicum*, *S. bovis* and *S. haematobium* in calves and sheep and the levels of immunity produced by homologous and heterologous systems were compared. Immunization with *O. turkestanicum* produced a high degree of resistance in mice against challenge with *S. bovis*, *S. haematobium*, or *S. mansoni*, and calves, exposed to the relatively non-pathogenic *O. turkestanicum*, developed a marked resistance to a heavy challenge infection with *S. bovis*, and vice versa. The results with *S. haematobium* were even more striking, as this parasite was generally regarded as being non-infective to livestock. Exposure of the calves to large doses of *S. haematobium* showed that a considerable number of worms developed to maturity (between 1% and 3% of the cercariae applied) but no viable eggs were produced. Nevertheless, calves were partially resistant to heterologous challenge with *S. bovis* or *O. turkestanicum* after exposure to *S. haematobium* cercariae. The results of the experiments in sheep were less dramatic, but again some protection was demonstrated.

It was concluded that the simultaneous transmission of *S. bovis* and *S. haematobium* in the same area may be to mutual benefit of man and his livestock in reducing the effects of schistosomiasis. It was further recommended that field studies should be carried out in the Middle East and Africa to assess the epidemiological significance of heterologous resistance, and that experiments should be carried out to develop optimal immunization procedures in cattle and sheep with a view to field testing in areas where schistosomiasis was an important veterinary problem.

Heterologous and homologous immunity to *S. mansoni* compared in baboons

Because of the encouraging results obtained by immunizing rhesus monkeys with heterologous cercariae, Nelson and his colleagues next sought to determine whether heterologous immunization of baboons against *S. mansoni* was possible, using either schistosomes naturally non-pathogenic in the baboon, e.g. *S. rodhaini* and *S. bovis*, or genetically altered hybrid forms (Taylor *et al.* 1973a). It had previously been found that *S. rodhaini* and *S. bovis* do not mature in baboons and that the two hybrids to be used were of relatively low pathogenicity in baboons, even though mature adult worms were produced (Taylor *et al.* 1973b). It had already been shown that rhesus monkeys could be partially immunized against *S. mansoni* challenge by immunization with *S. bovis* or *S. mattheei* cercariae (Amin *et al.* 1968, Eveland *et al.* 1969), but it was not known how other non-human primate species would react. The baboon, a natural host of *S. mansoni* and *S. haematobium* in Africa (Nelson *et al.* 1962) was chosen for this purpose because it appeared to be one of the best experimental models of human schistosomiasis (Sadun *et al.* 1966, Jordan *et al.* 1967, Webbe *et al.* 1972, Damian *et al.* 1972). Furthermore, preliminary observations by Newsome (1956) and McMahon (1967) had already suggested that baboons could develop partial acquired resistance to reinfection with *S. mansoni* and likewise, Webbe and James (1973) had reported that baboons could also become partially resistant to reinfection with *S. haematobium*. The large range of immunizing agents used was also expected to provide indirect evidence of the possible importance of heterologous acquired immunity in the epidemiology of human schistosomiasis. For comparison, parallel attempts were also made to immunize baboons with the homologous parasite, *S. mansoni*. It was considered of great interest to determine whether 'concomitant' immunity, as described in the rhesus monkey, also occurred in other species of primates, particularly those that had host–parasite relationships more similar to that between man and *S. mansoni*.

It was found that baboons with established *S. mansoni* infections were indeed partially immune to reinfection with the homologous parasite, as shown by their faecal egg counts, worm recoveries, and tissue egg counts. Baboons could also be partially protected against *S. mansoni* by exposure to heterologous cercariae of *S. rodhaini* and *S. bovis* that presumably died before maturing, since no adults or eggs were found. Unexpectedly, less resistance was stimulated by the hybrids, even though substantial numbers of hybrid adult worms were present. Thus in the baboon model, the presence of an adult worm infection did not necessarily engender much resistance to challenge with *S. mansoni*; more resistance could be stimulated by the short-lived schistosomula stages. This suggested that further investigations would be worthwhile on the development of live

S. mansoni vaccines incorporating attenuated early stages of the parasites, or heterologous infections that failed to mature, and that the development of 'defined antigen' vaccines based on stage-specific schistosomula antigens might eventually be feasible.

In subsequent experiments, Nelson and colleagues reinvestigated the epidemiologically important question of whether heterologous resistance develops between *S. mansoni* and *S. haematobium*. Earlier experiments in mice (Nelson *et al.* 1968) had shown an absence of heterologous immunity, but as the mouse was acknowledged to be a poor model for *S. haematobium* infections, further experiments in more 'natural' host species were considered necessary. For these, hamsters and later baboons were used.

In contrast to the mouse, a high proportion of *S. haematobium* cercariae mature and produce viable eggs in hamsters and acquired homologous resistance to both *S. haematobium* and *S. mansoni* occurs in hamsters (Purnell, 1966, Smith *et al.* 1976). In the heterologous resistance experiments, it was found that an eight-week-old primary infection with *S. mansoni* induced a strong heterologous resistance to *S. haematobium* and vice versa (Smith *et al.* 1976). Subsequently, experiments were carried out to determine whether a primary infection in the baboon with *S. haematobium* could stimulate significant acquired immunity to *S. mansoni* (Webbe *et al.* 1979). Baboons immunized with *S. haematobium* were first shown to have developed a strong resistance to the homologous infection before challenge with the heterologous parasite. It was then shown that heterologous immunity was also present: this reduced not only the proportion of adult worms that developed, but also egg excretion, and the pathological reactions to the eggs in the tissues. In studies on the possible mechanisms involved in this resistance, it was shown that serum from all the immunized baboons had antibodies that were cytotoxic for *S. mansoni* schistosomula *in vitro* in the presence of unfractionated granulocytes. The range of schistosome species tested as heterologous immunizing agents was extended by Eveland *et al.* (1969), who demonstrated reciprocal heterologous immunity between *S. japonicum* and *S. mansoni* in rhesus monkeys, and also confirmed the Amin *et al.* (1968) observations that monkeys could be partially protected against *S. mansoni* challenge by immunization with *S. bovis*. No heterologous immunity was observed, but in monkeys immunized with *S. bovis* and challenged with *S. japonicum* it was suggested that the degree of heterologous immunity was related to the phylogenetic closeness between the schistosome species involved.

In cattle, further evidence of heterologous immunity was obtained, using *S. mansoni* to immunize against *S. bovis*, though this time the immunizing cercariae (*S. mansoni*) were first attenuated by irradiation (Hussein & Bushara 1976). This was considered necessary following further evidence that not only cattle but also sheep (Adam & Magzoub

1976, 1977) and goats (Gar el Nabi, in preparation) could develop patent *S. mansoni* infections experimentally and that this parasite could lead to prominent portal vascular lesions in sheep (Saeed & Nelson 1974). Following irradiation, the *S. mansoni* cercariae indeed failed to attain sexual maturity in cattle, yet they were still capable of eliciting partial resistance to *S. bovis*. A further confirmation was thus provided in support of the view that neither the presence of adult worms nor their eggs was essential for the acquisition of resistance. This study was also one of the several studies emphasizing the differences between animal models in their ability to mount resistance against bilharziasis, since, for example, parallel studies in sheep (Bickle *et al.* 1979a) showed that although irradiated *S. mattheei* schistosomula induced strong homologous resistance to *S. mattheei* cercariae challenge, no resistance to heterologous challenge with *S. bovis* cercariae was present; neither did vaccination with irradiated *S. mansoni* schistosomula protect sheep against *S. mattheei* cercariae challenge.

Heterologous immunity between schistosomes and liver flukes – implications for understanding the immune mechanisms involved

In spite of the repeated demonstrations of heterologous immunity in experimental schistosomiasis, surprisingly little work has been carried out on the mechanisms involved. However, there have been several investigations on heterologous immunity between schistosomes and a range of other types of organisms, particularly the liver flukes *Fasciola hepatica* and *F. gigantica*, some of which shed light on the mechanisms involved. Some of these studies involved cattle and sheep and they are of special interest since these ruminants are natural hosts of schistosomes and liver flukes, both of which often co-exist in the same habitat under field conditions in Africa. For example, Siraj *et al.* (1981) showed that calves undergoing primary, patent infection with *S. bovis* resisted challenge with *F. hepatica*; they regarded the liver fibrosis elicited by the immunizing infection as insufficient to account for this heterologous resistance. Heterologous immunity between these two parasites was also demonstrated in sheep (Monrad *et al.* 1981) immunized by either immature, or newly patent, schistosome infections whereas, unexpectedly, chronic *S. bovis* infections were non-protective against the liver fluke. These results again suggested that resistance to heterologous challenge was unlikely to be caused by pathology resulting from the immunizing infections. More recently, and partly through the facilities of the bovine bilharziasis project which Nelson started in the Sudan, a group of Nelson's former students and their colleagues investigated heterologous immunity between bilharziasis and fascioliasis in ruminants. This time, heterologous immunity between Sudanese strains of *S. bovis* and the larger, tropical liver fluke, *F. gigantica*, was investigated in Sudanese zebu calves by Yagi *et al.* (1986). They found very high levels of resistance to challenge with *S. bovis* in calves infected with *F. gigantica* for eight weeks, and similarly high resistance in calves primarily infected with *S. bovis* and challenged, eight weeks later, with *F. gigantica*. Liver enzyme tests showed that, in both cases, the primary infections had caused some liver damage, suggesting the possible contribution of liver damage to the expression of resistance to challenge.

More detailed investigations on immune mechanisms have been possible in rodent models. In mice, a marked, reciprocal resistance has been demonstrated between *S. mansoni* and *F. hepatica*. However, although some immunity to *S. mansoni* was detected in mice simultaneously infected with *S. mansoni* and *F. hepatica*, neither single-sex, nor prepatent *S. mansoni*, nor immature *F. hepatica* infections induced heterologous immunity. These results suggested that 'heterologous immunity' may in fact be dependent upon pathological changes induced by the primary infections. Ford *et al.* (1987a) consequently carried out detailed investigations into the mechanisms involved; for this, the rat was chosen as a more suitable host species than the mouse because of the severe pathology caused in mice by even light *F. hepatica* infections and because mice fail to develop specific immunity to reinfection with *F. hepatica*. In the rat, on the other hand, there was general agreement that the immune system was effective in limiting infections of *F. hepatica*, mainly due to the consistent success of passive transfer experiments. The rat was also amenable to immunological studies on *S. mansoni*, as strong resistance developed, which was known to be at least partially mediated by specific immune mechanisms. In order to avoid the possible complications of pathology, attempts were made to induce heterologous immunity by using irradiated infections, implantation of flukes in diffusion chambers, and immune serum transfers. In this regard, the only similar studies involving ruminants are those of Yagi *et al.* (1986) on calves. Although these authors had shown a strong, reciprocal heterologous immunity between *F. gigantica* and *S. bovis*, that immunity was attainable only after immunization with normal cercariae; only a much lower degree of resistance could be stimulated in these animals when the immunizing larvae were attenuated by irradiation at 3 Krad. It was suggested, therefore, that heterologous immunity in this case was either stimulated by stage-specific adult worm antigens or that some liver damage by the immunizing infections was required for the stimulation of resistance.

High levels of heterologous immunity to *S. mansoni* challenge were detected in Fischer rats infected four weeks previously with *F. hepatica*, but serum from *F. hepatica* infected rats or rabbits was unable to passively transfer resistance against *S. mansoni*, suggesting that humoral factors were not responsible for this heterologous resistance, particularly since the same sera were effective against homologous challenge (Ford *et al.* 1987a). Furthermore, no immunity to *S. mansoni* followed intraperitoneal implantation of liver flukes, although this did stimulate homologous immunity to *F. hepatica*. Next the possibility was investigated that, rather than being killed by specific cytotoxic mechanisms, the challenge *S. mansoni* schistosomula in *F. hepatica*-infected rats were simply shunted from the hepatic portal vein via portal-systemic shunts, thus escaping sequestration in the liver, as suggested by Wilson *et al.* (1983) to explain the 'pathology mediated' resistance of mice to homologous reinfection with *S. mansoni*. Significant shunting of 15 μm microspheres (about the diameter of lung schistosomula) was indeed observed in rats exposed to *F. hepatica*. These experiments therefore implied that, although there was strong heterologous immunity to *S. mansoni* in *F. hepatica*-infected rats, this was probably not mediated by specific immune mechanisms (see also Ford *et al.* 1987b). The relative contributions of specific and non-specific mechanisms in inter-specific schistosome immunity have as yet not been formally investigated.

Pathogenesis of bovine bilharziasis

Studies on the pathogenesis of bovine (and other ruminant) bilharziasis proceeded concurrently with early vaccine development studies, the objectives being to elucidate the mechanisms of the disease and to utilize this knowledge in evaluating vaccines. Most investigations were made with reference to *S. bovis* and *S. mattheei* and many of the studies concerning the former species were undertaken as part of the London–Khartoum bovine bilharziasis project in the Sudan. These investigations, together with the studies of J. M. Preston and his colleagues from Glasgow University and J. A. Lawrence in Zimbabwe form the basis of the present understanding of the pathogenesis of ruminant bilharziasis.

Knowledge on the pathology caused by infection with *S. bovis* and *S. mattheei* in animals in the mid-1960s was fragmentary but some information was available on *S. mattheei* in sheep. Le Roux (1929) described *S. mattheei* as a new bilharzia species (Veglia & Le Roux 1929) and also drew attention for the first time to the seriousness of animal bilharziasis in Africa, a finding later confirmed by many other investigators (Dinnik & Dinnik 1965, Eisa 1966, Hussein 1969, Reinecke 1970, Van Wyk *et al.* 1974). It was realized that, although epizootics such as those reported by these workers in different parts of Africa were likely to reappear from time to time as a result of water conservation schemes and changing methods of husbandry, the greater economic impact was probably due to the long-term effects of more insidious types of infection. This was because the chronic disease was far more widespread among African domestic ruminants than the epizoonotic form and was often exacerbated under field conditions by concurrent diseases, poor nutrition, adverse climatic factors and physical fatigue (Dargie 1980). For this reason, it was necessary to investigate the disease process under different conditions of exposure both in experimentally and naturally infected animals.

Initially, studies focused on lesions induced by *S. bovis* and *S. mattheei* in experimentally infected calves (Hussein 1971) and, subsequently, further studies were made on the lesions induced by either or both of these species in cattle (Hussein *et al.* 1975, 1976, Lawrence 1977a, 1977b, 1977c, 1978a, 1978b, Saad *et al.* 1984a), sheep (Hussein *et al.* 1976, 1984, Saad *et al.* 1984b) and goats (Saad *et al.* 1984b). The lesions produced in these different ruminants were essentially similar and are described together. Following a prepatent period of 6–7 weeks, during which only mild, transient lesions were seen, the animals developed significant pathological lesions and clinical signs. The two parasites, being inhabitants of the portal and mesenteric veins, usually affected mainly the liver, intestines and associated lymph nodes and less frequently the lungs and pancreas, while in severe cases other organs could be involved, especially the abomasum and the rumen (Fig. 9.1). The lesions had several features in common with bilharziasis of other hosts, including man, but there were also some important differences. For instance, whereas granulomata, periportal fibrosis and cellular infiltrations were found consistently in the livers of infected cattle, no lesions typical of Symmer's clay-pipe stem fibrosis of man were found in these animals and they also showed no evidence of portal hypertension or its complications such as congestive splenomegaly and gastro-oesophageal varicosities (Hussein 1971, Hussein *et al.* 1975). It has been suggested, therefore, that cattle either possess or can develop spontaneously efficient collateral channels and thereby succeed in bypassing the portal obstruction (Hussein 1971). A striking pathological feature in bovine bilharziasis, on the other hand, was found to be the occurrence of massive medial hyperplasia of the portal radicles. This was often associated with, and probably arose secondarily to, severe intimal hyperplasia and subintimal eosinophilic infiltration, especially in chronically infected and resistant animals. In old-standing infections, these changes coupled with portal thrombosis and fibroplasia led to marked distortion of the portal fields and gave them a peculiar 'knotted' appearance, which Sobrero (1960) earlier designated 'sclerosi epatica nodosa' or 'knotted hepatic fibrosis'.

Another unusual finding in bovine bilharziasis was the development of huge lymphoid nodules around dead parasites (Fig. 9.2) and occasionally around pre-existing granulomas. These lesions, which became evident from around the third month post-infection, by which time some worms were being destroyed in the liver and other sites (Hussein 1971), were not dissimilar to the 'milk spots' of porcine ascariasis and the lymphoid of lungworm in sheep (Beresford-Jones 1964). They had been met with

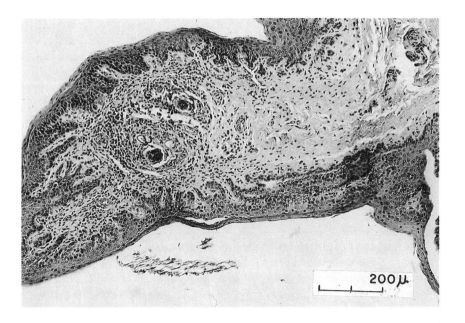

Figure 9.1 *S. bovis* granulomata in the rumen.

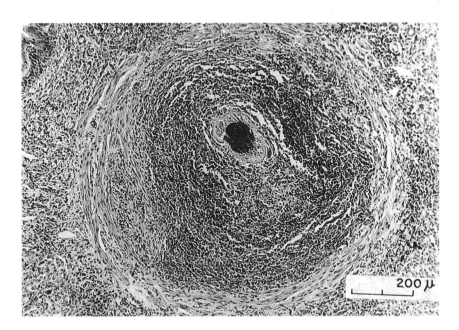

Figure 9.2 Intravascular lymphoid nodule around dead *S. bovis* worm in the liver of a sheep.

more often in calves immunized with irradiated bilharzia vaccines than in non-immunized calves, denoting the increased rate of worm destruction in the former case (Bushara *et al.* 1978). Intestinal lesions during early stages of the infection were more marked in the small, compared to the large, intestine in both bovine and ovine bilharziasis (Hussein 1971, Saad *et al.* 1980, 1984b) whereas in goats, the small and the large intestines appeared to be equally affected (Saad *et al.* 1984b). Haemorrhages in the small intestine during the early acute stages of the disease accounted for the most serious clinical consequences of bovine bilharziasis, namely anaemia and hypoalbuminemia. During more chronic stages, these haemorrhages subsided partly and the worms moved to the veins of the large intestine and became sterile. No rectal or colonic polypi like those occurring in 'intestinal bilharziasis' of man were seen in cattle or other ruminants.

The lungs were sometimes riddled with granulomas showing endarteritic lesions, perivascular cuffing, bronchiolar hypertrophy and worm embolization into the pulmonary arteries. Pancreatic involvement was denoted by the presence of worms, eggs and granulomas, coupled with interstitial inflammatory infiltration and fat necrosis.

Vesical lesions have been reported in cattle infected with *S. mattheei* (Condy 1960, McCully & Kruger 1969) but not in animals infected with *S. bovis* (Hussein 1971, Hussein *et al.* 1975). Urinary manifestations are rare in both *S. bovis* and *S. mattheei* infections and are probably seen only as aberrant conditions during superinfection.

Practical evaluation of live vaccines – sheep and cattle

Following the experiments demonstrating the successful protection of baboons against *S. mansoni* with heterologous cercariae of *S. bovis* and *S. rodhaini* and the series of experiments showing that sheep and cattle could be partially protected against their own species of schistosomes by heterologous immunization, experiments were carried out to determine the effectiveness of homologous, irradiation-attenuated infections and to compare the degree of resistance induced with that stimulated by heterologous immunization with normal cercariae, with a view to evaluating the practical potential of these live vaccines. To do this, parallel studies on vaccination against *S. mansoni* in baboons and against the domestic animal parasites *S. mattheei* and *S. bovis* in their natural hosts were carried out. It was of course realized at the outset that no practical application of these types of live vaccines was going to be possible, as long as they consisted of short-lived aquatic suspensions of cercariae. To develop a more practical form of organism for the vaccine, various techniques for the transformation of cercariae into schistosomula-like organisms were tested and James and colleagues (1986) later went on to develop effective cryopreservation techniques for schistosomula.

A number of laboratory and field studies have been carried out to test live vaccines incorporating irradiated *S. bovis* and *S. mattheei* schistosomula in cattle and sheep (reviewed by Taylor 1987). The results confirmed that vaccination was highly efficacious in reducing worm and egg burdens, that the immunity was long-lived and that no significant vaccinal pathology was encountered in these large animals. The economic benefits of vaccination were evident from improved weight gains and better carcass quality, other favourable returns in vaccinated compared to non-vaccinated animals and by cost-effectiveness studies in the field (McCauley *et al.* 1984). The studies demonstrated that faecal egg counts are a more useful parasitological criterion of protection than is the number of adult parasites, or even tissue egg counts.

Independent studies confirming the effectiveness of the irradiated bovine bilharziasis vaccines were soon forthcoming from other cattle/parasite systems, when Hsu *et al.* (1984) showed that calves immunized with highly irradiated schistosomula of a Chinese strain of *S. japonicum* developed a significant immunity against homologous challenge, under both laboratory and field conditions, especially with repeated immunizations. These authors concluded that vaccination was highly beneficial in protecting cattle against the effects of *S. japonicum* and furthermore, that the protection of this reservoir of the parasite might have a considerable impact on public health.

Current research on the irradiated *S. bovis* vaccine for cattle aims to test whether improved levels of immunity can be obtained by varying the irradiation dose, number of organisms per dose, and the route of injection of the vaccine (Bushara *et al.*, unpublished). Preliminary results on the migration of irradiated-attenuated *S. bovis* schistosomula in mice showed that a dose of 20 Krad prevented migration past the lungs, whereas a dose of 50 Krad prevented the parasites from leaving the muscle injection site. In cattle experiments, 3 Krad-irradiated organisms allow a complete migration of some of the irradiated schistosomula. In recent studies calves were immunized with a single dose of 10 000 larvae and challenged with normal cercariae eight weeks later. Immunized groups were vaccinated with: 3 Krad-irradiated schistosomula injected i.m.; 20 Krad schistosomula i.m; 20 Krad schistosomula injected i.d; and 50 Krad schistosomula injected i.m. Significant levels of protection were stimulated by vaccination with 3 Krad, i.m., or 20 Krad i.d. larvae, but less resistance was induced by 20 Krad, i.m. or 50 Krad i.m. vaccination. The first two treatments (3 Krad, i.m. and 20 Krad i.d.) were selected for further study, and groups of calves were vaccinated with between 5000 and 50 000 irradiated cercariae or schistosomula: the vaccines were administered in one dose, or in three doses over a four-week period. 17–20 weeks after the single immunization, or the first of the multiple doses, all the calves were challenged with 10 000 normal *S. bovis* cercariae and perfused 19 weeks later. The results showed that each of the vaccination schedules was protective, with no significant differences between the groups. These immunization schedules are currently undergoing field trials.

Practical evaluation of live vaccines in baboons

Experiments with S. mansoni

In parallel with the early vaccination experiments on *S. mattheei* in sheep attempts have been made to vaccinate baboons against *S. mansoni* using irradiated *S. mansoni* cercariae and schistosomula (Taylor *et al.* 1976). In a direct comparison of irradiated 'homologous' cercariae with normal 'heterologous' cercariae, groups of baboons were exposed to three doses of 5000 6 Krad-irradiated *S. mansoni* cercariae or to similar numbers of normal *S. rodhaini* cercariae and challenged 15 weeks later with 500 normal *S. mansoni* cercariae. Faecal egg counts, worm and tissue egg counts and histological examination showed that neither of the immunizing schedules induced significant protection. In another experiment baboons were injected intramuscularly with 31 000 schistosomula of *S. mansoni* in four doses using an irradiation dose of 2.1–2.4 Krad. Challenge with 3500 normal *S. mansoni* cercariae 21 weeks after the first immunizing dose again showed no significant protection.

These results were clearly much less encouraging than those obtained with ruminant schistosomiasis and a series of experiments was therefore carried out using irradiated *S. mansoni* vaccines in the mouse model in an attempt to find methods of boosting the degree of resistance induced by irradiated vaccines, prior to retesting in baboons (Bickle *et al.* 1979b). Single intramuscular injections of 500 schistosomula exposed to irradiation doses in the range 2.3–160 Krad all resulted in significant protection (as assessed by reduced worm burdens) against a challenge infection with normal *S. mansoni* cercariae administered at intervals from 3–24 weeks post vaccination. However, schistosomula irradiated with 20 Krad consistently resulted in better protection than those exposed to either higher or lower doses, despite the persistence of stunted adult worms from the infections irradiated with 2.3 Krad. Varying the number of irradiated schistosomula, the frequency and route of their administration, the site of challenge and the strain of host all failed to enhance the level of resistance. Percutaneously applied cercariae were found to be more effective in stimulating resistance than intramuscularly injected irradiated schistosomula.

These findings did not immediately suggest a way of dramatically increasing the effectiveness of irradiated *S. mansoni* vaccines in baboons, although they suggested that somewhat better results might be obtained by using a higher irradiation dose, such as 20 Krad, than the dose of 6 Krad or 2.1–2.4 Krad used in the previous baboon experiments. Indeed, more encouraging results were later obtained in baboons by other workers, using highly irradiated *S. mansoni* cercariae, baboons immunized with 56 Krad irradiated *S. mansoni* cercariae being protected from 70% of the challenge infection (Stek *et al.* 1981), although other similar experiments produced much lower levels of resistance (Damian *et al.* 1985, James *et al.* 1986).

The latter studies demonstrated that while such immunizations had undoubtedly stimulated marked immune responses to *S. mansoni* (as judged by both ELISA and ADCC assays), they failed to induce protection in that the vaccinated baboons were infected with similar worm burdens to the unvaccinated challenge control animals. Further research in this area is required, especially since the baboon is the only experimental model in schistosomiasis research which exhibits significant resistance to reinfection after a primary unattenuated infection but whose protection to challenge infection after immunization with highly irradiated cercariae is in some doubt.

Experiments with S. haematobium

In terms of resistance, the host–parasite relationship of baboons infected or immunized with *S. haematobium* is more straightforward than that for *S. mansoni*-infected baboons. Thus, baboons sensitized to *S. haematobium* infection exhibit a relatively greater degree of resistance to homologous infection than baboons sensitized to *S. mansoni* infection. Nevertheless, as previously described (Jordan *et al.* 1967, Webbe *et al.* 1974, Cheever *et al.* 1974), the baboon is fully susceptible to primary *S. haematobium* infection and exhibits many of the characteristics of human infection, with the proviso that, in baboons, more tissue-bound eggs are recovered from the large intestine and less from the urogenital system than in humans. The distribution of tissue-bound eggs reflects the sites of residence of the adult worms and hence illustrates the major difference between the dual involvement of the urogenital and intestinal system in baboons as opposed to the predominantly vesical plexus involvement in humans.

Baboons with primary infections of 1000 *S. haematobium* cercariae showed significant resistance to a challenge with 10 000 cercariae (Webbe & James 1973, Webbe *et al.* 1976). In the latter study the degree of resistance was associated with the chronicity of the primary infection in that animals with 73-week infections were more resistant to reinfection (80% of challenge larvae failed to survive to maturity) than baboons with 27-week infections (37% of challenge-derived worms recovered at perfusion). Other manifestations of immunity in this model were depression of worm fecundity, reduction in pathology and the absence of any increase in faecal or urine egg output following challenge infection.

In investigations of the mechanisms of this resistance, it was shown that 'fecundity-suppressed' adult worms resumed full egg production upon surgical transplantation to naive recipients and that some resistance to challenge could be conferred upon naive recipients by mesenteric vein transplantation of adult worms. With regard to possible immunological correlates of resistance, a positive association between antibody titres in an ADCC assay and resistance was observed (Webbe *et al.* 1976), whereas ELISA titres did not correlate with resistance. An attempt to enhance the level of resistance to reinfection in *S. haematobium*-infected baboons by non-specific stimulation of cytotoxic cells just prior to challenge infection (through injection of Cord factor-CF or muramyl dipeptide-MDP, which Mahmoud *et al.* (1977) and Olds *et al.* (1980) had found to elevate resistance in *S. mansoni* infected mice) was vitiated by the fact that the infected control animals were nearly 100% resistant to the challenge infection (Sturrock *et al.* 1985). This experiment differed from most other resistance to reinfection studies in that the challenge infection (3000 cercariae per animal) consisted of fewer cercariae than the primary infection (5000 cercariae per baboon) and this may have accounted for the unprecedented high levels of protection observed.

It is evident from the above that baboons infected with *S. haematobium* are more resistant to homologous challenge infections than any other fully permissive hosts infected with either *S. haematobium* or *S. mansoni*. This has led some researchers to question the validity of the baboon as a susceptible host to *S. haematobium* and by inference its value as an appropriate model for the disease in man (Damian 1984, Damian *et al.* 1985). Long-lasting vaccines which, if they are to be derived from the parasites themselves, can only be developed by using a model which exhibits high protection. Since *S. haematobium*-infected baboons suffer extreme egg-related pathology, the relative roles of specific immune responses and of 'pathology mediated resistance' are unclear. If protection in this model could be induced by less pathogenic immunizing procedures, such as the use of radiation-attenuated larvae, definition of the 'protective' antigens may be more straightforward. Baboons immunized with irradiated *S. haematobium* larvae have been shown to be protected from 90% of a homologous challenge infection (Webbe *et al.* 1982). This was the first time that an immunized, permissive host species had been protected from challenge infection to a degree which exceeded the levels of protection normally obtained from hosts with unattenuated infections. It was also demonstrated that, as with irradiated *S. mansoni* vaccines in mice, irradiated, percutaneously applied cercariae induced higher levels of protection than mechanically transformed schistosomula and that larvae irradiated at the 20 Krad level induced higher levels of resistance (cercariae 89%, schistosomula 70%) than larvae irradiated at the 3 Krad level (cercariae 71%, schistosomula 64%). The reduction in excreta egg output and tissue-lodged eggs reflected the size of the worm burden, i.e. there was no evidence worm fecundity had been reduced. The size of the individual granulomata in the vaccinated baboons was no different from that in the non-immunized challenge control animals, although the gross pathology in the latter group was markedly greater than in the vaccinated animals, owing to the increased tissue egg burden. Gross pathology associated with injection of schistosomula was not evident. The results of this study prompted a series of experiments designed to optimize the performance of the live-attenuated vaccine against *S. haematobium*, with particular emphasis on increasing the level of protection and utilizing schistosomula instead of cercariae, with a view towards reducing the pathology and obviating practical constraints associated with percutaneous exposures to cercariae.

Baboons immunized with 20 Krad *S. haematobium* schistosomula were more highly protected against challenge infection than baboons immunized with 60 Krad (Harrison *et al.* unpublished). Furthermore, evidence was obtained to indicate that multiple immunizations induced higher levels of protection than a single immunization with the same number of schistosomula. In this experiment, levels of protection using irradiated schistosomula were equal to, and in some groups exceeded, those observed with irradiated cercariae in the previous study by Webbe *et al.* (1982). The result can probably be attributed to the greater number of larvae comprising the immunization and the greater number of immunizing doses used in the more recent experiment. In addition to the reduction in worm burden, protection was manifested by a reduction in faecal and urine egg counts and tissue egg counts. The reduction in egg production was also reflected in a reduction in gross pathology. Results of this study also indicated an inverse association between the degree of resistance and the interval between the last immunization and the challenge infection. Thus, baboons given multiple immunizations of 20 Krad schistosomula and challenged eight weeks later were 91% resistant, whereas baboons receiving the same vaccinations but challenged 28 weeks later were only 60% resistant. There was a concomitant decrease of antibody titres to *S. haematobium* larvae with time as assessed by ELISA using cercariae antigens and by the degree of reactivity to radiolabelled schistosomula surface antigens. Furthermore, sera from animals vaccinated with 60 Krad larvae exhibited significantly lower anti-cercariae antibody titres immediately prior to challenge than sera from baboons immunized with the more highly protective 20 Krad larvae. Sera from baboons vaccinated with either 20 Krad or 60 Krad reacted to the same schistosomula surface antigens as did sera from chronically infected baboons and humans, thereby preventing the immediate identification of antigens responsible for protection. An accompanying study was performed to examine the pathology associated with vaccination of baboons with live irradiated *S. haematobium* schistosomula, to assess their feasibility for future human vaccination (Byram *et al.*, unpublished). Baboons were immunized (with a total of 18 000 irradiated larvae in 1 ml of media injected into the thigh muscle) in ways which were considered to be likely scenarios in the event of human vaccination and then necropsied at various intervals after immunization. The animals whose vaccination lesions were most pronounced (a palpable mass 1–2 cm in length, lasting 2–4 weeks) had either received a previous vaccination or had a previous unattenuated infection which had been treated with praziquantel just prior to immunization. Animals with non-treated, patent *S. haematobium* infections and vaccinated a year after infection exhibited milder lesions at the site of immunization and animals receiving a single vaccination alone showed least reaction to the immunization. On the grounds that (a) the human population most in need of protection against schistosomiasis haematobia live in endemic areas and therefore would be highly sensitized, (b) infected people would be chemotherapeutically cured prior to vaccination and, (c) the human response to vaccination might be similar to that of experimental baboons, it was concluded that this form of immunization would be inadvisable in humans.

Subsequent experiments demonstrated that cryopreserved, irradiated *S. haematobium* schistosomula induced only about 20% protection to challenge infection and, in a separate experiment, that baboons immunized with non-cryopreserved schistosomula were increasingly less resistant as the interval between immunization and challenge infection was lengthened, thereby confirming earlier observations (Harrison *et al.* unpublished). These results, taken in conjunction with the results of the pathogenicity study led to the conclusion that alternative methodologies of vaccinating humans against *S. haematobium* should be examined and that, in particular, attempts to develop experimental 'defined antigen' *S. haematobium* vaccines should be made. This type of approach, which is already far advanced in *S. mansoni* vaccine research, employs a range of investigative methods, including studies on the mechanism of immunity and attempts to characterize the 'protective antigens' by the use of monoclonal and polyclonal antibody probes, with the aim of their large-scale production by recombinant DNA technology or by chemical synthesis. But that is another story!

References

Adam, S. E. I. & M. Magzoub 1976. Susceptibility of desert sheep to infection with *Schistosoma mansoni* of Northern Sudan. *Vet. Path.* **13**, 211–5.

Adam, S. E. I. & M. Magzoub 1977. Clinical-pathological changes associated with experimental *Schistosoma mansoni* infection in the goat. *Brit. Vet. J.* **133**, 201–10.

Amin, M. A. & G. S. Nelson 1969. Studies on heterologous immunity in schistosomiasis. 3. Further observations on heterologous immunity in mice. *Bull. Wld Hlth Org.* **41**, 225–32.

Amin, M. A., G. S. Nelson & M. F. A. Saoud 1968. Studies on heterologous immunity in schistosomiasis. 2. Heterologous schistosome immunity in rhesus monkeys. *Bull. Wld Hlth Org.* **38**, 19–27.

Beresford-Jones, W. P. 1964. The pathology and parasitology of sheep lungworm infections in Great Britain with special reference to *Meullerius capillaris* (Müller 1889). PhD thesis. University of London.

Bickle, Q. D., M. G. Taylor, E. R. James, G. S. Nelson *et al.* 1979a. Further observations on immunisation of sheep against *Schistosoma mattheei* and *S. bovis* using irradiation-attenuated schistosomula of homologous and heterologous species. *Parasitology.* **78**, 185–93.

Bickle, Q. D., M. G. Taylor, M. J. Doenhoff & G. S. Nelson 1979b. Immunisation of mice with gamma-irradiated, intramuscularly-injected schistosomula of *Schistosoma mansoni*. *Parasitology* **79**, 209–22.

Bushara, H. O., M. F. Hussein, A. M. Saad, M. G. Taylor *et al.* 1978. Immunisation of calves against *Schistosoma bovis* using irradiated cercariae or schistosomula of *S. bovis*. *Parasitology* **77**, 303–11.

Bushara, H. O., A. A. Gameel, B. Y. A. Bajid, I. Khitma *et al.* 1983. Observations on cattle schistosomiasis in the Sudan, a study in comparative medicine. VI. Demonstration of resistance to *Schistosoma bovis* challenge after a single exposure to normal cercariae or to transplanted adult worms. *Am. J. Trop. Med. Hyg.* **32**, 1375–80.

Cheever, A. W., D. G. Erikson, E. H. Sadun & V. von Lichtenberg 1974. *Schistosoma japonicum* in monkeys and baboons: parasitological and pathological findings. *Am. J. Trop. Med. Hyg.* **23**, 51–64.

Condy, J. B. 1960. Bovine schistosomiasis in Southern Rhodesia. *Cent. Afr. J. Med.* **6**, 381–4

Damian, R. T. 1984. Immunity in schistosomiasis: an holistic view. In *Immunology of parasites and parasitic infection.* J. J. Marchanolis (ed.), *Contemporary topics in immunology,* **12**, 359–420. New York: Plenum Press.

Damian, R. T., N. D. Greene & K. Fitzgerald 1972. Schistosomiasis mansoni in baboons. The effect of surgical transfer of adult *Schistosoma mansoni* upon subsequent challenge infection. *Am. J. Trop. Med. Hyg.* **21**, 951–8.

Damian, R. T., M. R. Powell, M. L. Roberts, J. D. Clark *et al.* 1985. *Schistosoma mansoni*: parasitology and immunology of baboons vaccinated with irradiated cryopreserved schistosomula. *Int. J. Parasit.* **15**, 333–44.

Dargie, J. D., 1980. The pathogenesis of *Schistosoma bovis* infection in Sudanese cattle. *Trans. R. Soc. Trop. Med. Hyg.* **74**, 560–2.

Dinnik, J. A. & N. N. Dinnik 1965. The schistosomes of domestic ruminants in Eastern Africa. *Bull. Epiz. Dis. Afr.* **13**, 341–59.

Eisa, A. M. 1966. Parasitism – a challenge to animal health in the Sudan. *Sudan J. Vet. Sci. Anim. Husb.* **7**, 85–94.

Eveland, L. K., S. Y. L. Hsu & H. F. Hsu 1969. Cross-immunity of *Schistosoma japonicum*, *S. mansoni* and *S. bovis* in rhesus monkeys. *J. Parasitol.* **55**, 279–88.

Ford, M. J., M. G. Taylor, S. M. McHugh, R. A. Wilson *et al.* 1987a. Studies of heterologous resistance between *Schistosoma mansoni* and *Fasciola hepatica* in inbred rats. *Parasitology* **94**, 55–67.

Ford, M. J., M. G. Taylor & Q. D. Bickle 1987b. Re-evaluation of the potential of *Fasciola hepatica* antigens for immunisation against *Schistosoma mansoni* infection. *Parasitology* **94**, 327–36.

Harrison, R. A., Q. D. Bickle & M. J. Doenhoff 1982. Factors affecting the acquisition of resistance against *Schistosoma mansoni* in the mouse. Evidence that the mechanisms which mediate resistance during early patent infections may lack immunological specificity. *Parasitology* **84**, 93–110.

Hsu, S. Y. L. & H. F. Hsu 1961. New approach to immunisation against *Schistosoma japonicum*. *Science* **133**, 766.

Hsu, S. Y. L. & H. F. Hsu 1963. Further studies on rhesus monkeys immunised against *Schistosoma japonicum* by administration of cercariae of the Formosan strain. *Z. ParasitKde* **14**, 506–11.

Hsu, S. Y. L., J. R. Davis & H. F. Hsu 1965. Histopathology in rhesus monkeys infected four times with the Formosan strain of *Schistosoma japonicum*. *Z. Tropenmed. Parasitol.* **16**, 297–304.

Hsu, S. Y. L., H. F. Hsu, K. Y. Chu, C. T. Tsai *et al.* 1966. Immunisation against *Schistosoma haematobium* in rhesus monkeys by administration of cercariae of *Schistosoma bovis*. *Z. Tropenmed. Parasitol.* **17**, 407–12.

Hsu, S. Y. L., H. F. Hsu, S. T. Xu, F. H. Shi *et al.* 1983. Vaccination against bovine schistosomiasis japonica with highly X-irradiated schistosomula. *Am. J. Trop. Med. Hyg.* **32**, 367–70.

Hsu, S. Y. L., S. T. Xu, Y. X. He, F. H. Shi *et al.* 1984. Vaccination of bovines against schistosomiasis japonica with highly irradiated schistosomula in China. *Am. J. Trop. Med. Hyg.* **33**, 891–8.

Hussein, M. F. 1969. The pathology of spontaneous and experimental bovine schistosomiasis. PhD thesis. University of London.

Hussein, M. F. 1971. The pathology of experimental schistosomiasis in calves. *Res. Vet. Sci.* **12**, 246–52.

Hussein, M. F. & H. O. Bushara 1976. Investigations on the development of an irradiated vaccine for animal schistosomiasis. In *Nuclear techniques in animal production and health*, 421–31. Vienna: International Atomic Energy Agency.

Hussein, M. F., A. A. Saeed & G. S. Nelson 1970. Studies on heterologous immunity in schistosomiasis. 4. Heterologous schistosome immunity in cattle. *Bull. Wld Hlth Org.* **42**, 745–9.

Hussein, M. F., G. Tartour, S. E. Imbabi & K. E. Ali 1975. The pathology of naturally occurring bovine schistosomiasis in the Sudan. *Ann. Trop. Med. Parasit.* **69**, 217–25.

Hussein, M. F., H. O. Bushara & K. E. Ali 1976. The pathology of experimental *Schistosoma bovis* infection in sheep. *J. Helminthol.* **50**, 235–41.

Hussein, M. F., S. M. Basmaeil & H. A. R. Hassan 1984. Urinary schistosomiasis in a sheep; a case report. *Ann. Trop. Med. Parasitol.* **78**, 439–41.

James, E. R., M. Otieno, R. Harrison, A. R. Dobinson *et al.* 1986. Partial protection of baboons against *Schistosoma mansoni* using radiation-attenuated, cryopreserved schistosomula. *Trans. R. Soc. Trop. Med. Hyg.* **80**, 378–84.

Jordan, P., F. von Lichtenberg & K. D. Goatly 1967. Experimental schistosomiasis in primates in Tanzania. Preliminary observations on the susceptibility of the baboon (*Papio anubis*) to *Schistosoma mansoni* and *S. haematobium*. *Bull. Wld Hlth Org.* **37**, 393–403.

Lawrence, J. A. 1977a. *Schistosoma mattheei* in the ox. The clinical pathological observations. *Res. Vet. Sci.* **23**, 280–7.

Lawrence, J. A. 1977b. *Schistosoma mattheei* in the ox, observations on the parasite. *Vet. Parasitol.* **3**, 291–303.

Lawrence, J. A. 1977c. *Schistosoma mattheei* in the ox. The intestinal syndrome. *J. South Afr. Vet. Assoc.* **48**, 55–8.

Lawrence, J. A. 1978a. The pathology of *Schistosoma mattheei* infection in the ox. Lesions attributable to the eggs. *J. Comp. Pathol.* **88**, 1–14.

Lawrence, J. A. 1978b. The pathology of *Schistosoma mattheei* infection in the ox. Lesions attributable to the adult parasite. *J. Comp. Pathol.* **88**, 15–29.

Le Roux, P. L. 1929. Remarks on the habits and pathogenesis of *Schistosoma mattheei*, together with notes on the pathological lesions observed in infested sheep. *Fifteenth Annual Report of the Director of Veterinary Services, Union of South Africa*, October 1929. 347–406.

Le Roux, P. L. 1961. Some problems in bilharziasis in Africa and in the adjoining countries. *J. Helminthol. R.T. Leiper Supplement*, 117–26.

Mahmoud, A. A. F., R. H. Civil, E. Lederer & L. Chedid 1977. Enhancement of resistance against a multicellular helminth (*Schistosoma mansoni*) by cord factor (CF) (Abstract). *Clin. Res.* **25**, 380A.

Massoud, J. & G. S. Nelson 1972. Studies on heterologous immunity in schistosomiasis. 6. Observations on cross-immunity to *Ornithobilharzia turkestanicum*, *Schistosoma bovis*, *S. mansoni*, and *S. haematobium* in mice, sheep and cattle in Iran. *Bull. Wld Hlth Org.* **47**, 591–600.

McCauley, E. H., A. A. Majid & A. Tayeb 1984. Economic evaluation of the production impact of bovine schistosomiasis and vaccination in the Sudan. *Prev. Vet. Med.* **2**, 735–54.

McCully, R. M. & S. P. Kruger 1969. Observations on bilharziasis of domestic ruminants in South Africa. *Onder. J. Vet. Res.* **36**, 129–61.

McMahon, J. E. 1967. The effect of suppressive therapy of immunity in baboons (*Papio anubis*) infected with *Schistosoma mansoni*. *E. Afr. Med. J.* **44**, 250–5.

Meleney, J. E. & D. V. Moore 1954. Observations on immunity to superinfection with *Schistosoma mansoni* and *S. haematobium* in monkeys. *Exp. Parasitol.* **3**, 128–239.

Monrad, J., N. O. Christensen, P. Nansen & F. Frandsen 1981. Resistance to *Fasciola hepatica* in sheep harboring primary *Schistosoma bovis* infections. *J. Helminthol.* **55**, 261–71.

Nelson, G. S., C. Teesdale & R. B. Highton 1962. The role of animals as reservoirs of bilharziasis in Africa. In *CIBA Foundation Symposium on Bilharziasis*. G. E. W. Wolstenholme & M. O'Connor (eds), 127–49. London: Churchill.

Nelson, G. S., M. A. Amin, M. F. A. Saoud & C. Teesdale 1968. Studies on heterologous immunity in schistosomiasis. I. Heterologous schistosome immunity in mice. *Bull. Wld Hlth Org.* **38**, 9–17.

Newsome, J. 1956. Problems of fluke immunity: with special reference to schistosomiasis. *Trans. R. Soc. Trop. Med. Hyg.* **50**, 258–74.

Olds, G. R., L. Chedid, E. Lederer & A. A. F. Mahmoud 1980. Induction of resistance to *Schistosoma mansoni* by natural cord factor and synthetic lower homologues. *J. Infect. Dis.* **141**, 473–8.

Preston, J. M., G. S. Nelson & A. A. Saeed 1972. Studies on heterologous immunity in schistosomiasis. 5. Heterologous schistosome immunity in sheep. *Bull. Wld Hlth Org.* **47**, 587–90.

Purnell, R. E. 1966. Host–parasite relationships in schistosomiasis. II The effects of age and sex on the infection of mice and hamsters with cercariae of *Schistosoma mansoni* and of hamsters with cercariae of *Schistosoma haematobium*. *Ann. Trop. Med. Parasitol.* **60**, 94–9.

Reinecke, R. K. 1970. The epizootology of an outbreak of bilharziasis in Zululand. *Cent. Afr. J. Med.* **16**, 10–12.

Saad, A. M., M. F. Hussein, J. D. Dargie *et al.* 1980. *Schistosoma bovis* in calves: the development and clinical pathology of primary infections. *Res. Vet. Sci.* **28**, 105–11.

Saad, A. M., M. F. Hussein, J. D. Dargie & M. G. Taylor 1984a. Erythrokinetics and albumin metabolism in primary experimental *Schistosoma bovis* infection in Zebu cattle. *J. Comp. Pathol.* **94**, 249–62.

Saad, A. M., M. F. Hussein, J. D. Dargie & M. G. Taylor 1984b. The pathogenesis of experimental *Schistosoma bovis* infections in sheep and goats. *J. Comp. Pathol.* **94**, 371–85.

Sadun, E. H., F. von Lichtenberg & J. I. Bruce 1966. Susceptibility and comparative pathology of ten species of primates exposed to infections with *Schistosoma mansoni*. *Am. J. Trop. Med. Hyg.* **15**, 715–18.

Saeed, A. A. & G. S. Nelson 1974. Experimental *Schistosoma mansoni* in sheep. *Trop. Anim. Hlth Product.* **6**, 45–52.

Siraj, S. B., N. O. Christensen, P. Nansen & F. Frandsen 1981. Resistance to *Fasciola hepatica* in calves harbouring primary patent *Schistosoma bovis* infections. *J. Helminthol.* **55**, 63–70.

Smith, M. A., J. A. Clegg & G. Webbe 1976. Cross-immunity to *Schistosoma mansoni* and *S. haematobium* in the hamster. *Parasitology* **73**, 53–64.

Smithers, S. R. 1968. Immunity to blood helminths. In *Immunity to parasites*. A. E. R. Taylor. (ed.), 55–66. Oxford: Blackwell Scientific Publications.

Smithers, S. R. & R. J. Terry 1965. Naturally acquired resistance to experimental infections of *Schistosoma mansoni* in the rhesus monkey (*Macaca mulatta*). *Parasitology* **55**, 701–10.

Smithers, S. R. & R. J. Terry 1967. Resistance to experimental infection with *Schistosoma mansoni* in rhesus monkeys induced by the transfer of adult worms. *Trans. R. Soc. Trop. Med. Hyg.* **61**, 517–33.

Sobrero, R. 1960. Animili domistici ospiti naturali di *Schistosoma bovis* in Somalia. *Rivista Parassitol.* **21**, 125–30.

Stek, M., P. Minard, D. A. Dean & J. E. Hall 1981. Immunisation of baboons with *Schistosoma mansoni* cercariae attenuated by gamma irradiation. *Science* **213**, 1518–20.

Sturrock, R. F., A. E. Butterworth, V. Houba, B. J. Cottrell et al. 1985. Attempts to manipulate specific responses to induce resistance to *Schistosoma mansoni* in Kenyan baboons (*Papio anubis*). *J. Helminthol.* **59**, 175–86.

Taylor, M. G. 1970. Hybridization experiments on five species of African schistosomes. *J. Helminthol.* **44**, 253–314.

Taylor, M. G. 1987. Schistosomes of domestic animals: *Schistosoma bovis* and other animal forms. In *Immune responses in parasitic infections: immunology, immunopathology and immunoprophylaxis*. E. J. L. Soulsby. (ed.), 50–90. Boca Raton, Florida: CRC Press.

Taylor, M. G., M. A. Amin & G. S. Nelson 1969. 'Parthenogenesis' in *Schistosoma mattheei*, *J. Helminthol.* **43**, 197–206.

Taylor, M. G., G. S. Nelson, M. Smith & B. J. Andrews 1973a. Studies on heterologous immunity in schistosomiasis. 7. Observations on the development of acquired homologous and heterologous immunity to *Schistosoma mansoni* in baboons. *Bull. Wld Hlth Org.* **49**, 57–65.

Taylor, M. G., G. S. Nelson, M. Smith & B. J. Andrews 1973b. Comparison of the infectivity and pathogenicity of six species of African schistosomes and their hybrids. 2. Baboons. *J. Helminthol.* **47**, 455–85.

Taylor, M. G., E. R. James, G. S. Nelson, Q. D. Bickle et al. 1976. Immunisation of baboons against *Schistosoma mansoni* using irradiated *S. mansoni* cercariae. *J. Helminthol.* **50**, 215–21.

Van Wyk, J. A., R. C. Bartsch, L. J. Van Rensburg, L. P. Hamilton et al. 1974. Studies on schistosomiasis. 6. A field outbreak of bilharzia in cattle. *Onder. J. Vet. Res.* **41**, 39–49.

Veglia, F. & P. L. Le Roux 1929. On the morphology of *Schistosoma mattheei* (sp. nov.) from the sheep in the Cape Province. In *Fifteenth Annual Report of the Director of Veterinary Services, Union of South Africa*, 335–46.

Vogel, H. & W. Minning 1953. Uber die erworkbenre. Resistenz von *Macacus rhesus* gegenuber *Schistosoma japonicum*. *Z. Tropenmed. Parasitol.* **4**, 418–505.

Webbe, G. & C. James 1973. Acquired resistance to *Schistosoma haematobium* in the baboon (*Papio anubis*). *Trans. R. Soc. Trop. Med. Hyg.* **67**, 151–2.

Webbe, G., C. James & G. S. Nelson 1974. *Schistosoma haematobium* in the baboon (*Papio anubis*). *Ann. Trop. Med. Parasitol.* **68**, 187–203.

Webbe, G., R. F. Sturrock, E. R. James & C. James 1982. *Schistosoma haematobium* in the baboon (*Papio anubis*): effect of vaccination with irradiated larvae on the subsequent infection with percutaneously applied cercariae. *Trans. R. Soc. Trop. Med. Hyg.* **76**, 354–60.

Webbe, G., C. James, G. S. Nelson, M.M. Ismail *et al.* 1979. Cross resistance between *Schistosoma haematobium* and *S. mansoni* in the baboon. *Trans. R. Soc. Trop. Med. Hyg.* **73**, 42–54.

Webbe, G., C. James, G. S. Nelson, S. R. Smithers *et al.* 1976. Acquired resistance to *Schistosoma haematobium* in the baboon (*Papio anubis*) after cercarial exposure and adult worm transfer. *Trans. R. Soc. Trop. Med. Hyg.* **70**, 411–24.

Webbe, G., G. S. Nelson, C. James, H. Furse *et al.* 1972. Intravenous pyelograms and associated histopathology of baboons infected with *Schistosoma haematobium* (Nigerian strain). *Trans. R. Soc. Trop. Med. Hyg.* **66**, 15–16.

Wilson, E. R. A., P. S. Calcine & S. M. McHugh 1983. A significant part of 'concomitant immunity' in mice to *Schistosoma mansoni* is the consequence of a leaky hepatic portal system, not of immune killing. *Parasite Immunol.* **5**, 595–601.

Yagi, A. I., S. A. Younis, E. M. Haroun, A. A. Gameel *et al.* 1986. Studies on heterologous resistance between *Schistosoma bovis* and *Fasciola gigantica* in Sudanese cattle. *J. Helminthol.* **60**, 55–9.

10 Zoonotic helminths of wild and domestic animals in Africa

Introduction

Zoonoses have assumed a greater importance in Africa due to increased contact between humans and animals. In the last 50 years, wars of independence, territorial disputes, civil wars and famines resulted in massive population movements exposing people to abnormal environmental conditions and introducing exotic diseases. Shortages of food forced people to consume lower forms of life. Facilities for proper cooking were also reduced and food was eaten either raw or insufficiently cooked. Infective stages of parasites survived and completed their life-cycles. Lack of hygiene, local customs and primitive medicines provided opportunities for parasites to get established and multiply.

The joint FAO/WHO Expert Committees in 1959 and 1967 defined zoonoses as 'those diseases and infections which are naturally transmitted between vertebrate animals and man'. Consequently parasitic zoonoses are parasitic diseases and infections which are naturally transmitted between vertebrate animals and man. Parasitic zoonoses include obligate zoonoses which are transmitted from vertebrates to humans and facultative zoonoses which will generally be transmitted from person to person or animal to animal respectively but may infect human beings occasionally.

The concern of this chapter is with infections that are transmitted from domestic and wild animals in Africa to humans. Some of these infections, such as schistosomiasis, hydatidosis, trichinosis and others have already been mentioned elsewhere in this volume and will not be dealt with here. Some of those not previously mentioned are now reported, but no attempt is made to include them all, the chapter presents a selection of some of the important ones.

Trematode zoonoses

Paragonimiasis

Paragonimiasis is caused by lung flukes of the genus *Paragonimus*. Several species of this genus infect wild carnivores and small mammals in

various parts of the world. Only two species belonging to this genus have
so far been reported from Africa. *P. africanus* was found in the mongoose
Crossarchus obscurus in Cameroon and later from the swamp mongoose
Atilax paludinosus, the mandril *Mandrillus leucephalus*, the potto
Perodictus potto and the civet cat *Viverra civetta*. Several hosts, including
monkeys, dogs and cats were experimentally infected. *P. uterobilateralis*
was found in the march mouse *Malcomys edwardsi* and the shrew
Crocidura flavescens in Liberia and later from the civet cat *V. civetta* in
Nigeria (Voelker *et al.* 1975). Two species of freshwater crabs,
Sudanonotus africanus and *S. aubryi* serve as second intermediate hosts in
Nigeria (Voelker *et al.* 1975). Humans become infected after eating raw
or insufficiently cooked crabs.

No adult worms of *P. africanus* have so far been recovered from
humans but on the basis of measurements of eggs in human sputum,
human cases were suspected in Cameroon. Adult worms identified as
P. uterobilateralis were found in an abdominal cyst in an eight-year-old
girl in Nigeria and other human cases have been identified on the basis of
egg measurements in the sputum of several patients in Nigeria.
Pulmonary paragonimiasis in Africa has been so far restricted to West
Africa. Human cases have been reported from Liberia, Gabon,
Cameroon, Guinea and Nigeria. Previous cases reported as *P. westermani*
were wrongly identified. Cases have been reported by Craig & Faust
(1937), Zahra (1952), Nnochiri (1968), Sachs *et al.* (1983) and others.
Nwokolo (1964) observed only four human cases of pulmonary paragonimiasis in Enugu, Nigeria, between 1959 and 1962 but after the civil war
he (1972) reported more than 100 cases in the same area in ten months.
Due to shortage of proteins, inhabitants in the famine-stricken area
increased their consumption of water crabs and this led to the sudden
increase in cases of paragonimiases. Udonsi (1987) reported an outbreak
of paragonimiasis due to *P. uterobilateralis* in the same area in Nigeria.
In a 24-month survey, 332 (16.8%) out of 1973 individuals examined were
found to be infected. There was evidence of an increasing annual
prevalence rate from 15% in 1983 to 18.7% in 1985. Infection was present
in all age groups with a prevalence of 18.9% in males and 14.5% in
females.

Heterophyidiasis

Heterophyidiasis is caused by a number of species of small digeneans of
the family Heterophyidae. Adults of this family are parasitic in
piscivorous birds and mammals. Their life-cycle involves a snail as first
intermediate host and several species of fresh and brackish water fishes as
second intermediate hosts in which the cercariae encyst.

About 10 heterophyid species have been found in humans, mainly in
the Far and Middle East. In Africa the most common heterophyid species

infecting humans is *Heterophyes heterophyes* which is found particularly in Egypt. Dogs and cats are the primary hosts but a number of other mammals as well as birds are also infected and are responsible for the dissemination of eggs of this species. The parasite is specific to the molluscan first intermediate host, *Pironella*, but not to either the fish second intermediate host or the final host. Numerous species of fish act as intermediate hosts but mullets of the genus *Mugil* and cichlid species such as *Tilapia* spp. are most abundantly infected. The brackish lagoons of the Nile Delta in Egypt are favourite areas for the breeding of these species of fish as well as the snail intermediate host. Humans become infected by consuming raw or undercooked fish containing metacercaria. In Egypt salted fish 'fessikh' is regarded as a delicacy and is probably the main source of human infection with *H. heterophyes* (Wells & Randall 1955, Elias *et al.* 1969).

Several studies were carried out on *H. heterophyes* and its prevalence in humans in certain areas of Egypt. There is an indication, however, that the disease is gradually decreasing. Khalil (1933) reported an infection rate of 88% in people examined at Mataria; Nagaty & Khalil (1964) recorded a rate of infection of 10.5% in inhabitants of the same area while Rifaat *et al.* (1980) recorded a rate of 5%. Youssef *et al.* (1987) attribute this decrease to the current high price and scarcity of mullets, due to the alteration in water nutrients brought about by the construction of the Aswan High Dam.

Several other heterophyids occur in Egypt, some such as *Heterophyes dispar* and *H. equalis* have been reported in humans. *H. heterophyes* has also been found in natural infection in cats in Tunisia (Baloget & Callot 1939) and a closely related species *H. pleomorphis* was found in dogs and cats in Uganda. As far as I can find, no human cases have so far been reported from Tunisia or Uganda.

Dicrocoeliasis

The lancet liver flukes of the genus *Dicrocoelium* are mainly parasites of sheep, goats and cattle occurring in these hosts with extreme densities and incidences of infection. More than 60 other species of mammals, including hares, rabbits, antelopes and rodents have also been found naturally infected but the coincidences and intensities of infection in these hosts are relatively low. Humans and monkeys have also been found occasionally infected (Berghe & Denecke 1938). The completion of the life-cycle of species of *Dicrocoelium* requires a land snail and an ant as intermediate hosts.

D. lanceolatum is widely distributed and occurs in Asia, Europe, North and South America and probably North Africa. *D. hospes* is restricted to Africa and has been reported from the Sudan, Chad, Ghana, Uganda, Tanzania, Sierra Leone, Zaire and other countries. Cases of human infection with *D. hospes* have been reported from Ghana and Sierra Leone. However, the appearance of eggs of *Dicrocoelium* in human faeces is not an indication of a true human infection. In endemic areas where infested livers are eaten by people, the liver tissues and the contained digeneans are completely digested but the eggs are only partially digested. The undigested eggs may appear occasionally in human faeces but the presence of eggs in repeated examinations of human faeces, particularly after abstention from eating liver meat, may indicate a true infection. Serological tests and adequate clinical examinations are added confirmations.

D. *lanceolatum* was reported from two Europeans who had resided in Zaire (Berghe & Denecke 1938). They also found eggs of this species in the faeces of three chimpanzees (*Pan satyrus*), six baboons (*Papio* sp.), five Mangabey (*Cercocebus* sp.) and three vervet monkeys (*Cercopithecus* sp.), and recovered adult worms from the duodenum of *Papio* sp. and *Cercocebus* sp. No eggs were present in the bile duct.

Fascioliasis

Fascioliasis is the disease caused by liver flukes of the genus *Fasciola*. It is one of the most important diseases of ruminants throughout the world and causes severe economic losses in domestic animals due to mortality, reduction in milk and meat production and liver condemnation. Adult parasites generally live in the bile duct and lay non-embryonated eggs which pass with bile into the intestine and are eventually evacuated in faeces. Lymnaeid snails serve as intermediate hosts and the cercariae encyst on various plants and occasionally on the surface of water enclosing small bubbles of air which allow them to float. The final host becomes infected after ingesting metacercaria encysted on plants or floating in water.

In Africa *Fasciola gigantica* is the most prevalent species of the genus. It occurs in domestic animals and has been reported from at least ten different species of wild animals (Round 1968). *F. hepatica* is also present in Africa but its distribution is restricted to certain areas where its snail intermediate host is available. *F. nyanzae* occurs in the hippopotamus in East Africa, and *F. tragelaphi*, reported from the sitatunga *Tragelaphus spekei* in Zimbabwe, has been moved to the genus *Tenuifasciola* (Yamaguti 1971).

Fascioliasis is a common disease of domestic animals in many parts of the world and human cases have also been reported. Most cases of human fascioliasis have been attributed to *F. hepatica* and only a few to *F. gigantica*. Hammond (1974) discussed records of human infections with *F. gigantica* and suggested that the infection may be more common than was thought because it may have been overlooked. As with dicrocoeliasis, it is worth noting that in localities where people are accustomed to eat livers of infected animals, undigested eggs may appear in the faeces which may lead to false diagnosis. The presence of eggs in repeated examinations, particularly after abstention from eating liver may confirm a true infection. No cases of human infection with *F. nyanzae* or *F. tragelaphi* have been reported.

In Africa human cases of *F. hepatica* have been reported in Morocco, Algeria, Tunisia, Egypt and Malaŵi. *F. hepatica* has been established for a long time in North Africa, having been found in Egyptian mummies (Curry *et al.* 1979).

Human cases of *F. gigantica* have been reported from Rwanda, Burundi, Uganda, Madagascar, Cameroon, Zimbabwe, Zambia and Egypt. An unusual case of ectopic infection has been reported in Egypt (Ghawabi *et al.* 1978) in a boy presenting with a disabling abscess of the left ankle. Incision of the abscess revealed a folded, apparently flattened worm which on sectioning proved to be an immature *F. gigantica*.

Poikilorchiasis

Fain & Vandepitte (1957) described a new fluke, *Poikilorchis congolensis*, which they recovered from a subcutaneous cyst in the retro-auricular region of a native from Kasai province in Zaire. They proposed a new genus for this worm which they attached to the family Achillurbainiidae (Dollfus 1939). The eggs of this species, found in the pus of the cyst, resemble eggs of *Paragonimus* but are smaller. They reported four human cases of similar cysts containing the same eggs and they believe that the case reported by Yarwood & Elmes (1943) in Nigeria and attributed to *Paragonimus* is probably due to *Poikilorchis*. None of these patients with *Poikilorchis* had eggs in their sputum or any traces of pulmonary distomiasis.

Although the natural host of species of *P. congolensis* is not known it is likely to be a rodent or a similar host. Species of related genera of the family Achillurbainiidae were found in rodents and leopards in the Far East and in opossums in Brazil. A human case of a related species was reported from a retro-auricular abscess in a ten-year-old girl in China. The life-cycle of this species is not known.

Cestode zoonoses

Coenuriasis

The term coenurus refers to a bladder-like metacestode belonging to a number of species of tapeworms of the genus *Multiceps* which is related to the genus *Taenia*. In fact some workers argue that the two genera are synonymous as they cannot be differentiated on the basis of the morphology of the adult tapeworm. The coenurus has many scolices and differs from the cysticercus metacestodes; workers attribute these differences to host rather than to parasite properties.

Adult worms occur normally in carnivores, and the larval forms occur in subcutaneous tissues, or tissues of the central nervous system of a number of vertebrate hosts. Various species of domestic and wild herbivores, rabbits, rodents, porcupines and others are intermediate hosts and humans are accidentally infected. In both natural and accidental infections oncospheres hatch in the lumen of the intestine, actively penetrate the intestinal wall, enter the circulatory system and are carried to subcutaneous tissues, somatic muscles, to the central nervous system or ocular tissues. In one of these sites the oncosphere develops into a coenurus; preference for a particular site depends in most cases on the species of the parasite.

Coenuriasis of animals is well known in central and southern Africa. A number of human cases have also been reported. The first human case in Africa, described by Turner & Leiper (1919), was a tumour containing a coenurus cyst excised from the intercostal muscle of a patient from northern Nigeria. Subsequently several human cases have been reported from various parts in Africa including Zaire, Uganda, Kenya, Rwanda, Ghana and South Africa. Templeton (1968) reviewed the cases up to that time and reported 14 cases from Uganda. Wilson et al. (1972), reporting the first case of human coenuriasis in Ghana, presented a critical review of the literature and discussed the difficulty of species identification within the genus *Multiceps* based on morphological, site and epidemiological differences.

Kaminsky et al. (1978) reported four cases of coenuriasis; the cysts were removed from the muscles of three children and one adolescent from different localities in Kenya.

Sparganosis

Sparganosis is the term used to describe a tissue infection caused by the plerocercoid stage of pseudophyllidean tapeworms of the genus *Spirometra*. The validity of this genus is disputed and some regard it as a subgenus of *Diphyllobothrium*. The group-name *Sparganum* was introduced to include all unidentified plerocercoid larvae. Later this was reserved for plerocercoids of *Diphyllobothrium* and *Spirometra*.

The life history of *Spirometra* involves copepods of the genus *Cyclops* as first intermediate host where the procercoid develops. The plerocercoid (= sparganum) develops in a wide range of second intermediate hosts including amphibians, snakes, mammals and birds. The adult worms develop in domestic and wild carnivores. Humans are occasionally found to harbour spargana and it is presumed that they become infected accidentally. The mechanisms of human infection in nature is not clear and it is probable that humans acquire infection by ingesting infected *Cyclops* with drinking water, by ingesting raw or imperfectly cooked flesh of an infected second intermediate host, or by the local application of the flesh of an infected second intermediate host to the eye or to an ulcerated skin as a poultice.

Two adult species of *Spirometra* have so far been reported from Africa; *S. pretoriensis* from wild canids and *S. theileri* in wild felids. Only a few other records of spargana have been reported from animals in Africa and these include monkeys, baboons, genets, okapi, rodents and warthogs. A number of human cases with spargana have also been reported in Africa. Sambon (1907) reported the first case from an abscess in the leg of a Maasai in Kenya. Fain and Piraux (1959) gave a summary of human cases in Africa and since then human cases have been reported from a number of countries including Uganda, Tanzania, Botswana, Zaire, Madagascar, Liberia, Rwanda and Mozambique. Nelson *et al.* (1965) believed that sparganosis in humans was probably much more prevalent in Africa than these records indicated.

Other cestode infections

Species of the anoplocephalid cestode genus *Inermicapsifer* are very common in Africa, mainly in rodents and hyrax. *I. madagascariensis* is widespread in animals and a number of human infections with this parasite have been reported. Human cases have been reported from Kenya, Zimbabwe, Zambia, South Africa, Zaire, Rwanda, Madagascar and Mauritius. Nelson *et al.* (1965) also suspected that this parasite was more common in Africa than such records indicated.

Species of another anoplocephalid genus *Bertiella* are also prevalent in Africa. *B. studeri* is widespread in non-human primates. A number of human cases with this parasite have been reported.

The dwarf tapeworm *Hymenolepis nana* has a cosmopolitan distribution in rodents and humans. Although the human and rodent strains are morphologically identical, they are physiologically different. Because of the unusual life-cycle of this species where no intermediate host is essential, humans are probably the main source of human infection. The low incidence of infection in humans indicates that there is little inter-human infection. Human infections have been reported from Algeria, Egypt, Sudan, Burkina Faso, Senegal and South Africa.

H. diminuta on the other hand is a common cestode of rodents and an occasional parasite of humans. Various insects, including fleas, are obligatory intermediate hosts. Human cases have been reported from Zimbabwe (Goldsmid 1973).

Nematode zoonoses

Angiostrongyliasis

The lungworm species of the genus *Angiostrongylus* are of considerable public health importance in certain parts of the world because some species can be transmitted to humans. *A. cantonensis* and *A. malayensis* are associated with eosinophilic meningitis in humans caused by cerebral angiostrongyliasis. *A. costaricensis* causes abdominal angiostrongyliasis. The definitive hosts of these species are rodents; humans are accidental hosts. Khalil (1986) has provided an account of these parasites and their rodent hosts. Certain molluscs serve as intermediate hosts in which the parasite develops. The infective larvae invade and partially develop in the brain and the central nervous system of the final host before travelling to the lungs where they reach sexual maturity.

A. cantonensis is the most important and widespread of the three species. Recent surveys have shown that this parasite is prevalent in tropical and subtropical belts, in areas conducive for the propagation of rodents and molluscan intermediate hosts. The disease in humans is spreading through infected stowaway rodents in ships. It also appears that the dissemination of the disease has been assisted by the introduction of snail intermediate hosts, particularly the giant African snail *Achatina fulca* which is a very suitable intermediate host and is consumed as a delicacy in certain parts of the world.

Humans become infected when raw or undercooked molluscs containing infective larvae are ingested, sometimes accidentally. Paratenic hosts, such as prawns and fish, harbouring infective larvae after eating infected molluscs, are sometimes important sources of human infection. Although the parasite does not develop to maturity in humans, a high proportion of the ingested third stage larvae reach the central nervous system, producing eosinophilic meningitis. In Africa the parasite has been reported in rats in Madagascar (Brygoo & Chabaud 1964) and from man in the Ivory Coast where a living pre-adult *A. cantonensis* was found in the cerebrospinal fluid of a ten-year-old boy (Assi-Adou *et al.* 1980). In Egypt this parasite was found in rodents and snail intermediate hosts where *Rattus norvegicus* and *Lanistes carinatus* were established as the definitive and intermediate hosts respectively (Yousif *et al.* 1980). It is likely that this parasite is much more prevalent in Africa that these records indicate. The cryptic location of the parasite in the lungs of rodents, organs which are not routinely examined, may have contributed to the overlooking and under-reporting of *A. cantonensis*. The fact that this parasite can utilize a wide range of intermediate hosts which are abundant in Africa, and the cosmopolitan distribution, high mobility and adaptability of rodents favours its spread.

Trichostrongyliasis

Trichostrongyliasis is caused by nematodes of the family Trichostrongylidae (see also Chapter 5). These are small hair-like worms which generally inhabit the stomach and small intestine, primarily of domestic and wild ruminants. They have a direct life cycle where no intermediate host is involved. The eggs are eliminated with the faeces of the host. Under favourable conditions of warmth and moisture these eggs quickly hatch, producing first stage larvae. In the soil or on pastures the larvae undergo two moults and are finally transformed to the infective third stage larvae. On being swallowed by a suitable final host the third stage larvae are capable of developing into sexually mature adults. Trichostrongylid nematodes are widely spread throughout the world. Heavy infections of livestock cause great economic losses.

It was previously thought that trichostrongyliasis was fairly rare in humans but recent surveys have shown that in certain foci in the world the disease is fairly prevalent. Species of the genus *Trichostrongylus* are the ones that generally infect humans and so far several species have been incriminated. Stoll (1947) estimated that 5.5 million people were infected. Wattrelos (1981) reviewed all the known human cases worldwide and it seems that the disease is more prevalent than was thought but its distribution is very patchy. Humans are most frequently affected in rural areas where contamination of the soil with animal excrement takes place. Humans generally become infected after eating unwashed vegetables or plant material contaminated with infective larvae.

Africa has a rich and varied number of herbivorous domestic and wild animal species and consequently a large number of trichostrongylid species are found there (Gibbons 1977). Four different species of *Trichostrongylus* have been found in humans in Africa reported from 25 different countries. The disease is most prevalent in Egypt where it is considered endemic in certain areas. Lawless *et al.* (1956) found that 70% of the adult population they examined in an Egyptian village were infected. Chandler (1954) found an incidence of 29% and 40% in two Egyptian villages reflecting the close association between the people and their animals in fields and at home in that country.

Oesophagostomiasis

Several species of the nematode genus *Oesophagostomum* have been recorded in domestic and wild animals. Some species are parasites of primates and at least three different species have been found in humans. It is assumed that people become infected after swallowing infective stages and the possibility of infection through the skin has also been considered (Haaf & Van Soest 1964).

Numerous records of human infections have been reported from

Africa. Some of these reports were based on finding eggs during faecal examinations. This is probably an unreliable criterion as eggs of *Oesophagostomum* are not easily distinguishable from eggs of hookworms. Confirmation, however, occurred in a number of cases by laparotomy when worms were recovered from nodules or seen in sections (see also Ch. 5).

Human cases in Africa have been reported from Ethiopia, Nigeria, Uganda, Sudan, Ghana and Kenya. Leiper (1913) reported a 4% incidence among prisoners in Nigeria. Barrowclough & Crome (1979) have reviewed some of the reported cases.

Capillariasis

Capillariasis in humans is caused by species of the genus *Capillaria*. Although several species of the genus affect humans, the commonest and best known species are *C. philippinensis* and *C. hepatica*. *C. philippinensis* has so far not been found in Africa. *C. hepatica* is a very common parasite of rodents all over the world and has been reported from other hosts such as carnivores, ungulates, primates and humans.

The adult worms of *C. hepatica* lodge in the hepatic parenchyma and produce eggs which do not pass out in the faeces but accumulate in the liver and form large whitish spots under the surface. When infected livers of rodents are eaten through cannibalism by other rodents or through predation by carnivores, eggs escape and appear in the faeces of these animals, complete their development and become infective.

There are few records of human infection with *C. hepatica*. Because eggs are rarely seen in human faeces, light infections have escaped detection. Only very heavy infections where the disease has been fatal or cases complicated by a noncommitant disease have come to light. Prominent signs of the disease are hepatomegaly, morning fever and vomiting. Ingestion of eggs through dirt or contamination of food appears to be the route of human infection. Attah *et al.* (1983) reported a *C. hepatica* infection in a 27-year-old Nigerian woman. The parasite is very prevalent in rodents in Africa and it is likely that numerous human infections occur.

Future prospects

In this review only a few selected examples of some of the important zoonotic infections are discussed but there are several other helminth parasites that are naturally transmitted between animals and humans in Africa. Normally most of these parasites produce mild infections that have little health significance and are consequently disregarded. The recent emergence and rapid spread of the Acquired Immunodeficiency syndrome (AIDS) may however change the importance of these infections. Parasites causing latent or mild infections in immunocompetent hosts may induce severe or even lethal diseases in immunocompromised patients. A new type of zoonoses, termed 'Aids-related zoonoses', is recognized and the parasites are regarded as producing opportunistic infections.

This changed situation should give an impetus for urgent studies of all infections in Africa. Progress in controlling such infections will only be achieved after thorough studies of the epidemiology, immunodiagnosis and chemotherapy of these helminth infections. In other fields such as bacteriology, virology and protozoology significant progress has already been achieved. This success should demonstrate to responsible authorities and funding agencies that support for research and development could lead to progress in the control of zoonotic infections.

References

Assi Adou, J., J. K. Kouame, J. Moreau, M. A. Timite *et al.* 1980. L'angiostrongyloidose. Mise au point à propos d'un cas. *Med. Afr. Noire* **27**, 421–5.

Attah, E. B., S. Nagarajan, E. N. Obineche & S. C. Gera 1983. Hepatic capillariasis. *Am. J. Clin. Path.* **79**, 127–30.

Baloget, L. & J. Callot 1939. Trématodes de Tunise. 3 Superfamille Hétérophyoidea. *Arch. l'Inst. Pasteur Tunise*, **28**, 34–63.

Barrowclough, H. & L. Crome 1979. *Oesophagostomum* in man. *Trop. Geogr. Med.* **31**, 133–8.

Berghe, L. & K. Denecke 1938. *Dicrocoelium dendriticum* (*Fasciola lanceolata*) chez l'homme et les singes au Congo Belge. *Ann. Soc. Belge. Med. Trop.* **18**, 509–14.

Brygoo, E. R. & A. G. Chabaud 1964. Présence d'*Angiostrongylus cantonensis* (chen) à Madagascar. *Ann. Parasit. Hum. Comp.* **39**, 793.

Chandler, A. C. 1954. A comparison of helminthic and protozoan infections in two Egyptian villages two years after the installation of sanitary improvements in one of them. *Am. J. Trop. Med. Hyg.* **3**, 59–73.

Craig, C. F. & E. C. Faust 1937. *Clinical Parasitology.* 733. Philadelphia: Lea & Febiger.

Curry, A., C. Anfield & E. Tapp 1979. Electron microscopy of the Manchester mummies. In *Manchester Museum mummy project.* Multidisciplinary research on ancient Egyptian mummified remains. A. R. David (ed.), 103–11. Manchester: Manchester Museum.

Dollfus, R. P. 1939. Distome d'un abcès palpébro-orbitaire chez une panthère. Possibilité d'affinités lointaines entre ce distome et les Paragonimidae. *Ann. Parasit. Hum. Comp.* **17**, 209–35.

Elias, A., M. G. E. Ahmed, F. A. El-Wakeil & I. O. Foda 1969. Fish salting with reference to *Heterophyes* infection. *J. Vet. Sci. Unit. Arab. Rep.* **6**, 75–87.

Fain, A. & A. Piraux 1959. Sparganose chez l'homme et les animaux au Rwanda-Burundi. *Bull. Soc. Pathol. Exot.* **52**, 804–8.

Fain, A. & J. Vandepitte 1957. Description du nouveau stome vivant dans des kystes au abcès rulaires chez l'homme au Congo Belge. *Ann. Soc. Belge. Med. Trop.* **37**, 251–8.

Ghawabi, M. H., S. A. Sale & M. E. Azab 1978. A case of ectopic fascioliasis. *J. Egypt. Soc. Parasitol.* **8**, 141–6.

Gibbons, L. M. 1977. Morphology and taxonomy of the superfamily Trichostrongyloidae (Nematoda), with special reference to those occurring in African Artiodactyla. PhD thesis, University of London.

Goldsmid, J. M. 1973. A note on the occurrence of *Hymenolepis diminuta* (Rudolphi, 1819) Blanchard, 1891 (Cestoda) in Rhodesia. *Cent. Afr. J. Med.* **19**, 51–2.

Haaf, E. & A. H. van Soest 1964. *Oesophagostomum* in man in North Ghana. *Trop. Geogr. Med.* **16**, 743–56.

Hammond, J. A. 1974. Human infection with the liver fluke *Fasciola gigantica. Trans. R. Soc. Trop. Med. Hyg.* **68**, 253–4.

Kaminsky, R. G., D. G. Gatei & R. R. Zimmermann 1978. Human coenuriasis from Kenya. *E. Afr. Med. J.* **55**, 355–9.

Khalil, L. F. 1986. The helminth parasites of rodents and their importance. In *Proceedings of the second symposium on recent advances in rodent control, Kuwait*, 141–9.

Khalil, M. 1933. The life history of the human trematode *H. heterophyes* in Egypt. *Lancet* **225**, 537.

Lawless, D. K., R. E. Kuntz & C. P. A. Strome 1956. Intestinal parasites in an Egyptian village of the Nile Valley with emphasis on the protozoa. *Am. J. Trop. Med. Hyg.* **5**, 1010–14.

Leiper, R. T. 1913. Observations on certain helminths of man. *Trans. R. Soc. Trop. Med. Hyg.* **6**, 265–97.

Nagaty, H. F. & H. M. Khalil 1964. Incidence of helminth infections among outpatients in the clinic of Mataria Collective Unit, Dakahlia Governorate. *J. Egypt. Med. Assoc.* **47**, 341–6.

Nelson, G. S., F. R. N. Pester & R. Rickman 1965. The significance of wild animals in the transmission of cestodes of medical importance in Kenya. *Trans. R. Soc. Trop. Med. Hyg.* **59**, 507–24.

Nnochiri, E. 1968. *Parasitic diseases and urbanization in a developing community*. Oxford: Oxford University Press.

Nwokolo, C. 1964. Paragonimiasis in Eastern Nigeria. *J. Trop. Med. Hyg.* **67**, 1–10.

Nwokolo, C. 1972. Endemic paragonimiasis in Eastern Nigeria. Clinical features and epidemiology of the recent outbreak following the Nigerian civil war. *Trop. Geogr. Med.* **24**, 138–47.

Rifaat, M. A., S. A. Salem, S. I. El-Kholy, M. M. Hegazi *et al.* 1980. Studies on the incidence of *Heterophyes heterophyes* in Dakahlia governorate. *J. Egypt. Soc. Parasitol.* **10**, 369–73.

Round, M. 1968. *Check list of the helminth parasites of African mammals*. Technical Communication No 38 of the Commonwealth Bureau of Helminthology, St Albans.

Sachs, R., P. Kern & J. Voelker 1983. Le *Paragonimus uterobilateralis* comme cause de trois cas de paragonimiase humaine au Gabon. *Tropenmed. Parasitol.* **34**, 105–8.

Sambon, L. W. 1907. Description of some new species of animal parasites. *Proc. Zool. Soc. London,* **19**, 282–3.

Stoll, N. R. 1947. This wormy world. *J. Parasitol.* **33**, 1–18.

Templeton, A. C. 1968. Human coenurus infection. A report of 14 cases from Uganda. *Trans. R. Soc. Trop. Med. Hyg.* **62,** 251–5.

Turner, M. & R. T. Leiper 1919. On the occurrence of *Coenurus glomeratus* in man in West Africa. *Trans. R. Soc. Trop. Med. Hyg.* **13,** 23–4.

Udonsi, J. K. 1987. Endemic *Paragonimus* infection in Upper Igwun Basin, Nigeria: A preliminary report on a renewed outbreak. *Ann. Trop. Med. Parasitol.* **81,** 57–62.

Voelker, H., R. Sachs, K. J. Volkmer & H. Braband 1975. Zur epidemiologie der Paragonimiasis bei Mensch und Tier in Nigeria, Westafrika. *Vet. Med. Nachrich.* No. 1/2, 158–72.

Wattrelos, P. 1981. Contribution a l'étude de la trichostrongylose humaine (A propos de 81 observations). Thèse pour le doctorat en médecine, Lille, France.

Wells, W. H. & B. H. Randall 1955. Salted mullet (fessikh) as a source of human infection with *Heterophyes heterophyes.* *J. Egypt. Publ. Hlth Assoc.* **30,** 83–6.

Wilson, V. C. L. C., D. M. Wayte & R. O. Addae 1972. Human coenurosis in the first reported case from Ghana. *Trans. R. Soc. Trop. Med. Hyg.* **66,** 611–23.

Yamaguti, S. 1971. *Synopsis of digenetic trematodes of vertebrates.* Tokyo: Keigaku Publishing.

Yarwood, G. R. & G. T. Elmes 1943. *Paragonimus* cyst in a West African native. *Trans. R. Soc. Trop. Med. Hyg.* **36,** 347–51.

Yousif, F., M. Roushdey & M. El-Emam 1980. The host–parasite relationships of *Angiostrongylus cantonensis* in Egypt. 1. Natural and experimental infection of the snail intermediate host *Lanistes carinatus.* *J. Egypt. Soc. Parasitol.* **10,** 399–412.

Youssef, M. A., N. S. Mansour, H. N. Awadalla, N. A. Hammouda *et al.* 1987. Heterophyid parasites of man from Idku, Maryut and Manzala lakes area in Egypt. *J. Egypt. Soc. Parasitol.* **17,** 475–9.

Zahra, A. 1952. Paragonimiasis in the Southern Cameroons. A preliminary report. *W. Afr. Med. J.* **1,** 74–82.

Index